Lecture Notes in Physics

Managing Editor

W. Beiglböck
Assisted by Mrs. Sabine Landgraf
c/o Springer-Verlag, Physics Editorial Department II
Tiergartenstrasse 17, D-69121 Heidelberg, Germany

The Editorial Policy for Proceedings

The series Lecture Notes in Physics reports new developments in physical research and teaching – quickly, informally, and at a high level. The proceedings to be considered for publication in this series should be limited to only a few areas of research, and these should be closely related to each other. The contributions should be of a high standard and should avoid lengthy redraftings of papers already published or about to be published elsewhere. As a whole, the proceedings should aim for a balanced presentation of the theme of the conference including a description of the techniques used and enough motivation for a broad readership. It should not be assumed that the published proceedings must reflect the conference in its entirety. (A listing or abstracts of papers presented at the meeting but not included in the proceedings could be added as an appendix.)

When applying for publication in the series Lecture Notes in Physics the volume's editor(s) should submit sufficient material to enable the series editors and their referees to make a fairly accurate evaluation (e.g. a complete list of speakers and titles of papers to be presented and abstracts). If, based on this information, the proceedings are (tentatively) accepted, the volume's editor(s), whose name(s) will appear on the title pages, should select the papers suitable for publication and have them refereed (as for a journal) when appropriate. As a rule discussions will not be accepted. The series editors and Springer-Verlag will normally not interfere with the detailed editing except in fairly obvious cases or on technical matters.

Final acceptance is expressed by the series editor in charge, in consultation with Springer-Verlag only after receiving the complete manuscript. It might help to send a copy of the authors' manuscripts in advance to the editor in charge to discuss possible revisions with him. As a general rule, the series editor will confirm his tentative acceptance if the final manuscript corresponds to the original concept discussed, if the quality of the contribution meets the requirements of the series, and if the final size of the manuscript does not greatly exceed the number of pages originally agreed upon. The manuscript should be forwarded to Springer-Verlag shortly after the meeting. In cases of extreme delay (more than six months after the conference) the series editors will check once more the timeliness of the papers. Therefore, the volume's editor(s) should establish strict deadlines, or collect the articles during the conference and have them revised on the spot. If a delay is unavoidable, one should encourage the authors to update their contributions if appropriate. The editors of proceedings are strongly advised to inform contributors about these points at an early stage.

The final manuscript should contain a table of contents and an informative introduction accessible also to readers not particularly familiar with the topic of the conference. The contributions should be in English. The volume's editor(s) should check the contributions for the correct use of language. At Springer-Verlag only the prefaces will be checked by a copy-editor for language and style. Grave linguistic or technical shortcomings may lead to the rejection of contributions by the series editors. A conference report should not exceed a total of 500 pages. Keeping the size within this bound should be achieved by a stricter selection of articles and not by imposing an upper limit to the length of the individual papers. Editors receive jointly 30 complimentary copies of their book. They are entitled to purchase further copies of their book at a reduced rate. As a rule no reprints of individual contributions can be supplied. No royalty is paid on Lecture Notes in Physics volumes. Commitment to publish is made by letter of interest rather than by signing a formal contract. Springer-Verlag secures the copyright for each volume.

The Production Process

The books are hardbound, and the publisher will select quality paper appropriate to the needs of the author(s). Publication time is about ten weeks. More than twenty years of experience guarantee authors the best possible service. To reach the goal of rapid publication at a low price the technique of photographic reproduction from a camera-ready manuscript was chosen. This process shifts the main responsibility for the technical quality considerably from the publisher to the authors. We therefore urge all authors and editors of proceedings to observe very carefully the essentials for the preparation of camera-ready manuscripts, which we will supply on request. This applies especially to the quality of figures and halftones submitted for publication. In addition, it might be useful to look at some of the volumes already published. As a special service, we offer free of charge LATEX and TEX macro packages to format the text according to Springer-Verlag's quality requirements. We strongly recommend that you make use of this offer, since the result will be a book of considerably improved technical quality. To avoid mistakes and time-consuming correspondence during the production period the conference editors should request special instructions from the publisher well before the beginning of the conference. Manuscripts not meeting the technical standard of the series will have to be returned for improvement.

For further information please contact Springer-Verlag, Physics Editorial Department II, Tiergartenstrasse 17, D-69121 Heidelberg, Germany

John Buckmaster Tadao Takeno (Eds.)

Modeling in Combustion Science

Proceedings of the US-Japan Seminar
Held in Kapaa, Kauai, Hawaii, 24-29 July 1994

Springer

Editors

John Buckmaster
Talbot Laboratory, Room 321
University of Illinois
104 South Wright Street
Urbana, IL 61801, USA

Tadao Takeno
Department of Mechanical Engineering
Nagoya University
Chikusa-ku, Nagoya 464-01, Japan

ISBN 978-3-662-14028-4   ISBN 978-3-540-49226-9 (eBook)
DOI 10.1007/978-3-540-49226-9

CIP data applied for

© Springer-Verlag Berlin Heidelberg 1995

Originally published by Springer-Verlag Berlin Heidelberg New York in 1995
Softcover reprint of the hardcover 1ts edition 1995

Typesetting: Camera-ready by the editors
SPIN: 10481119   55/3142-543210 - Printed on acid-free paper

PREFACE

During the week of July 25, 1994, a workshop on combustion was held in Kapaa, Kauai, under the auspices of the National Science Foundation (USA) and the Japanese Society for the Promotion of Science (Japan). It was designed to bring together Japanese and American combustion scientists, together with a handful from other parts of the world, who share a common interest in the mutual interaction between mathematical modeling, numerical modeling, and experiment. Each participant was required to present a major lecture, and papers based on most of these talks are contained in this volume. In addition, there are a small number of extra papers describing subjects that were part of the workshop discussions. The papers cover a variety of issues that are of importance to combustion scientists.

The burden of organization was lightened by a number of people, and we are grateful for that. Particular thanks must go to Diane Kawamoto, of the Aston Kauai Beachboy Hotel, whose extra efforts on our behalf were important in defining a valuable and enjoyable meeting.

December 1994   J. Buckmaster, Hong Kong
                                 T. Takeno, Nagoya

CONTENTS

1. Turbulence - Premixed Flame

2. Turbulence Non-premixed Flame

3. Modeling

4. Flame - Pressure Interactions

5. Numerical Treatments

6. Combustion Waves

7. High Mach Numbers

1. Turbulence - Premixed Flame

Some Open Issues in Premixed Turbulent Combustion

Paul D. Ronney
Department of Mechanical Engineering
University of Southern California
Los Angeles, CA 90089-1453, USA

Abstract

The widely studied subject of premixed turbulent combustion is discussed, with particular attention to identification and review of controversial and unresolved fundamental issues having practical significance. Four such topics are discussed: (1) prediction of turbulent burning velocity (S_T) in the flamelet regime, (2) the role of the velocity spectrum on S_T, (3) the existence and properties of turbulent combustion with distributed reaction zones, and (4) quenching of flames by turbulence. Directions for future research are suggested.

Keywords: turbulence, combustion, burning velocity, extinction

1 Introduction

Turbulent combustion is employed practically all mobile and stationary power generation devices because turbulence increases the mass consumption rate of reactants to values much greater than laminar flames can achieve. This in turn increases the heat release rate and thus power available from a combustor or internal combustion engine of a given size. Few combustion engines would function without the increase in burning rates brought about by turbulence.

For devices employing premixed reactants, the mass consumption rate per unit cross-section area of flame front is given by $\rho_R S_T$, where ρ_R is the unburned gas density and, by definition, S_T is the turbulent burning velocity. Turbulence may increase S_T to values well above the laminar burning velocity (S_L) [1]. However, increasing turbulence levels beyond a certain value increases S_T very little if any (Fig. 1) and may lead to complete quenching of the flame [2]. This effect is particularly pronounced when S_L is small compared to the RMS velocity fluctuation (u'), e.g. for lean fuel-air mixtures. This indicates that the propagation rates of very lean mixtures cannot be increased *ad infinitum* merely by increasing u'. Thus lean mixtures, which thermodynamically promise higher thermal efficiencies and lower pollutant emissions, will exhibit unsatisfactory combustion rates in many practical systems. In addition to its long-standing relevance to automotive applications [3], lean premixed turbulent has recently been considered for gas turbine applications because of this potential for

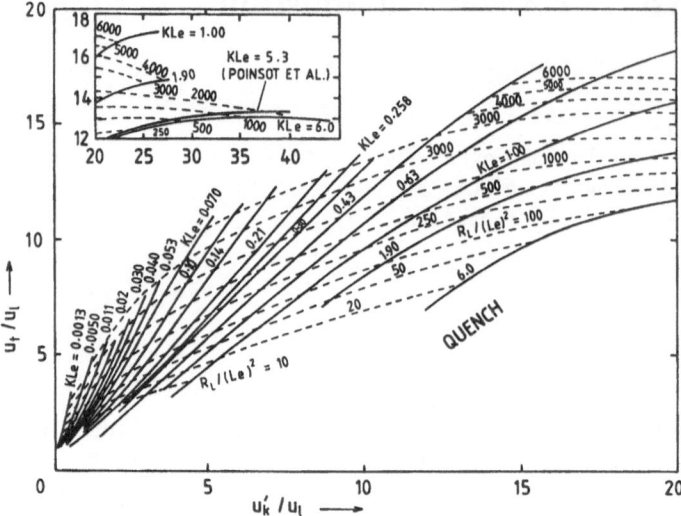

Figure 1. Effect of turbulent intensity and integral-scale Reynolds number on turbulent burning velocity [1]. This plot is obtained from "smoothing" an extensive compilation of experimental data.

reduced NO_x emissions [4]. Furthermore, studies of premixed turbulent combustion may be relevant to some initially nonpremixed systems as well since some partial premixing may occur near blowoff or extinction limits.

After many years of study, much has been learned about premixed turbulent combustion through theory, computation, and experiment. Nevertheless, there are numerous issues of practical importance about which little is known or where agreement between different theoretical predictions and experiments is poor. In this paper we focus on fundamental issues which are currently controversial and unresolved. The focus of this paper on idealized systems that, while quite different from practical flames, are necessary to understand as a prerequisite to solving problems of practical interest.

One important facet of premixed turbulent flames that affects virtually all important properties is the mode of combustion. There are at least three important time scales in turbulent premixed combustion processes: (1) the large-eddy turnover time scale (t_i) ~ L_I/u', where L_I is the integral length scale of turbulence; (2) the small-eddy cutoff time scale (e.g. the Kolmogorov time scale in inertial-range turbulence) (t_k) ~ $t_i Re_L^{-1/2}$, where $Re_L \equiv u'L_I/\nu$ is the turbulent Reynolds number and ν the kinematic viscosity; and (3) the chemical time scale (t_c) ~ D/S_L^2, where D is a characteristic molecular diffusivity. From these three time scales, two independent non-dimensional parameters can be constructed. Following the Leeds group studies led by Prof. Bradley, we choose Re_L ~ $(t_i/t_k)^2$ and the Karlovitz number (Ka) ~ t_c/t_k which is generally written as

$$Ka \equiv 0.157 \, Re_L^{-1/2}U^2Sc^{-1}; \quad U \equiv u'/S_L \, ; \, Sc \equiv \nu/D \qquad (1),$$

where Sc is the Schmidt number. The constant 0.157 is that recommended in [5].

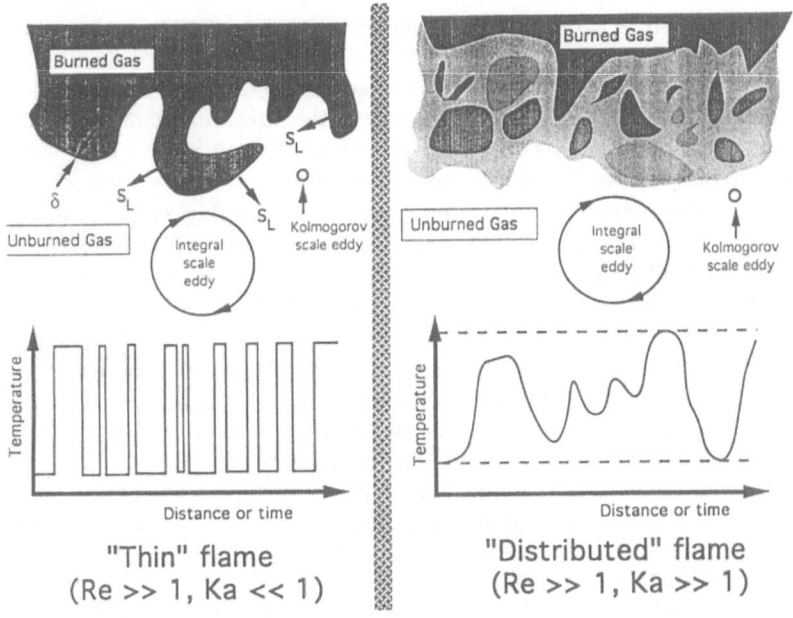

Figure 2. Schematic diagrams of flamelet and distributed combustion.

Various combustion regimes have been identified depending on the values of the two non-dimensional parameters [3, 6, 7]. For strongly turbulent flows, i.e. large Re_L, the two extreme cases are "flamelet" combustion at Ka<<1 and "distributed" combustion at Ka>>1 (see Fig. 2). According to [8], for gaseous turbulent flames the limit of flamelet-like behavior is Ka ≈ 0.15. When this condition is satisfied, t_c << t_k and thus the flame front is expected to be a continuous surface that propagates normal to itself with constant S_L. The role of turbulence is to wrinkle the front, thus increasing its area and thereby its propagation rate (see Eq. 2). There are only two types of fluid, reactant and product, with a sharp interface between them. In the other extreme when Ka is large, t_c >> t_k and thus in most locations the local turbulent strain rate is sufficient to extinguish the flame front, perhaps in a manner similar to that of strained laminar flames [6], thereby disrupting the flame front and causing the identity of the flamelets to be lost entirely. In this case turbulence acts to increase the effective D within the chemically active zone, which in turn increases the thickness (δ_T) of this zone. Consequently, the mechanism of acceleration of flame propagation is quite different from that of wrinkling in the flamelet regime. In the distributed combustion regime, unlike the flamelet regime, there should be significant probability of finding compositions or temperatures intermediate between those of pure reactants and pure products. This expectation has been confirmed experimentally by Yoshida [9] in studies on a specially constructed counterflow burner. Of course, the transition between flamelet and distributed combustion is not sharp, but rather occurs gradually over a range of Ka; Ronney and Yakhot [10] present an analysis of the intermediate regime.

Even in the widely-studied flamelet combustion mode, there are several important issues that are unresolved. For example, one of the most important goals of premixed turbulent flame models is to determine the effect of u' on S_T, yet there is no widely accepted model for predicting S_T, nor is it known whether a general relation exists. Indeed, the very existence of a unique turbulent burning velocity has not been established theoretically except for certain very restrictive cases [11]. Another parameter thought to affect S_T is the turbulent kinetic energy spectrum, however, existing models differ on the role of the energy spectrum on S_T.

The distributed combustion mode has received far less attention than the flamelet mode in existing literature. There are few models of S_T and very few experimental data relevant to this mode. Indeed, some writers (e.g. [3]) believe that all practical combustion devices operate in the flamelet mode, which would suggest that the distributed combustion mode is of no practical importance. However, it is advocated here that some mode of non-flamelet combustion must be relevant to some practical conditions. The reasoning is that in the flamelet mode the front is continuous and always propagating, hence there is no mechanism for extinguishment. Thus at a minimum the distributed combustion mode should have some bearing on flame quenching by turbulence. However, the mechanism of quenching is not well understood, and no model is able to predict the experimentally observed quenching behavior without employing empirical correlations.

Based on this discussion, four subjects have been identified for further discussion and are addressed in the following sections:

- Prediction of turbulent burning velocity in the flamelet regime
- The effect of the kinetic energy spectrum on turbulent burning velocity
- The existence and properties of turbulent combustion with distributed reaction zones
- The mechanism of flame quenching by turbulence

2 Prediction of turbulent burning velocity in the flamelet regime

2.1 Effect of turbulence intensity

Most models of S_T in the flamelet regime are based on Damköhler's [12] proposition that the flame front is wrinkled by turbulence but propagates normal to itself with constant normal velocity S_L (the *Huygens propagation* model). The area increase caused by front wrinkling leads to the relation

$$S_T/S_L = A_T/A_L \qquad (2)$$

where A_T is the surface area of the wrinkled front and A_L is the cross-section area projected in the direction of front propagation. Thus, the objective of flamelet models is essentially to determine the surface area increase caused by wrinkling.

As an example of how daunting this problem can be, consider the special case U << 1. The Clavin-Williams relation [13] $U_T - 1 = U^2$, where $U_T \equiv S_T/S_L$, had been widely accepted since its introduction. Recently, this relation was challenged by

Kerstein and Ashurst [14], who showed that the Clavin-Williams relation is probably applicable only to periodic flows. Instead, they proposed $U_T - 1 \sim U^{4/3}$ for random flows. They also show that the distance which the front must propagate before reaching its steady value is proportional to $U^{-2/3}L_I$, thus at low U a very large experimental or computational domain is required to test their model.

A surprising feature of the experimental record can be seen in Fig. 1, where a very sharp rise in U_T is seen at small U, followed by a rather abrupt leveling off at larger U. This is found even for low values of Ka, where Huygens propagation is expected to prevail. For example, for $Re_L = 3000$, U_T increases from 1 to 7 as U increases from 0 to 2 then U_T increases only to 10 as U increases from 2 to 5. On first glance this behavior that would seem difficult to predict based on Eq. (2) or similar relations.

Fig. 3 shows some predictions of the effect of U on U_T made by flamelet models based on Eq. (2), along with "smoothed" experimental data. As might have been expected based on the above discussion, the experiments do not agree closely with any of the proposed theories, nor do the theories agree with each other. Note that there is no consensus on the quantitative relationship between U_T and U, whether heat release increases or decreases U_T, nor whether the plots have constant slope or bend towards the horizontal at high U. The only point of agreement is that U_T should increase as U increases. (Ironically, Fig. 1 shows that under some circumstances experimental values of U_T can actually *decrease* with increasing U).

Given the complex nature of this problem, even for the "simple" case U << 1, how

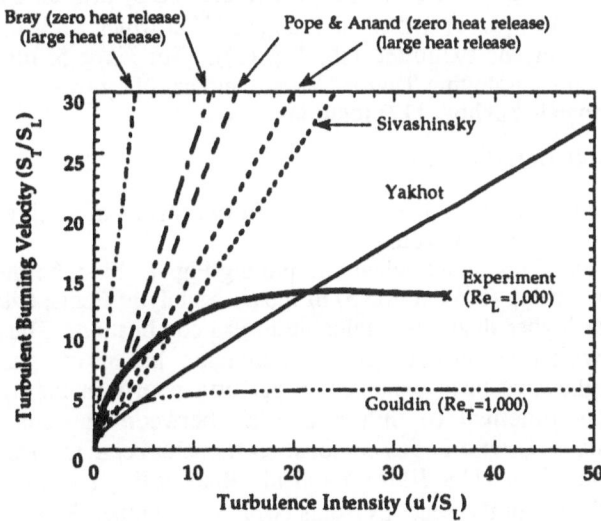

Figure 3. Predicted effect of turbulence intensity (u'/S_L) on turbulent burning velocity (S_T/S_L) from "thin-flame" theories: Bray [15] with zero heat release and large (density ratio = 7) heat release; Anand and Pope [16] with zero and infinite heat release; Yakhot [17]; Sivashinsky [18]; Gouldin [19] with $Re_L = 1,000$; experimental values from Bradley [1] for $Re_L = 1,000$. Where Re_L is not specified, predictions are independent of Re_L.

is one to chose between conflicting models? The "obvious" method would be through direct numerical simulation (DNS) based on the so-called "G equation" [20]

$$\frac{\partial G}{\partial t} + \mathbf{V} \cdot \nabla G = S_L |\nabla G| \tag{3},$$

which is an exact kinematic relation for Huygens propagation in a flow field $\mathbf{V}$; here G is a scalar, any level surface (G = constant) of which can be interpreted as a flame surface. However, practically all DNS studies have been conducted at low U, where the model predictions are similar, and extension to high U is difficult because as U increases, the self-propagation velocity (S_L) becomes small in comparison to the convection velocity $|\mathbf{V}|$ and may be overwhelmed by numerical noise. This is somewhat analogous to the problem of maintaining the accuracy of diffusion of momentum (viscous effects) in high Reynolds number flows where the convection of momentum is large. Turning to conventional experiments for an answer is equally unsatisfying because of the wide scatter in the data (see, e.g., [5]). Even if one is to accept some smoothed data set as accurate, it is unlikely that such data can be directly compared to models because of differences between model assumptions and experimental reality, e.g. constant density and transport properties, adiabatic combustion, and homogeneous and isotropic turbulence.

Perhaps the experimental results best suited for comparison with the proposed relations are those employing aqueous autocatalytic reactions that produce propagating fronts analogous to premixed gas flame fronts [21]. The advantages of these systems for scrutinizing the various relationships for U_T are that the density and transport properties are constant across the front, S_L is not affected by heat loss, and their large Sc (≈ 500, versus ≈ 1 in gases) extends the range of U at which Ka is low enough that flamelet behavior may be exhibited (see Eq. (1)). For Ka < 5, results from two separate experiments employing Taylor-Couette and capillary wave apparatuses [22] (Fig. 4) seem to match Yakhot's [17] relation

$$U_T = exp(U^2/U_T^2) \tag{4}.$$

more closely that the others shown in Fig. 3. Initial results in a vibrating-grid flow [23] also fit Eq. (4) reasonably well.

Several points should be noted when interpreting Fig. 4. First, because of the high Sc and lack of quenching (see Section 5) of the autocatalytic fronts, data are obtained at values of U far higher than any attainable in gas combustion. Thus, some of the cited models, which employ empirical constants based on gas combustion experiments, might have been expected to perform poorly in the high-U regime. Second, it is not practical to obtain overlap between gaseous and aqueous experiments. This is because in gas combustion flamelet behavior occurs only at Ka ≤ 0.15 [8], which indicates U ≤ 10 for practical values of Re_L (see Eq. 1). However, to obtain the same U with the aqueous fronts (with much higher Sc) requires values of Re_L so low that the flow is not turbulent. Third, the experimental results may indicate some general validity of Eq. (4) because they include three types of flows in which u' and S_T were measured in entirely different ways. On the surface, Eq. (4) is appealing because it is free of adjustable parameters and is based on a solution of the G-equation, whereas the others shown in Fig. 3 (except for Sivashinsky [18]) are not. However, Yakhot's result is based on an approximate, not exact, solution obtained by applying a renormalized perturbation expansion procedure to Eq. (3). This solution

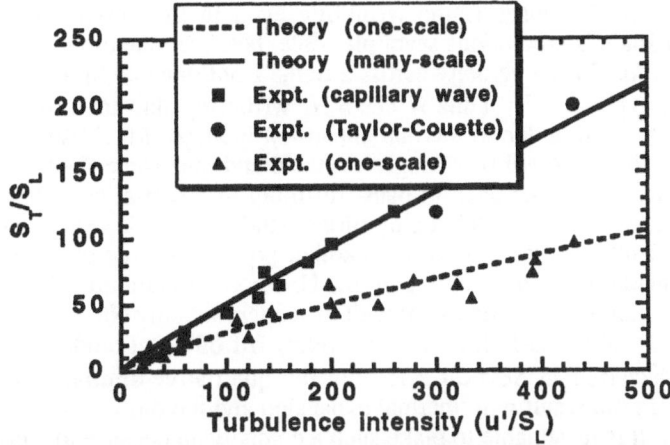

Figure 4. Measured effect of velocity disturbance intensity (u'/S_L) on the front propagation velocity (S_T/S_L) for two multiple-scale flows [22] and a one-scale flow [24] along with corresponding theoretical estimates [17, 24].

has been questioned by some, for example Pocheau [25], who proposes that for scale-invariant regimes the form

$$U_T = (1 + \beta U^\alpha)^{1/\alpha}; \quad \alpha, \beta \text{ constants} \tag{5}$$

must be satisfied. Note that, for $U \gg 1$, Eq. (4) can be written as $U_T \approx U/\sqrt{\ln(U)}$, whereas (5) becomes $U_T \approx \beta^{1/\alpha}U$, the fundamental difference being a weak $\sqrt{\ln(U)}$ deviation from linearity, sometimes called "bending" in the literature. Most other models also predict a linear relation at high U (see Fig. 3). Even as far back as 1943 Shchelkin [26] proposed a relation identical to (5) with $\alpha = 2$. However, physical or numerical experiments aimed at distinguishing between a linear relation and $\sqrt{\ln(U)}$ bending would need to be extremely accurate and be performed over a very wide range of U - a challenging proposition, even for the exotic experiments employing aqueous autocatalytic chemical reactions mentioned above. Indeed, the experimental data shown in Fig. 4 are not sufficiently accurate to delineate between linear and $\sqrt{\ln(U)}$ bending behavior.

2.2 Effect of thermal expansion

Another open issue is the effect of thermal expansion on U_T. As discussed above, some models (e.g. Bray [15]) predict thermal expansion increases U_T at constant U while others (e.g. Anand and Pope [16]) predict the opposite. For the special case $U \ll 1$, Aldredge and Williams [27] predict a strong increase in U_T with increasing density ratio. In a computational study, Ashurst and Barr [28] have also found that U_T is moderately larger with thermal expansion than at constant density. In a semi-analytical study, Cambray and Joulin [29] (Fig. 5) predict thermal expansion increases U_T, but the increase is less at higher U. (It should be noted that Cambray and Joulin do not employ a turbulent flow model, but rather an oscillating flow containing only 1 or 4 independent modes.) Furthermore, the Cambray-Joulin model does not predict $U_T \to 1$ as $U \to 0$, because thermal expansion effects, i.e. the Darrieus-Landau instability (see [6]), causes flame wrinkling even in the absence of turbulence. Cambray and Joulin's predictions seem plausible, because, due to mass conservation, the increase in local flow velocity across a flame front must be S_L multiplied by the density ratio, typically 7. Thus if U >> 7, wrinkling due to turbulence should outweigh any wrinkling due to thermal expansion induced flow. Even at smaller U, i.e. 1 < U < 7, the front will be strongly wrinkled and thus some flame elements will be oriented face-to-face or back-to-back, resulting in a partial cancellation of their induced pressure fields. If validated by further studies, this discussion and Cambray and Joulin's results would suggest that another possible source of "bending" is the thermal-expansion induced wrinkling at low U, whose effect diminishes at higher U. This would be qualitatively consistent with the experimentally observed behavior of a sharp rise in U_T at low U followed by a leveling off of U_T at moderate U shown in Fig. 1 and discussed in Section 2.1; note the qualitative similarity between the Cambray and Joulin result with thermal expansion and the dashed curves of constant Re_L in Fig. 1. It is reasonable to make such a comparison because the dashed curves correspond to constant Re_L, so that the properties of the flow field ahead of the front are constant along these curves; this is consistent with Cambray and Joulin's model.

The constant-density autocatalytic chemically reacting fronts would make an ideal candidate for comparison with gaseous flame fronts to study thermal expansion effects. However, as discussed in section 2.1, there is essentially no overlap between the values of U attainable with the two systems, so that direct comparison is not possible. However, if it is supposed that Eq. (4) predict U_T for small U as well as large U in the constant-density case, then the results of [22] would support the assertion that thermal expansion increases U_T at constant U, since Eq. (4) lies close to the lowest values of U_T for the variable-density results shown in Fig. 1.

2.3 Flame stretch effects

The models shown in Fig. 3 presume constant S_L for use in Eq. (2). Some models [15, 30] additionally consider the effect of curvature or strain ("flame stretch") on the local S_L and its modification of S_T. Since for a wrinkled flame that is statistically planar, the *mean* curvature is zero, the effects of curvature on positively and

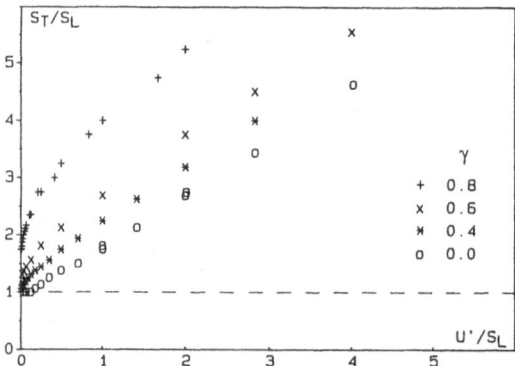

Figure 5. Calculated effect of flow disturbance intensity on turbulent burning velocity for a steady one-scale flow with varying degrees of thermal expansion [29]; $\gamma \equiv 1 - \rho_P/\rho_R$.

negatively curved flame elements might be expected to cancel; DNS results [31] support this expectation. The mean strain along the flame front is not necessarily zero because the flame front may preferentially align itself in a particular way with respect to the local principal axes of strain. DNS results show that as a result, the mean strain is generally found to be biased toward positive values [31]. Since, except for weak strain in mixtures with low Lewis numbers, the effect of strain is to decrease S_L [32], turbulent combustion models that incorporate flame stretch effects predict increasing flame stretch (i.e. increasing Ka) decreases S_T. Additionally, according to this mechanism one would expect that at fixed U, U_T would be higher in mixtures with low Le, an expectation that is confirmed experimentally [33]. Thus, the effects of flame stretch on S_T seems to be one case where theory is consistent with experimental findings (Fig. 1). Moreover, this decrease in the mean or effect S_L at high Ka due to the flame stretch provides yet another possible mechanism for the "bending" seen experimentally.

3 The effect of the kinetic energy spectrum on turbulent burning velocity

Another unresolved issue concerning the turbulent burning velocity, and one of practical importance, is for a given turbulent kinetic energy (or u'), what distribution of this energy in wavenumber space would produce the largest S_T? The energy spectrum is established by Re_L and the type of flow (shear flow, jet, channel flow, etc.) Renormalization-group theories [17, 18] predict no effect of Re_L on U_T if the spectrum is sufficiently broad that the renormalization procedure is valid. Fractal models [19] presuming a Kolmogorov or other spectrum predict an important effect of Re_L. Theories employing probability-density functions [16] and second-moment closure schemes [15] do not explicitly predict Re_L effects but assumptions about the spectrum are built into the empirical constants employed in these models. The experimental data in Fig. 1 show that at fixed U, U_T increases, roughly according to $U_T \sim Re_L^{0.15}$ at low Ka. Of course, higher Re_L yields a broader range of scales, which suggests that a broader range of scales increases U_T at fixed U.

Some additional insight into the role of the velocity spectrum can be obtained from a comparison of the experimental results of Shy *et al* [24] and Haslam *et al* [22]. Both studies employed liquid-phase autocatalytic chemical reactions that produce constant-density propagating fronts and both studies employed Taylor-Couette flow in the annulus between two rotating concentric cylinders. In [24], only the inner cylinder was rotated, in which case the flow is characterized primarily by single-scale toroidal vortex pairs that fill the annulus [34]. In this case the measured values (Fig. 4) of U_T were considerably lower than the predictions of Eq. (4). In [22], both cylinders were rotated in such as way as to produce a "featureless turbulence" [34]. In this case, Fig. 4 shows that the measured values of U_T are close to the predictions of Eq. 4. Additional experiments in a capillary-wave apparatus that produces a turbulent-like flow yield similar results (Fig. 4). These data also suggest that a broader range of scales increases U_T at fixed U. Furthermore, the data suggest that for a sufficiently broad range of scales U_T may be independent of the details of the energy spectrum.

The difference between the two types of Taylor-Couette flows may be that pockets of reactants were consistently observed in the single-scale flow studied by Shy *et al*

[24] but infrequently in the broad-scale flows studied by Haslam *et al* [22]. Thus it seems that when too much kinetic energy is concentrated in a narrow range of scales, islands of reactants are formed, which decreases amount of flame surface area that is growing exponentially with time and thereby decreases U_T. (It is clear on simple physical principles that islands of products cannot form so long as Huygens propagation applies. To obtain an island of product, two elements of flame front that share a region of products must come arbitrarily close to one another and then merge, but if they share a product region and come arbitrarily close to one another, they must propagate away from each other, since propagation is from products toward reactants, and therefore they cannot merge.)

The following physical interpretation of the influence of island formation is proposed. Batchelor [35] showed that the length a material line or the area of a material surface in a turbulent flow increases exponentially with time. If this result can be extrapolated from a material line or surface ($S_L = 0$) to a propagating front ($S_L \neq 0$), then it suggests that the A_T will grow exponentially with time. However, if islands of reactants are formed, the amount of area available for this exponential growth will decrease, and thus the propagation rate will be less than if no islands are formed. Evidence of the validity of the concept of exponential growth can be seen in Kerstein [36], who used this hypothesis in an alternative derivation of Eq. 4. Evidence of the validity of concept of loss of area due to island formation can be seen in the results of Shy *et al* [24], who used the exponential growth hypothesis to derive a relationship for U_T in a array of single-scale vortices. They presumed the front started as a flat front on one side of the vortex and grew only until the front had consumed the vortex, at which time the front was restored to its original flat state. Hence, this model essentially presumes a loss of front area due to island formation. Their result

$$U_T \approx exp(U/U_T) \tag{6}$$

agrees well with their experiments (Fig. 4) and numerical simulations [37, 38]. Furthermore, this concept of the role of island formation on U_T is supported by the analysis of Joulin and Sivashinsky [39]. On the other hand, a recent study by Aldredge [40] in a one-scale, two dimensional array of vortices produced a U_T comparable to that predicted by Eq. 4, rather than Eq. 6 as would be expected based on this discussion.

The experiments of Haslam *et al* [22] and Shy *et al* [24] suggest that Eqs. 2 and 3 may constitute asymptotic limits for the rate of front propagation in flows with very broad and very narrow ranges of flow scales, respectively. However, it has not been established what criteria delineates narrow from broad ranges of flow scales and how this might depend on U. Based on the discussion in the previous paragraph, one might hypothesize that "broad" construes broad enough that islands do not form. Recently Ashurst [41] has found that island formation in a single-scale flow begins at $u'/S_L \approx C$, where C is a constant whose value is about 1.8. One might then hypothesize that for the range of flow scales to be considered "broad" (i.e. to avoid islands), the portion of u' contributed by scales between size λ and $A\lambda$ (where A is another constant) must be less than $CS_T(\lambda)$, where $S_T(\lambda)$ is the turbulent burning velocity due to scales between size 0 and λ. If we further assume $S_T \approx u'$ in a broad-scale flow, as many of the aforementioned theories predict, and a velocity spectrum having $u'(\lambda) = u'(\lambda/L_I)^{[-(\alpha+1)/2]}$ ($\alpha = -5/3$ corresponds to a Kolmogorov spectrum) then the criterion for absence of islands is given by

$$u'(A\lambda) - u'(\lambda) < C\, S_T(\lambda)$$

$$\Rightarrow\; u'(A\lambda/L_I)^{[-(\alpha+1)/2]} - u'(\lambda/L_I)^{[-(\alpha+1)/2]} < C(\lambda/L_I)^{[-(\alpha+1)/2]}$$

$$\Rightarrow\; \alpha > -\left(\frac{2\ln(C+1)}{\ln(A)} + 1\right) \tag{7}.$$

Alternatively, Eq. (7) stated as the range of maximum (L_I) to minimum (λ_{min}) flow scales is given by $L_I/\lambda_{min} \approx (U/C)^{[-2/(\alpha+1)]}$. For representative values $A = 2$ and $C = 1.8$, Eq. (7) predicts $\alpha > -4$ or $L_I/\lambda_{min} > 1.5\, U^{2/3}$. Thus, it seems that the spectrum may be much "steeper" than a Kolmogorov spectrum without islands of reactants becoming a significant feature. This might explain why Manzaras et al [42] did not find islands in their turbulent IC engine flames, even at $U > 20$. Of course, because of intermittency, some islands could be observed for any spectrum. Also, there is considerable uncertainty about the critical value of α, as Wu and Driscoll [38] found that $U > 7$ to 8 was required for an isolated vortex pair to form islands, which would imply the criterion $\alpha > -7.2$ or $L_I/\lambda_{min} > 0.54\, U^{0.31}$.

As an alternative to this description of the role of the energy spectrum, Ashurst [41] proposes that some flows field may be modeled as having regions of intense vorticity interspersed with "quiet zones" of much lesser vorticity. By successive cascading of a semi-empirical relation over several independent scales, a relationship similar to Eq. (6) was obtained.

4 Distributed combustion

In the distributed combustion regime, Damköhler [12] proposed that the mean chemical reaction rate (ω_T) might be the same as in a laminar flame (ω_L), so by analogy with laminar flames, where $S_L \sim (D \cdot \omega)^{1/2}$,

$$U_T \equiv \frac{S_T}{S_L} = \sqrt{\frac{D_T}{D_L}} = \sqrt{\frac{\nu_T}{\nu_L}}\sqrt{\frac{Sc_L}{Sc_T}} \tag{8},$$

where the subscripts L and T denote laminar and turbulent conditions, respectively. Using the relations from turbulence theory [43] that $\nu_T/\nu_L \approx 0.061\, Re_L$ and $Sc_T \rightarrow 0.72$ as $Re_L \rightarrow \infty$, along with a typical $Sc_L \approx 0.7$ for gases, we obtain the prediction

$$U_T \approx 0.25\,\sqrt{Re_L} \quad (Ka \gg 1, Re_L \gg 1) \tag{9}$$

Note that U_T is independent of U at fixed Re_L, which is qualitatively consistent with the experimental data shown in Fig. 1 at high Ka (near extinction).

The predictions of Eq. (9) are plotted in Fig. 6. Figure 6 also shows that the experimental data, which suggest the relation $U_T \approx 5.5\, Re_L^{0.13}$, compare poorly with Eq. 9 except at the highest relevant values of Re_L. This lack of agreement is not surprising because Damköhler's hypothesis is probably too simplistic for gas combustion. This is a consequence of the strong sensitivity of ω to temperature for most gaseous flames. This causes small changes in temperature due to turbulent

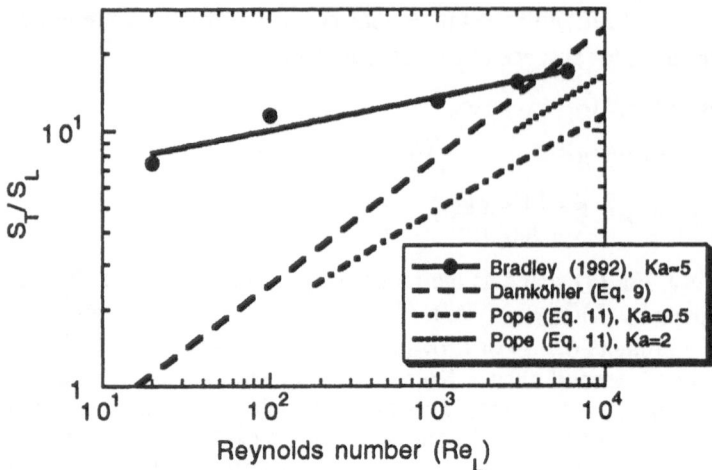

Figure 6. Comparison of theoretical predictions of turbulent burning velocity in the distributed combustion regime to experimental data taken from Fig. 1 at Ka ≈ 5. For plots of Eq. 11, only values of Re_L satisfying Eq. 12 are shown.

fluctuations to result in highly nonlinear changes in ω. This sensitivity is quantified by the Zeldovich number (Ze), defined as

$$Ze \equiv \frac{T_P - T_R}{T_P} \left(\frac{\partial \ln \omega}{\partial \ln T} \right)_{T=T_P} \tag{10},$$

where T indicates temperature and the subscripts P and R denote products and reactants, respectively. From this definition it is clear that for T close to T_P, $\omega \sim T^{Ze}$. Since typically Ze ≈ 10 for gaseous flames because of the turbulence-induced temperature fluctuations, it is then unreasonable to expect $\omega_T \approx \omega_L$, and thus the applicability of Eq. 8 or 9 is doubtful. In this context, Williams writes, "...Although the models [such as Eq. 8] cannot be right, the complexity of the problems makes it difficult to ascertain the most significant ways in which they are wrong."

Of course, computational studies can in principle overcome these analytical difficulties, but computations suffer from numerical "stiffness" problems due to the high Ze. The only detailed computational study of distributed combustion known to the author for high Ze from which values of U_T were predicted is that of Pope and Anand [44]. They assumed a constant-density flame with a one-step chemical reaction having Ze ≈ 12.3. Their results can be expressed as (see [10])

$$U_T \approx 0.3 \, U \, [0.64 + \log_{10}(Re_L) - 2 \log_{10}(U)] \tag{11}$$

which, on first glance, seems quite different from either Eq. 9 or the experimental fit. However, Fig. 6 shows that for the relatively limited range of Ka that is physically meaningful for distributed combustion in gases, the difference between Eqs. (9) and (11) is not particularly large.

In addition to theoretical difficulties, problems arise when comparing Eq. 9 or 11 directly to gaseous combustion experiments because complete flame quenching is observed at Ka ≈ 6 (Fig. 1), thus the range of Ka or U where distributed combustion might be observed is very limited. For distributed combustion to apply, δ_T should be larger than L_I so that all turbulent length scales cause increased transport within the reaction zone and no scale are large enough to wrinkle this thickened front. Following Ronney and Yakhot [10], who proposed $\delta_T \approx 6 \, \alpha_T/S_T$, and using Eq. (9) to estimate S_T, the criterion for distributed combustion becomes

$$Ka \geq 0.037 \sqrt{Re_L} \tag{12}$$

Experiments by Abdel-Gayed et al [8] suggest that $Ka \geq 0.3$, independent of Re_L, there is significant flame quenching within the reaction zone, which might correspond to the onset of distributed combustion. Equation (12) suggests that the minimum Ka for distributed combustion varies from 0.37 at $Re_L = 100$ to 3.7 at $Re_L = 10,000$. Thus, while the range of applicability of Eq. (9) or (11) is uncertain, it is probably only for conditions close to quenching.

In contrast, it may more valid to compare the predictions of Eq. 8 to experiments employing the aqueous autocatalytic systems described in Section 2.1 because $Ze \approx 0.05$ for this system and because there does not appear to be the same problem with quenching via turbulence. Since $Sc_L \approx 490$ for this system, Eq. (9) becomes $U_T \approx 6.5 \sqrt{Re_L}$. A comparison of the experimental results by Haslam *et al* [22] with this expression is shown in Fig. 7. The comparison is favorable for all data at Ka > 10. Thus is would appear that the principle upon which Eq. 8 is based may be valid, but perhaps only for systems with low Ze for the reasons discussed above.

Even if U_T in the distributed combustion cannot be modeled quantitatively, it is expected that U_T will increase less rapidly with U at high Ka than at low Ka. This is because most flamelet models (Fig. 3) predict $S_T \sim u'$, whereas for distributed combustion Eq. 8 predicts $S_T \sim \sqrt{u'}$. Thus, transition from flamelet to distributed combustion is yet another possible mechanism for the "bending" seen experimentally at high U.

5 Flame quenching by turbulence

All of the issues discussed above relate primarily to *quantitative* uncertainties in modeling turbulent flames and comparing these results to experiments. However, the mechanism of flame quenching by turbulence is not fully understood even in a *qualitative* way. Some authors [2] suggest that global extinguishment results from mass extinguishment of individual flamelets. That is, if Ka is sufficiently high, the probability of local strain rates that are strong enough to extinguish the flame locally is high enough that most of the flamelet surface will be extinguished, then the entire flame will cease to exist. This hypothesis might be stated as "*flamelet quenching causes flame quenching.*"

This hypothesis is challenged here because it overlooks the possibility of chemical reaction occurring in a mode other than that of thin fronts. In this context, Williams (1985) states that, "... for open turbulent flames in which residence times are permitted to approach infinity, the turbulent reactant mixture *must* be able to find a way to burn." To illustrate this point, consider the model system (Fig. 8) of a flame propagating in an adiabatic, chemically inert duct with arbitrarily high u'. The first law of thermodynamics dictates that the mixture *must* eventually reach the adiabatic flame temperature (T_{ad}). The second law dictates that heat *must* be transferred from the burned gas to the unburned gas. Then if Ze >> 1, a characteristic of all practical fuels, then $\omega(T_P)$ >> $\omega(T_R)$, so that reaction is much faster in mixture that is preheated by the burned gas, there is no mechanism to prevent self-sustaining flame propagation into the fresh mixture. (If Ze is not large, then it is possible that homogeneous reaction rather than flame propagation would occur, but even in this case eventually all of the reactants will be consumed.) Of course, this argument does not prescribe the type of ignition source required, whether propagation should be steady or unsteady, nor whether thin fronts would exist, but it does dictate that propagation *must* occur. Indeed, this argument also applies to laminar flames. Certainly heat losses or finite residence time (e.g. in a stagnation-point flow) may lead to extinction, but empirical correlations of quenching by turbulence [2] do not include heat loss or residence time factors, and thus may be missing the physics of quenching.

As evidence for the above assertion, Poinsot *et al* [45] numerically studied flame fronts interacting with isolated vortex pairs in the presence of an artificial heat loss and found that without heat losses, the vortices only suppress flame propagation temporarily, i.e. the flame always re-ignites eventually. With heat loss, the flame front can be permanently suppressed, but only when artificial heat loss is almost sufficient to extinguish the undisturbed plane flame. Also, Giovangigli and Smooke [46] have shown that there is no intrinsic flammability limit for planar, steady, *laminar* flames - arbitrarily dilute mixtures will burn, albeit very slowly and with a very large flame thickness. On the experimental side, Shy et al [24] and Haslam *et al* [22] conducted high-Ka experiments using aqueous autocatalytic reactions and found no evidence of front extinguishment by turbulence, even at highest experimentally-attainable values of Ka ($\approx$ 2,500) - at least *400 times* larger than values of Ka that extinguish gaseous flames. In these studies, the difference in quenching behavior between gas combustion and the aqueous reactions was proposed to be due to the absence of heat losses effects in the latter because of their low Ze ($\approx$ 0.05).

This discussion motivates a test of heat losses as a possible mechanism for flame

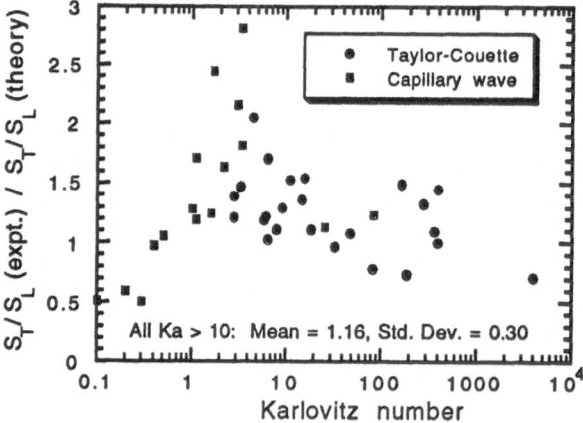

Figure 7. Comparison of experimental data on U_T in aqueous autocatalytic chemically reacting fronts [22] to theoretical prediction for distributed combustion with Sc_L = 490, i.e. $U_T = 6.5 \sqrt{Re_L}$.

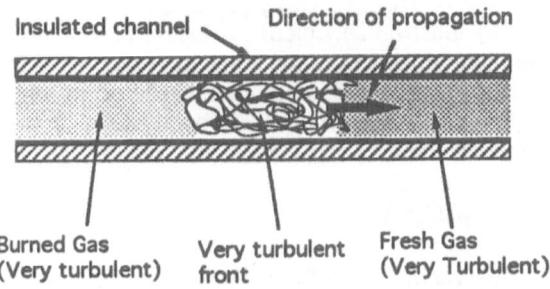

Figure 8. Schematic of idealized turbulent combustion apparatus.

extinguishment by heat losses. In laminar flames, heat losses due to conduction to walls [47] and band radiation from gases [48] are known to lead to flame quenching. In turbulent combustion, however, the situation is less clear. In many cases, e.g. fan-stirred bombs, the walls are probably too remote for conduction to walls to be significant. Radiant losses are probably not significant in the flamelet regime except in mixtures with extremely low S_L ($\approx$ 2 cm/sec) [48]. However, for distributed combustion, the front thickness may increase greatly (cf. Fig. 2). It has been proposed [10] that this thickened front is more susceptible to extinguishment by heat losses. Consider the following estimate for radiative loss. By analogy with laminar flame theory [6], extinction is expected to occur when the ratio of heat loss per unit volume within the flame front (Q_{loss}) to heat generation (Q_{gen}) is of the order 1/Ze at extinction. We estimate Q_{gen} = (heat generation rate)/(flame front volume) $\approx$ $\rho_R S_T A_L C_p \Delta T / \delta_T A_L$, where C_p the constant-pressure specific heat. The data in Fig. 1 indicate that near extinction, $U_T \approx 5.5 \, Re_L^{0.13}$. Then assuming $\delta_T \approx 6\alpha_T/S_T$ as suggested in [10],

$$\frac{Q_{loss}}{Q_{gen}} \approx 0.026 \, Re_L^{-0.76} \, Ka \, \frac{Q_{loss} L_I^2}{k\Delta T} \tag{13}.$$

where k is the thermal conductivity, evaluated at $T = T_R$. Assuming $Q_{loss}/Q_{gen} \approx$ 1/Ze at extinction and using the values for the combustion products of stoichiometric hydrocarbons in air, namely $Q_{loss} \approx 2 \times 10^6$ W/m^3 (see [49]), k = 0.026 W/mK, ΔT = 1900K and Ze $\approx$ 10, with L_I = 0.05 m as in Abdel-Gayed et al [5], we obtain prediction of Ka at quenching (Ka_q):

$$Ka_q \approx 0.38 \, Re_L^{0.76} \tag{14}.$$

Eq. 14 is plotted in Fig. 9 along with experimental data [1,2]. The model is bracketed by the experimental results for $Re_L \leq 800$. The order-of-magnitude agreement suggests that radiant heat loss could be an important factor in flame extinction by turbulence, at least at low and moderate Re_L. However, at high Re_L this mechanism probably is not operative. Spherically expanding turbulent flames in a fan-stirred bomb, ignited by an electric spark of $\approx$ 10J energy, are employed in the Leeds experiments led by Prof. Bradley and quoted above. Because of this configuration, it is possible that the ignition energy may constitute a limiting factor in flame propagation. Lewis and von Elbe [50] have shown that for laminar flames, an ignition criteria can be constructed by presuming that for successful ignition, the energy deposited (E) must be sufficient to raise a ball of hot gas whose radius is comparable to the flame thickness to the adiabatic flame temperature. In view of the possibly large increase in flame thickness that could occur when the transition from flamelet to distributed combustion occurs, by analogy with laminar flames, the criterion for ignition becomes

$$E \geq \frac{4}{3} \pi \, \delta_T^3 \rho_P C_P \Delta T \tag{15}.$$

Again employing $\delta_T \approx 6\alpha_T/S_T$ [10], Eq. (15) can be written as

$$Ka \leq 6.6 \left(\frac{E}{\rho_P C_p \, \Delta T \, L_I^3} \right)^{2/3} Re_L^{-0.24} \tag{16}$$

Using $E \approx 10$ J as in the Leeds experiments, $C_p \approx 1200$ J/kgK and $\rho_P \approx 0.17$ kg/m³ for hydrocarbon-air combustion products at 1 atm, with other parameters as above, one obtains an ignitability criterion

$$Ka_q \approx 2.3 \, Re_L^{-0.24} \tag{17}.$$

This criterion is plotted in Fig. 9. The order-of-magnitude agreement at high Re_L suggests that ignition energy could be an important factor in flame extinction by turbulence, at least at high Re_L. Also, in conjunction with the radiant loss model, it may describe the existence of two different quenching regimes proposed by Abdel-Gayed and Bradley [2] and shown in Fig. 9.

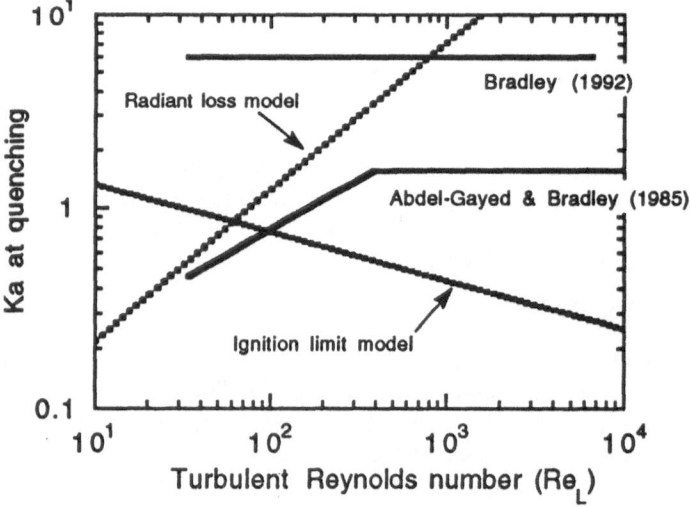

Figure 9. Comparison of empirical correlations of flame quenching by turbulence for Lewis numbers close to unity with theoretical estimates for radiant heat loss and ignition limit mechanisms.

6. Concluding remarks

6.1 Summary

Despite many years of study, there are many aspects of premixed turbulent combustion that are not well understood. For example, current models of turbulent burning velocity (S_T) do not agree with gaseous combustion experiments nor with each other. There is no generally accepted relationship between U and U_T, nor is it known whether one can exist in principle. However, recent experiments intended to mimic the common theoretical assumptions more closely than is possible in gas combustion systems may provide some preliminary indication that a universal relation exists, which, perhaps coincidentally, is reasonably well described by Eq. 4.

Experimental results generally show that at high turbulence intensity, the turbulent burning velocity increases slower than linearly, the so-called "bending" effect. It is possible that some bending could be a purely kinematical effect within the Huygens propagation regime without a topological change in the flame front due to "island" formation (Eq. 4), but the effect is too weak to explain the bulk of the experimentally observed bending. Other factors which are likely to contribute to bending include the formation of islands, Karlovitz number or flame stretch effects, thermal expansion, and transition from flamelet to distributed combustion, though it is unclear which of these factors is most important in a given experiment.

The properties of turbulent flames in the distributed combustion regime are not well understood, particularly for systems with large Zeldovich number (Ze), though some qualitative insight may be obtained by applying Damköhler's [12] original hypothesis concerning this regime.

Flame quenching by turbulence may be a consequence of the transition from flamelet-like to distributed like combustion at high Ka leads to a thickening of the active chemical reaction zone. When this thickening occurs, the flame is more susceptible to failure due to radiant heat losses, insufficient ignition energy, and perhaps other factors. Very simplistic approximate models can provide some guidance as to what effects are likely to be important in a particular range of Reynolds numbers.

6.2 Future directions

To clarify the effect of U on U_T in the Huygens propagation regime, direct numerical simulations of the G-equation in Navier-Stokes turbulence or artificial flows that produce flow statistics similar to Navier-Stokes turbulence [51] are warranted. In addition to determination of U_T, the existence of bending, and the effect of the velocity spectrum and island formation on U_T could be established in this way for comparison to theoretical models. While some studies of this type have been performed, e.g. Collins [52], they have been limited to low U and have not specifically addressed the issues mentioned here other than the effect of U on U_T. An initial study should employ the constant-density model because of the simplicity afforded by the fact that there is no coupling between the front propagation on the flow field in this case. A follow-on study should consider thermal expansion effects, which is of course far more difficult because of this coupling. In either case, the problem becomes increasingly more difficult at higher U as discussed in section 2.1.

It would be interesting to devise an experimental test of the Kerstein and Ashurst [14] prediction that at $U \ll 1$, $U_T - 1 \sim U^{4/3}$. One possible approach would be using a Taylor-Couette apparatus operating in the "featureless turbulence" regime as in Haslam *et al* [22] but with gaseous premixed flames with large S_L to obtain low U. This would enable observation of front propagation over many integral scales in the direction of propagation, limited only by the length of the cylinders, as Kerstein and Ashurst's analysis indicates is necessary. Also, in the Taylor-Couette flow the mean velocity is zero in the direction of propagation. This feature is critical so that one can distinguish between front propagation relative to the mean flow and apparent front propagation due to convection by the mean flow, which by definition does not contribute to S_T.

More detailed modeling of flame propagation in the distributed combustion regime is needed to clarify the conditions at which distributed combustion may occur, the effect on bending, and the role of distribution combustion in flame quenching. In addition to determining the effects of U and Re_L on U_T in this regime, a suitable model of the flame thickness δ_T is needed because of its influence on flame quenching (see Section 5). DNS studies employing a one-step chemical reaction with high Ze are needed. Some studies of this type have been conducted, e.g. Rutland *et al* [53], but these studies have focused on the low-Ka flamelet regimes rather than the high-Ka distributed combustion regime. Also, the addition of a volumetric heat-loss term could be used to test the possibility of radiative loss as a mechanism of quenching.

To test experimentally the revised mechanisms of flame quenching by turbulence proposed here, the scalings proposed here and by Abdel-Gayed and Bradley [2] can be compared. In particular, it may be useful to study the effects of pressure (P) on Ka_q, since very few of these data are reported on the literature. Note that for lower values of Re_L, Eq. 14 predicts $Ka_q \sim P^{-1}$, and at higher Re_L, Eq. 17 predicts $Ka_q \sim P^{-2/3}$, whereas the published shown in Fig. 9 do not predict such pressure effects. Also, if radiative loss is important, a means to modify the radiative loss should cause a change in the extinction boundaries. In this context, the effect of the addition of small quantities of inert, solid radiating particles may be useful. Such studies have provide useful for studying the effect of radiative losses on laminar flames [54]. Also, the possible influence of the ignition energy on quenching conditions can be tested by employing ignition sources of variable energy; according to Eq. 16, $Ka_q \sim E^{2/3}$.

Acknowledgments

The author's work in the area of premixed turbulent combustion has been supported by the National Science Foundation Presidential Young Investigator Program under Grant CBT 86-56228, the NASA Lewis Research Center under Grants NAG3-1242 and NAG3-1523, and the Gas Research Institute under Grants 5088-260-1688 and 5092-260-2486. The author thanks Drs. W. T. Ashurst and M. S. Wu for helpful comments on this manuscript.

References

1. Bradley, D. *Twenty-Fourth Symposium (International) on Combustion*, Combustion Institute, 1992, p. 247.
2. Abdel-Gayed, R. G. and Bradley, D. *Combust. Flame* 62, 61 (1985).

3. Bracco, F. V., Combust. Sci. Tech. 58, 209 (1988).
4. Correa, S., *Combust. Sci. Tech.* 87, 327 (1992).
5. Abdel-Gayed, R. G., Bradley, D. and Lawes, M., *Proc. Roy. Soc. (London)* A414, 389 (1987).
6. Williams, F. A.,*Combustion Theory*, 2nd ed., Benjamin-Cummins, 1985.
7. Peters, N., *Twenty-First Symposium (International) on Combustion*, Combustion Institute, 1986, p. 1231.
8. Abdel-Gayed, R. G., Bradley, D. and Lung, F. K.-K. *Combust. Flame* 76, 213 (1989).
9. Yoshida, A., *Twenty Second Symposium (International) on Combustion*, Combustion Institute, 1988, p. 1471.
10. Ronney, P. D. and Yakhot, V., *Combust. Sci. Tech.* 86, 31 (1992).
11. Xin, J. X., *Arch. Rational Mech. and Anal.* 121, 205, (1992).
12. Damköhler, G., *Z. Elektrochem. angew. phys. Chem* 46, 601 (1940).
13. Clavin, P. and Williams, F. A., *J. Fluid Mech.* 90, 589 (1979).
14. Kerstein, A. R. and Ashurst, W. T., *Phys. Rev. Lett.* 68, 934 (1992)
15. Bray, K. N. C., *Proc. Roy. Soc.(London)* A431, 315 (1990).
16. Anand, M. S. and Pope, S. B., *Combust. Flame* 67, 127 (1987).
17. Yakhot, V., *Combust. Sci. Tech.* 60, 191 (1988).
18. Sivashinsky, G. I., in: *Dissipative Structures in Transport Processes and Combustion* (D. Meinköhn, ed.), Springer Series in Synergetics, Vol. 48, Springer-Verlag, Berlin, 1990, p. 30.
19. Gouldin, F. C., *Combust. Flame* 68, 249 (1987).
20. Kerstein, A., Ashurst, W. T. and Williams, F. A., *Phys. Rev. A.*, 37, 2728 (1988).
21. Hanna, A., Saul, A. and Showalter, K., *J. Am. Chem. Soc.* 104, 3838 (1982).
22. Haslam, B. G., Ronney, P. D. and Rhys, N. A., manuscript in preparation; see also Haslam, B. G. and Ronney, P. D. "Experimental Simulation of Premixed Turbulent Combustion in Flows with a Broad Range of Temporal and Spatial Scales," Fall Technical Meeting, Western States Section, Combustion Institute, Oct. 12-13, 1992, Berkeley, CA.
23. Shy, S. S., Jang, R. H. and Tang, C. Y. Manuscript submitted to *Combustion and Flame*.
24. Shy, S. S., Ronney, P. D., Buckley, S. G. and Yakhot, V., *Twenty-Fourth Symposium (International) on Combustion*, Combustion Institute, Pittsburgh, 1992, p. 543.
25. Pocheau, A., *Phys. Rev. E* 49, 1109 (1994).
26. Shchelkin, K. I., *Zhur. Tekhn. Fiz.* 13, 520 (1943).
27. Aldredge, R. C. and Williams, F. A., *J. Fluid Mech.* 228, 487 (1991).
28. Ashurst, W. T. and Barr, P. K., *Combust. Sci. Tech.* 34, 227 (1983).
29. Cambray, P. and Joulin, G., *Twenty-Second Symposium (International) on Combustion*, Combustion Institute, 1992, p. 61.
30. Peters, N., *J. Fluid Mech.* 242, 611 (1992).
31. Rutland, C. J. and Trouve, A., *Combust. Flame* 94, 41 (1993).
32. Buckmaster, J. D. and Mikolaitis, D., *Combust. Flame* 47, 191 (1982).
33. Wu, M. S., Kwon, S., Driscoll, J. F. and Faeth, G. M., *Combust. Sci. Tech.* 78, 69 (1991).
34. Andereck, C. D., Liu, S. S. and Swinney, H. L., *J. Fluid Mech.* 164, 155 (1986).

35. Batchelor, G., *Proc. Roy. Soc. (London)* A213, 349 (1952).
36. Kerstein, A. R., *Combust. Sci. Tech.* 60, 163 (1988).
37. Zhu, J. Y. and Ronney, P. D., to appear in *Combustion Science and Technology* (1994).
38. Wu, M. S. and Driscoll, J. F., *Combust. Flame* 91, 310 (1991).
39. Joulin, G. and Sivashinsky, G. I., *Combust. Sci. Tech.* 77, 329 (1991).
40. Aldredge, R. C., manuscript submitted to *Combustion and Flame.*
41. Ashurst, W. T., *Combust. Sci. Tech.* 92, 87 (1993).
42. Mantazaras, J., Felton, P. G. and Bracco, F. V., *SAE Paper No. 881635*, 1988.
43. Yakhot, V. and Orzag, S. A., *Phys. Rev. Lett.* 57, 1722 (1986).
44. Pope, S. B. and Anand, M. S., *Twentieth Symposium (International) on Combustion,* Combustion Institute, 1984, p. 403.
45. Poinsot, T., Veyante, D. and Candel, S., *Twenty-Third Symposium (International) on Combustion*, Combustion Institute, 1990, p. 613.
46. Giovangigli, V. and Smooke, M., *Combust. Sci. Tech.* 87, 241 (1992).
47. Jarosinsky, J., *Combust. Flame* 50, 167 (1983).
48. Ronney, P. D., *Twenty Second Symposium (International) on Combustion,* Combustion Institute, 1988, p. 1615.
49. Hubbard, G. L. and Tien, C. L., *J. Heat Trans.* 100, 235 (1978).
50. Lewis, B. and von Elbe, G., *Combustion, Flames and Explosions of Gases*, 3rd ed., Academic Press, 1987.
51. Kwon, S., Wu, M. S., Driscoll, J. F. and Faeth, G. M., *Combust. Flame* 88, 221, (1992).
52. Collins, L. R., manuscript submitted to *Computers and Fluids.*
53. Rutland, C. J., Ferziger, J. H. and El Tahry, S. H., *Twenty-Third Symposium (International) on Combustion*, Combustion Institute, 1990, p. 621.
54. Abbud-Madrid, A. and Ronney, P. D., *AIAA J.* 31, 2179, (1993).

The Scalar-Field Front Propagation Equation and its Applications

R. C. Aldredge

Mechanical & Aeronautical Engineering
University of California, Davis
Davis, CA 95616-5294

keywords: eikonal equation, flame stretch, front-propagation equation,
          premixed combustion, turbulent flame propagation

1. INTRODUCTION

In many practical situations combustion occurs locally in a thin reaction zone which propagates through a reactive mixture of fuel and oxidizer. Examples of these situations can be found in spark-ignition engines, turbojets, ramjets, afterburners and rockets [1], although these devices may exhibit nonlocal and nonpropagating combustion phenomena as well. Conditions which promote the regime of thin-flame propagation in premixed combustion are large chemical activation energies of the reactive mixture and large temporal and spatial scales of flow-field variation. When the flame thickness and the characteristic time for chemical reaction within the flame are much smaller than the length scales and time scales, respectively, characterizing fluctuations in the flow field, the flame may be modeled as an interface separating cold reactants from hot products. The interface propagates into the premixed reactants normal to itself locally at a speed S_n, a function of the local reaction rate of the flame which is influenced by the local flame curvature, reactant concentration and flow-field strain rate. As the flame sheet propagates it is wrinkled and possibly torn by fluctuations in the flow field. Fig. 1 describes schematically five ideal regimes of flame propagation. In each case a combustion zone, defined by the separation of pure reactants on the far left from pure products on the far right, propagates to the left at some, possibly time-dependent, speed U_T. In case (a) the flow field fluctuations are such that a weakly wrinkled flame sheet is established, the location of which may be described by a single-valued function of the transverse coordinates. In case (b) the flame is wrinkled to the extent that it folds back upon itself and forms pockets of unburned reactants. Local extinction occurs in case (c), caused by large strain rate, curvature or concentration stratification, and results in a noncontiguous flame front followed by multiple flame sheets. The distributed-reaction regime is illustrated in case (d), where combustion occurs in a broad zone, significantly larger than the characteristic thickness of an unstrained planar flame propagating through a uniform flow of the same reactive mixture. A combination of the canonical regimes illustrated in Figs. 1a-d might be practically relevant. For example, a turbulent flow of reactants having a wide range of spatial and temporal scales might cause distributed combustion and flame broadening by action of the small-scale fluctuations while causing flame wrinkling by action of the large-scale fluctuations. This situation is illustrated in Fig. 1e.

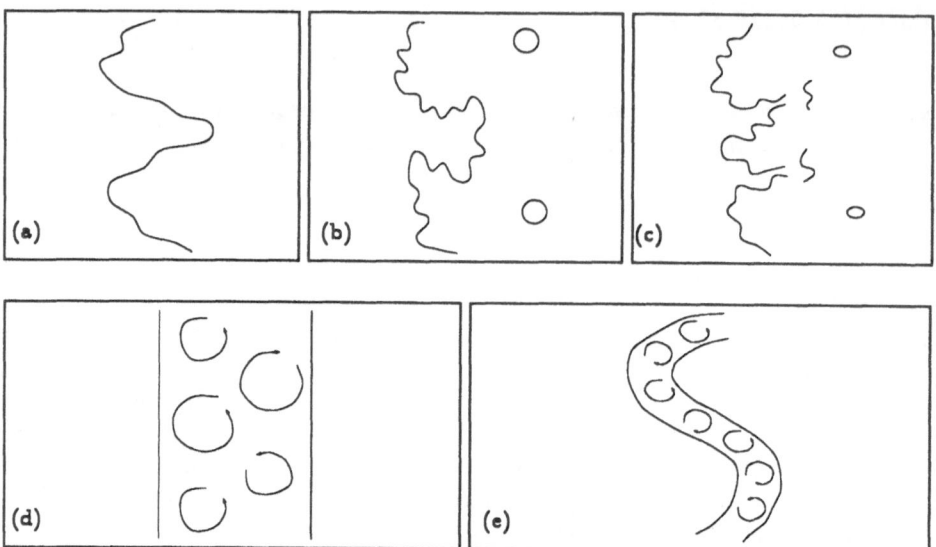

Figure 1: Regimes of turbulent flame propagation

In practical environments where thin-flame propagation is relevant the fuel consumption rate is governed by the evolution of flame structure and topography and the dynamics of the flow-field [2]. An approach toward accurate prediction of mean combustion rates in such cases is the calculation of the average mass flux of reactants across the total flame surface using direct numerical simulation. This is achieved by integration of the equations governing conservation of mass, momentum, energy and chemical-species throughout the computational domain of interest. Computational efficiency is generally improved by grid refinement at the flame surface, where temperature and species concentrations change most appreciably, and by solving only the mass and momentum equations in regions away from the flame surface. The cost of this worthwhile simplification is the necessity of tracking the flame surface so that the grid can be appropriately refined at each time step. The problem is then reduced to the solution of the full set of conservation equations on a refined, small subset of the total computational domain, the mass and momentum equations on a coarser grid, and a flame surface evolution equation. Front-propagation equations (FPE's), which describe the evolution of self-propagating interfaces, will be discussed further in the following sections.

2. WEAKLY WRINKLED FLAME REGIME

2.1. The Flame Surface Evolution Equation

For the regime of weakly wrinkled flame propagation the location of the flame surface may be written as a single-valued function of two space coordinates and time. When the flame is quasiplanar, having its local normal vector everywhere along the direction of mean flame propagation, on average, the two independent space variables are the transverse coordinates, perpendicular to the mean flame direction (c.f. Fig 2a). Weakly wrinkled spherical flames have their local normal vectors pointing everywhere along the radial direction, on average, so that the flame location is defined locally by the two orthogonal angular coordinates and time (c.f. Fig 2b).

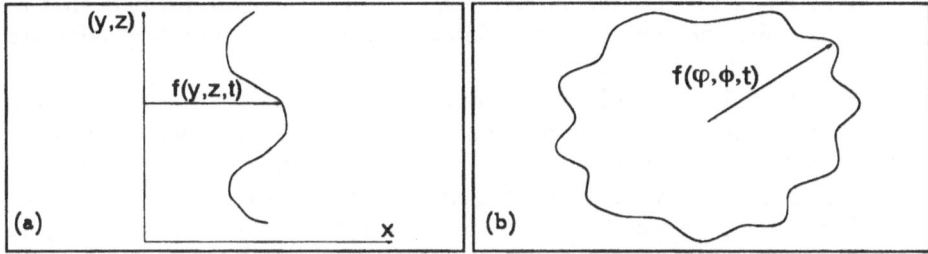

Figure 2: Weakly wrinkled flame configurations

The local response of the flame surface to fluctuations of the flow field may not in general be completely specified without numerical integration across the flame thickness, since this response must be consistent with the local structure of the flame. However, for the regime of weakly wrinkled flame propagation an FPE for the local flame position $f(y, z, t)$ may be derived from first principles by analysis of the flame structure, under assumptions of large activation energy, weak flame front curvature and weak flow-field strain [3-7]. Since the flow-field fluctuations are typically of small amplitude for this regime only the linear version of the equation is introduced here, namely,

$$\frac{\partial f}{\partial t} = u \mid_{x=f} + L \left(\nabla \cdot v \mid_{x=f} + S_0 \nabla^2 f \right). \tag{1}$$

In this equation (u, v) represents the three-dimensional flow-field velocity, to be evaluated on the reactant side of the flame surface, S_0 is the laminar burning velocity of a planar flame in a uniform flow of the same mixture, and L is the Markstein parameter associated with the influence of flame structure. It was Markstein [8] who suggested that the departure from S_0 of the local rate of normal flame propagation S_n be proportional to the flame surface curvature. This phenomenological model was later extended by Eckhaus [9] who accounted for the additional effect of flow-field strain. Equation (1) may be rewritten in terms of S_n, that is,

$$
\left.
\begin{aligned}
\frac{\partial f}{\partial t} &= u|_{x=f} - (S_n - S_0) \\
S_n &= S_0 - L\kappa \\
\kappa &= \nabla \cdot v|_{x=f} + S_0 \nabla^2 f
\end{aligned}
\right\} . \tag{2}
$$

In Eq. (2) κ denotes the linearized total stretch of the flame surface, the sum of flame surface curvature and flow-field strain rate. As the Markstein parameter is of the order of the flame thickness, which is much smaller than the length scale characterizing appreciable variations in f and in the velocity field (a prerequisite for the weakly wrinkled flame regime), the effect of stretch is a small correction to S_0 which damps (amplifies) flame wrinkling when $L \geq 0$ ($L \leq 0$). The sign of the Markstein parameter is determined by thermodynamic and transport properties of the reactive mixture, including the chemical heat release and the Lewis numbers of the reactive species. Equation (2) therefore predicts that at the leading order the velocity of the flame surface in the laboratory reference frame equals the flow-field velocity.

It is worthwhile to note here that Eq. (1) is based on the assumption that the effect of stretch on the flame structure occurs instantaneously. When the total volumetric stretch of the flame is considered, taking into account the rate of logarithmic increase of the flame thickness in addition to surface-area stretch, an FPE describing the evolution of *slowly varying flames* may be derived [10,11]. In this model flame structure modifications occur on a large characteristic time scale of the order the inverse of a nondimensional activation energy.

With the effect of flame structure modification on the evolution of f completely specified by Eq. (1), there is no further need for integration of the energy and species equations across the flame thickness. Equation (1) and the mass and momentum conservation equations for the reactant and product regions of the flow provide a complete description of the flame-surface and flow-field evolution for the weakly wrinkled flame regime. Local jump conditions on the velocity and pressure fields across the flame surface must be specified and are provided by the same analytical analysis used in the derivation of Eq. (1). At the leading order, the linearized jump conditions are

$$
\left.
\begin{aligned}
u_+ - u_- &= S_0(R - 1) + O(\epsilon) \\
v_+ - v_- &= -S_0(R - 1)\nabla f + O(\epsilon) \\
p_+ - p_- &= -\rho_- S_0^2 (R - 1) + O(\epsilon)
\end{aligned}
\right\} , \tag{3}
$$

where $R = \rho_-/\rho_+$ is the density ratio, ϵ is the ratio of the flame thickness to a characteristic length scale of the flow, and the subscripts (-) and (+) refer to the reactant and product regions, respectively. These jump conditions express the conservation of mass and tangential velocity components across a flame surface of zero thickness. The higher-order corrections of order ϵ which account for the influence of flame structure, resulting from a finite flame thickness, may be found in references [6,12-14] and will not be reproduced here for simplicity. In particular, reference [6] provides fully nonlinear jump conditions through $O(\epsilon)$ which are valid for order-unity velocity fluctuations and flame-surface gradients, but otherwise restricted to cases where the flame surface location may be written as single valued function of the transverse coordinates and time.

2.2. Flame-Dynamics and Flow-Field Coupling

The intrinsic instability of the problem defined by the leading order of Eq. (1) and the leading-order conditions given in Eq. (3), when R is not unity, was discovered independently by Darrieus [15] and Landau [16]. The reason for this instability is that the coupling between the dynamics of the flame surface and the flow field for the unforced problem is such that modifications to the velocity field are in phase with perturbations in f, locally, resulting in unbounded growth in gradients of f and in the total flame surface area, when no stabilizing influences are present [2]. The higher-order corrections to Eqs. (1) and (3) discussed above can be stabilizing to the growth of high-wavenumber wrinkles (for example, for $L \geq 0$ and near-unity Lewis numbers [17]), however another mechanism of stabilization is needed for small-wavenumber wrinkles. For downward propagating flames in open tubes the action of gravity is stabilizing for this range of wavenumbers if S_0 is not too large [12]. Another possible mechanism for this range might be the action of high intensity, large-frequency velocity fluctuations of a stationary flow.

The linearized Navier-Stokes equations in conjunction with Eq. (1) and flow-field jump conditions valid through $O(\epsilon)$ have been solved analytically [4] in order to examine the influence of flame structure on the rate of downward propagation of an intrinsically stable premixed flame through a large-scale, low-intensity turbulent flow. The modification of the turbulence properties of the flow, resulting from the coupling between dynamics of the flame surface and the flow field, was also examined. In that study a stationary excitation turbulence, characterized by an integral scale and an autocorrelation function, specified far upstream from the flame front was considered to be initially isotropic and homogeneous. Fig. 3 presents the predicted variation of the difference $u_T \equiv U_T/S_0 - 1$ with the ratio ϵ of the flame thickness to the integral scale of the excitation turbulence. Two autocorrelation functions were considered, $g_{u,1}$ which characterizes a flow with a turbulence Reynolds number of $O(1)$ and $g_{u,2}$ which characterizes a flow having a large turbulence Reynolds number. In both cases $R = 6$, the Prantl and Lewis numbers defined in the reactant flow are both unity, and the turbulence intensity relative to S_0 is 30%.

Certain wavelengths of the flame surface wrinkles were found to be selectively amplified by the turbulent flow, causing enhanced flame wrinkling and a peak in u_T about the value ϵ^* characterizing the least-stable wrinkles. Hence, the coupling between flame surface dynamics and the flow field can result in significant modification to U_T in low-intensity flows, even for intrinsically stable premixed flames. The character of the excitation flow, however, is also of significance. Since the kinetic energy of turbulence for the flow characterized by $g_{u,2}$ is more evenly distributed over a wide range of length scales [18] selective excitation occurs to a lesser extent for this flow than for the one represented by $g_{u,1}$, where the turbulence energy is concentrated primarily about the integral scale of turbulence [18].

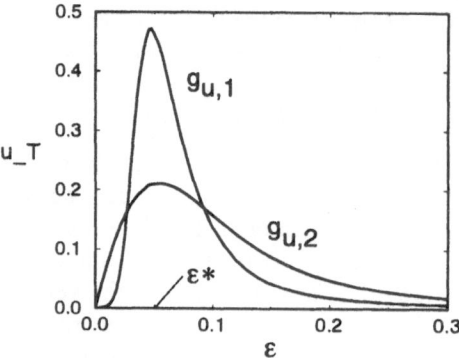

Figure 3: Turbulent burning velocity versus ratio of flame thickness to turbulence integral scale, with two different autocorrelation functions. Reprinted from reference [4].

The modification of turbulence properties by the presence of the flame is represented in Fig. 4. Figure 4a shows the variations in components of turbulence kinetic energy in both the reactant and product regions near the flame. The distance between the local instantaneous flame location and the mean flame location ($x=0$) is $f(y, z, t)$. Thus, f has an average of zero and its root-mean-square f' is a measure of the turbulent flame brush thickness. The turbulence energies are all normalized by the total energy of the isotropic excitation turbulence, so that upstream from the flame (to the left of $x=0$) the total turbulence kinetic energy approaches unity, while the energy associated with any one component of velocity fluctuation approaches a value of 1/3. The intensity of pressure fluctuations is normalized by ρS_0^2. Across the flame, the kinetic energy of transverse velocity fluctuations increases, while that of the longitudinal fluctuations remains unchanged. The flame therefore generates turbulence kinetic energy and causes anisotropy in the initially isotropic flow. However, downstream from the flame zone energy is transferred back from transverse fluctuations to longitudinal fluctuations, so that the final anisotropy in the product region exhibits higher longitudinal intensities than transverse intensities. The cause of the turbulence energy transfer downstream from the flame zone is the generation of the pressure intensity field in the hydrodynamic regions (c.f. Fig. 4a) and the correlation between this field and the flow-field strain. Fig. 4b shows variations over a larger distance in the product region only, where flame induced modifications are shown to occur over a length scale significantly shorter than the viscous-decay scale apparent in the figure.

3. LARGE-AMPLITUDE FLAME WRINKLING

3.1. The Scalar-Field Front Propagation Equation

When the fluctuations of the flow field are such that the flame surface is wrinkled to the extent depicted in Fig. 1b the formulation of the surface evolution in terms

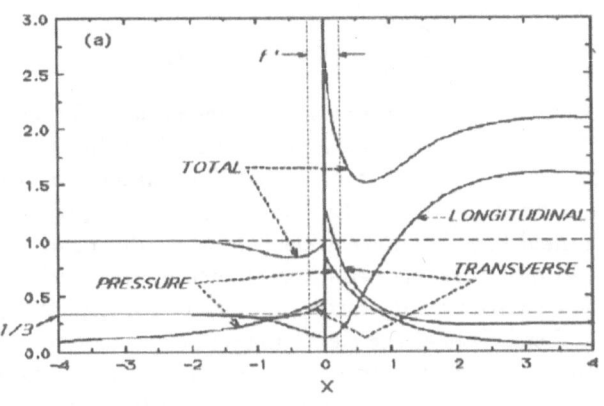

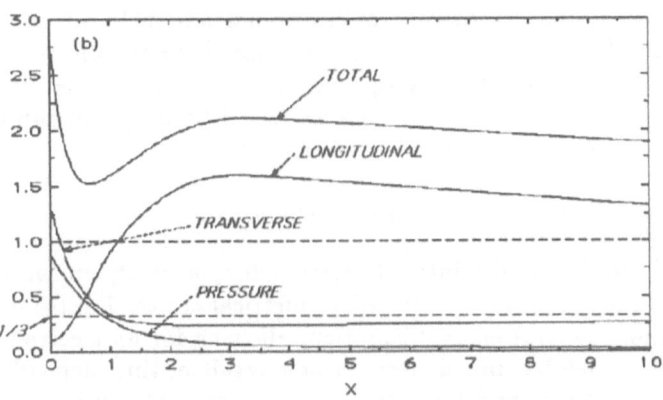

Figure 4: Turbulence kinetic energy modification by a weakly wrinkled premixed flame, through: (a) the upstream and downstream hydrodynamic zones. (b) the downstream hydrodynamic zone. Reprinted from reference [4].

of $f(y, z, t)$ is not appropriate, as this function does not remain single valued. Alternatively, a useful approach for this regime is the consideration of the flame surface as the locus of points in space defined by a constant level surface of a scalar field $G(x, y, z, t)$. On this level surface G remains constant and the local velocity of the flame surface in the laboratory reference frame, u, must satisfy the equation

$$\frac{\partial G}{\partial t} + \mathbf{u} \cdot \nabla G = 0. \tag{4}$$

The rate of propagation of the flame surface in the normal direction, S_n, is defined according to

$$\mathbf{S_n} \equiv (\mathbf{U} - \mathbf{u}) \cdot \mathbf{n}, \tag{5}$$

where U is the flow-field velocity and n is the unit normal $\nabla G / |\nabla G|$. Thus,

Eq. (4) becomes

$$\frac{\partial G}{\partial t} + \mathbf{U} \cdot \nabla G = S_n \, |\nabla G| \,. \tag{6}$$

This scalar-field evolution equation was introduced for the description of flame surface dynamics by Markstein [8], although earlier applications of this equation can be found in the field of geometrical optics, for the description of geometrical wave fronts. Within the approximations of geometrical optics, an electromagnetic field may be characterized by a single scalar function of space and time. Eq. (4) describes the evolution of such a scalar field whose constant-level surfaces define discontinuity surfaces in an electromagnetic field which propagate normal to themselves at the electromagnetic wave speed $c/\sqrt{\epsilon\mu}$, where c is the speed of light in a vacuum and ϵ and μ are the dielectric constant and magnetic permeability, respectively, of the quiescent medium of the electromagnetic field [19]. Although Hamilton [20] was the first to use scalar-field characteristic functions for the characterization of electromagnetic-field media in geometrical optics, Burns [21] later named certain similar characteristic functions that he independently considered *eikonals*, after the Greek word *ikon* meaning *image*. The equation obtained from Eq. (4) with $G = F(x, y, z) - ct$ and $\mathbf{u} \cdot \mathbf{n}$ equal to the electromagnetic wave speed $c/\sqrt{\epsilon\mu}$, namely,

$$|\nabla F|^2 = N(x, y, z)^2, \tag{7}$$

where $N \equiv \sqrt{\epsilon\mu}$ is the absolute refractive index, is widely known as the eikonal equation and is the basic equation of geometrical optics. Eq. (6) may be considered as a more general eikonal equation, allowing for an electromagnetic wave speed S_n that may be time dependent as a result of time-dependent properties (ϵ, μ) of the medium, and for a convection current $\mathbf{U}$ caused, for example, by nonstationary electric charges in the medium.

The influence of flame structure on the evolution of a flame surface, as described by Eq. (6), is accounted for by the dependence of S_n on properties of the velocity field and flame surface topography. In general, the complete set of conservation equations must be integrated across the thin flame for determination of the local value of S_n and the jump conditions to be satisfied by the hydrodynamic flow on either side of the flame. However, when the magnitude of flame stretch is weak, so that variations of properties within the flame along the direction normal to the reaction-zone surface occur over a much smaller scale than that characterizing variations along tangential directions, a linear relationship between S_n and the local flame stretch may be assumed. The validity of this weak-stretch approximation for flows with high-intensity velocity fluctuations may be questionable, however, because of the presence of large flame surface gradients which may make derivatives of fluid properties along directions tangential to the reaction zone relatively large.

The local flame-surface stretch is defined, following Klimov [22] and Williams [23], as

$$\kappa \equiv \frac{1}{\Delta} \frac{d\Delta}{dt}, \tag{8}$$

where Δ is the area of an infinitesimal local surface element. Exemplification of the right-hand side of this identity through purely kinematic arguments gives a nonlinear expression for κ in terms of the local flow-field properties and surface geometry [2,6,24],

$$\left. \begin{array}{l} \kappa = -\left(\mathbf{n} \cdot \mathbf{E} \cdot \mathbf{n} + S_n \nabla \cdot \mathbf{n}\right) \\ \mathbf{E} \equiv \frac{1}{2}\left[(\nabla \mathbf{U}) + (\nabla \mathbf{U})^{\mathrm{T}}\right] \end{array} \right\}, \tag{9}$$

where $\mathbf{E}$ is the symmetric rate-of-strain tensor of the flow field. An alternative form of this equation is obtained by elimination of the spatial velocity gradients in favor of the velocity field vector itself giving, by use of Eq. (6),

$$\left. \begin{array}{l} \kappa = -\frac{1}{|\nabla G|}\left[\nabla \cdot (S_n \nabla G) - \frac{D}{Dt}|\nabla G|\right] \\ \frac{D}{Dt} \equiv \frac{\delta}{\delta t} + \mathbf{U} \cdot \nabla \end{array} \right\}. \tag{10}$$

The weak-stretch approximation

$$S_n \equiv S_0 - L\kappa \tag{11}$$

then defines the local normal propagation speed of the flame and eliminates the need for integration through the reaction zone.

To complete the set of equations governing the evolution of the flow field and flame surface topography, jump conditions to be satisfied by the hydrodynamic flow must be specified. At the leading order these conditions, written in terms of the scalar-field normal $\mathbf{n}$, are

$$\left. \begin{array}{l} \mathbf{U}_+ - \mathbf{U}_- = S_0(R - 1)\mathbf{n} + \mathrm{O}(\epsilon) \\ p_+ - p_- = -\rho_- S_0^2(R - 1) + \mathrm{O}(\epsilon) \end{array} \right\}, \tag{12}$$

and reduce to the linear jump conditions given in Eq. (3) when the flame surface gradients are small. The complete set of governing equations, Eqs. (6,9,11,12) and the mass and momentum equations for the hydrodynamic flow on either side of the flame, are valid through order ϵ only when the higher-order corrections to Eq. (12) are specified. These may be provided analytically by integration across the flame thickness, under assumptions of large activation energies and weak stretch, but are expected to be only generalizations of the nonlinear conditions obtained by Matalon and Matkowsky [6] in terms of the single-valued variable f defined above.

3.2. Propagation in a High-Intensity Vortical Flow

Premixed-flame propagation through a large-scale, high-intensity vortical flow was studied numerically using the scalar-field front propagation equation assuming a constant normal flame propagation speed ($S_n = S_0$) [25]. The density change across the flame was neglected, resulting in continuity of the velocity and pressure fields at the flame surface (c.f. Eq. 12), so that the flow field could be decoupled from the flame surface dynamics and considered to be a known function of space and time. The parameters l, l/S_0, and S_0 were used for nondimensionalization of length, time and velocity units, where l is the length scale characterizing the two-dimensional monochromatic velocity field (u, v) given by

$$
\left.
\begin{array}{l}
u(x, y) = -2A\cos(2\pi x)\sin(2\pi y) \\
v(x, y) = 2A\sin(2\pi x)\cos(2\pi y)
\end{array}
\right\} .
\tag{13}
$$

This flow field was also considered by Ashurst and Sivashinsky [26]. In nondimensional variables and with $S_n = S_0$ Eq. (6) becomes

$$
\frac{\partial G}{\partial t} + U \cdot \nabla G = |\nabla G| .
\tag{14}
$$

This equation was solved with U given in Eq. (13) with jump periodic boundary conditions imposed,

$$
\left.
\begin{array}{l}
-0.5 \le x \le 0.5, -0.5 \le y \le 0.5 \\
G(0.5, y, t) = 1 + G(-0.5, y, t) \\
G(x, 0.5, t) = G(x, -0.5, t) \\
G(x, y, 0) = x
\end{array}
\right\} .
\tag{15}
$$

The initial condition on G corresponds to a flat flame aligned parallel to the y axis at each location x at time $t=0$. The evolution of the flame located initially at $x=0$ was tracked in time by determination of the locus of points in the x-y plane, at a given time t, at which $G(x, y, t)=0$.

Figure 5 illustrates the formation of unburnt pockets of reactant for the case $A=5$ as a result of the intense convolution of the flame surface by the vortical flow. In Fig. 6a the average speed of propagation of the flame in the $-x$ direction U_T is plotted versus the intensity A. The results of the computation with $S_n = S_0$ ($\epsilon=0$) agree well with the prediction of Yakhot [27],

$$
U_T = e^{(A/U_T)^2},
\tag{16}
$$

obtained analytically by renormalization group theory for $S_n = S_0$ and passive

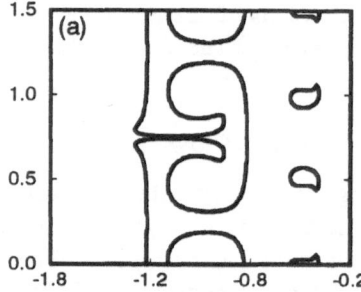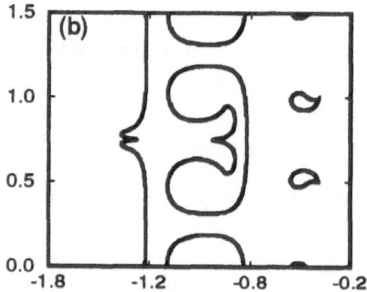

Figure 5: Wrinkled flame surface: (a) at the onset of pocket formation, $t=0.291$. (b) just after pocket formation, $t=0.300$. $A=5$ for both cases. Reprinted from reference [25].

flame propagation ($R=1$) in a homogeneous large-Reynolds-number turbulence.

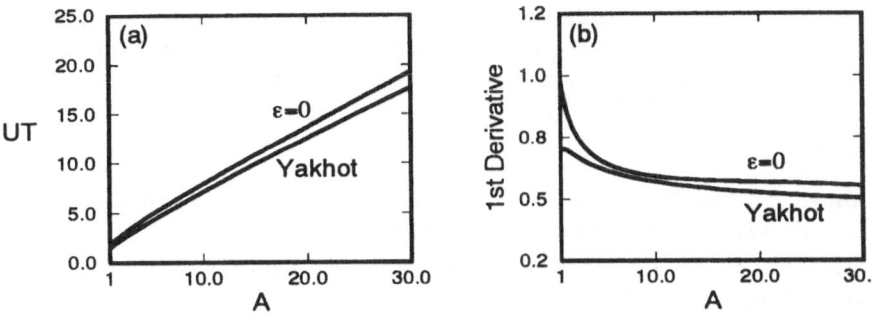

Figure 6: A comparison of the computational results without artificial viscosity ($\epsilon = 0$) to Yakhot's result: (a) Average flame speed versus velocity fluctuation intensity A. (b) The derivative dU_T/dA versus A. Reprinted from reference [25].

The computational results reveal that for the one-scale flow the rate of pocket formation controls the rate of flame surface-area increase with increasing intensity at large intensities. At the lower intensities, of the order of S_0, surface area growth is controlled by Lagrangian dispersion of the flame sheet. This explains the abrupt decrease in the slope dU_T/dA over a midrange of intensities (see Fig. 6b). The likelihood of pocket formation in a fully developed, high-intensity turbulent flow, however, is uncertain, as the distribution of energy among a wide range of spatial and temporal scales would tend to suppress pocket formation in this type of flow. A more complete discussion of the computational results can be found in reference [25].

4. SUMMARY AND CONCLUDING REMARKS

The use of a flame-front propagation equation in the calculation of the average propagation rate of a wrinkled flame, and for accurate description of the coupling between flame surface dynamics and flow-field fluctuations, has been discussed. For the regime of small-amplitude flame wrinkling a linear flame-surface evolution equation may be specified for tracking the motion of the flame surface. For the regime of large-amplitude flame wrinkling the eikonal scalar-field front-propagation equation, Eq. (6), provides for a more appropriate description of the flame surface evolution, while allowing naturally for the formation of multiple pockets of unburnt reactant.

Another method for tracking flame sheets, mentioned only briefly here, is the volume-of-fluid technique. In this technique the volume fraction of product in each cell of the computational domain is tracked in time. The flame sheet is constructed at each time step by the collocation of adjacent cells having product fractions in the range (0,1). An advantage of this method over other methods of front tracking is the ease with which fluxes of conserved quantities, such as density, and fluid volume can be handled in order to assure that all of the jump discontinuities of the fluid flow coincide with the front interface [28]. Its disadvantage for flame front tracking is that it does not easily allow for consideration of departures of S_n from S_0 resulting from flame-structure modification.

References

1. Liñán, A., and Williams, F. A., *Fundamental Aspects of Combustion*, Oxford University Press, New York, 1993, chap. 5.
2. Williams, F. A., *Combustion Theory*, 2nd ed., Addison-Wesley, Reading, MA, 1985, chapter 9.
3. Clavin, P., *Progress in Energy and Combustion Science*, 11:1-59 (1985).
4. Aldredge, R. C., and Williams, F. A., *J. Fluid Mechanics* 228:487-511 (1991).
5. Clavin, P., and Williams, F. A., *J. Fluid Mechanics* 116:251-282 (1982).
6. Matalon, M., and Matkowsky, B. J., *J. Fluid Mechanics* 124:239-259 (1982).
7. Clavin, P., and Joulin, G., *J. de Physique-Letters* 44:L-1 to L-12 (1983).
8. Markstein, G. H., *Nonsteady Flame Propagation*, Macmillan, 1964.
9. Eckhaus, W., *J. Fluid Mechanics*, 10:80 (1961).
10. Sivashinsky, G. I., *Acta Astronautica* 3:889 (1976).
11. Buckmaster, J. D., *Combustion and Flame* 28:225 (1977).
12. Pelce, P., and Clavin, P., *J. Fluid Mechanics* 124:219-237 (1982).
13. Clavin, P., and Garcia, P., *J. de Mécanique Théorique et Appliquée* 1:245-263 (1983).
14. Aldredge, R. C., Theory of Premixed Flame Propagation in Large-Scale Turbulence, Ph.D. dissertation, Princeton University (1990).
15. Darrieus, G. Propagation d'un front de flamme. Essai de théorie des vitesses anomales de déflagration par développement spontané de la turbulence. Unpublished manuscript, 1938; 6th International Congress of Applied Mechanics, Paris, 1946; La Mécanique des fluides. Quelques progrès récents (Reprints from La Technique Moderne, 1938-1941), pp. 15-16. Dunod, Paris, 1941.
16. Landau, L., *Acta Physicochim* (URSS) 19:77 (1944).
17. Joulin, G., and Clavin, P., *Combustion and Flame* 35:139-153 (1979).
18. Hinze, J. O., *Turbulence*, 2nd ed., McGraw Hill, New York, 1975.
19. Born, M., and Wolf, E., *Principles of Optics*, Pergamon Press, New York, 1959, chaps. 1-4 and pp. 760-763.
20. Hamilton, Sir W. R., *Trans. Roy. Irish Acad.* 15:69 (1828); *ibid.*, 16:1 (1830); *ibid.*, 16:93 (1831); *ibid.*, 17:1 (1837); Reprinted in "The Mathematical Papers of Sir W. R. Hamilton," Vol. I (*Geometrical Optics*), edited by A. W. Conway ad J. L. Synge, Cambridge University Press, 1931.
21. Bruns, H., *Abh. Kgl. Sächs. Ges. Wiss., math-phys. Kl.* 21:323 (1895).
22. Klimov, A. M., *Zhur. Prikl. Mekh. Tekhn. Fiz.* 3:49-58 (1963).

23. Williams, F. A., in *Analytical and Numerical Methods for Investigation of Flow Fields with Chemical Reactions, Especially Related to Combustion*, AGARD Conference Proceedings (Barrère, M., Ed.), AGARD, Paris, 1975, No. 164, pp. II1-1 to II1-25.

24. Buckmaster, J., *Acta Astronautica* 6:741-769 (1979).

25. Aldredge, R. C., *submitted for publication* (1994).

26. Ashurst, Wm., and Sivashinsky, G. I., *Combustion Science and Technology*, 80:159-164 (1991).

27. Yakhot, V., *Combustion Science and Technology* 60:191-214 (1988).

28. Pilliod, E. J., and Puckett, G. E., submitted for publication in *J. Computational Physics* (1994).

Effects of Local Flow Field on Chemical Reactions in Thin Reaction Zone of Premixed Flames

T. Takeno, M. Nishioka, X.L. Zhu and H. Yamashita
Department of Mechanical Engineering, Nagoya University
Chikusa-ku, Nagoya, 464-01 Japan

Keywords. turbulent premixed flame, numerical model, asymptotic flame structure

1 Introduction

Turbulent combustion is the most important subject of combustion research at present, and in accordance with the development of numerical calculation there have been many attempts to describe turbulent combustion in terms of numerical calculation. If we want to describe the turbulent flame behavior rigorously, however, the calculation required for direct numerical simulation becomes too large to be handled even by most sophisticated super-computers [1,2]. This is the reason why we need to develop a physical model to make the numerical calculation actually possible. The key to success of the model is how to separate the kinetics calculation from the flow calculation. Recently we have developed a method to combine the detailed chemical kinetics calculation of laminar flame with the turbulent fluctuation calculation to predict NO_x emission index of turbulent diffusion flames in laminar flamelet regime [3-5]. The details of the method and the calculated results obtained so far are presented in our companion paper [6]. The method is based on the asymptotic structure of laminar diffusion flames that in the thin reaction zone chemical reactions should be balanced by the molecular diffusion. Then the whole reaction process in the reaction zone must be governed by the available diffusion time evaluated at outer edge of the reaction zone. In the present paper we apply the same method to the numerical modeling of turbulent premixed flame in the laminar flamelet regime. In this paper we will describe the basic idea and the method how to make numerical calculation with the results obtained so far.

2 Method based on G-equation

In laminar flamelet regime, most of scales of flow fluctuations are much larger than the flame thickness, so that the flame can be regarded as a surface of zero thickness, which propagates normal to it with the laminar burning velocity S_u. Then the flame behavior can be predicted by solving so called *G-equation* [7], simultaneously with Navier-Stokes equation, as is shown in Fig. 1. The equation is given by

$$\frac{\partial G}{\partial t} + \mathbf{v} \cdot \nabla G = S_u |\nabla G|. \tag{1}$$

where $\mathbf{v}$ is velocity vector. Now, detailed kinetics can be decoupled from the fluctuation calculation, since the equation contains no terms representing effects of chemical reactions other than S_u. S_u is calculated for any laminar flames by adopting detailed kinetics, and is compiled as a function of some parameters, which represent effects of local flow field and of flame configuration. Then the problem is what are the best parameters to describe these effects. The perturbation theory, based on the simplified one-step kinetics, predicts that the effects can be expressed as a simple function of flame stretch and flame curvature, when the nonuniformity of flow is very small [7]. However, the nonuniformity in turbulent flow is not so small as can be treated by the perturbation theory, and the effects of detailed kinetics have to be taken into account. In addition, in most cases of turbulent combustion, Kolmogorov scale, the smallest scale of flow fluctuation, is not necessarily larger than the flame thickness. Therefore, we have to introduce another physical model to make more rigorous numerical calculation.

3 Physical Model Based on Asymptotic Flame Structure

Asymptotic analysis, based on one-step kinetics, of any premixed flames shows that the flame is composed of a relatively thick preheat zone and a thin reaction zone, as is shown in Fig. 2 [8]. δ is thickness of preheat zone and β is Zeldovich number, which is a large quantity as compared to unity. In the thick preheat zone, convection is balanced by molecular diffusion, whereas reaction is balanced by molecular diffusion

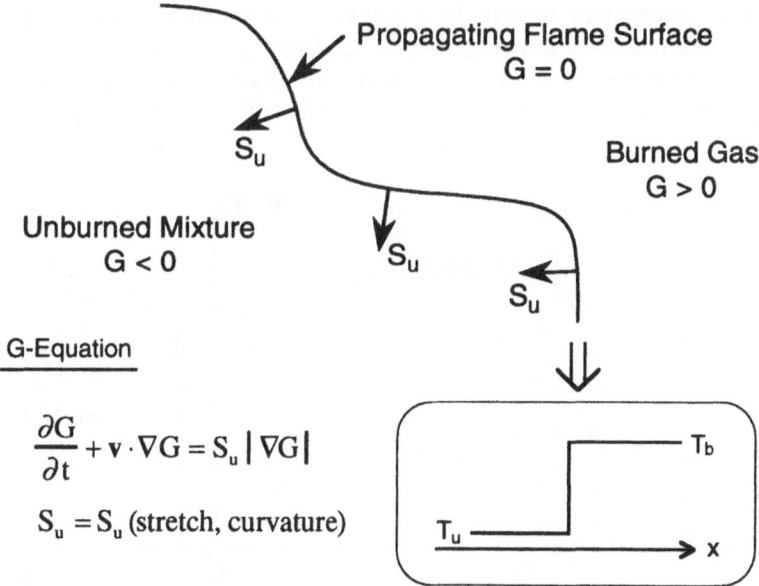

Fig. 1   Time-dependent behavior of propagating flame surface.

in the thin reaction zone. Then local flow field can affect the flame only through the preheat zone. In this zone convection will be altered, and through this molecular diffusion will be modified as well. Since the concentration of reactant, mass fraction Y, must be continuous at the boundary, the diffusion in the reaction zone is modified accordingly. In this way, local flow field will affect reactions in the reaction zone through molecular diffusion in the preheat zone. The effect can be evaluated in terms of the concentration gradient, or the scalar dissipation rate, at the downstream end of preheat zone. On the other hand, the relation between molecular diffusion and reaction in the reaction zone should be independent of local flow field.

In most cases, the reaction zone thickness is smaller than Kolmogorov scale, and the reaction zone can be approximated by an infinitely thin reaction sheet of zero thickness. The sheet propagates normal to it with the burning velocity $S_b = (\rho_u / \rho_b) S_u$, where ρ is density. Then time-dependent behavior of the sheet can be predicted by solving the following *G-equation*.

$$\frac{\partial G}{\partial t} + \mathbf{v} \cdot \nabla G = S_b |\nabla G| . \qquad (2)$$

On the other hand, the sheet accompanies the preheat zone of finite thickness in front of it, and all effects of flow and flame configurations manifest themselves in the structure of this zone. The structure can be predicted by solving an equation for a scalar Z.

$$\rho \frac{\partial Z}{\partial t} + \rho \mathbf{v} \cdot \nabla Z = \nabla \cdot (\rho D \nabla Z) . \qquad (3)$$

where D is diffusion coefficient, assumed equal to all components. Z is the normalized reactant concentration or temperature, and in the derivation of this equation assumptions of constant specific heat and unity Lewis number are adopted. The boundary conditions for Z are unity at upstream infinity and zero at the reaction sheet. The gradient of Z at the sheet gives the scalar dissipation rate, *SDR*, and the inverse of this quantity gives the representative molecular diffusion time τ_D in the preheat zone.

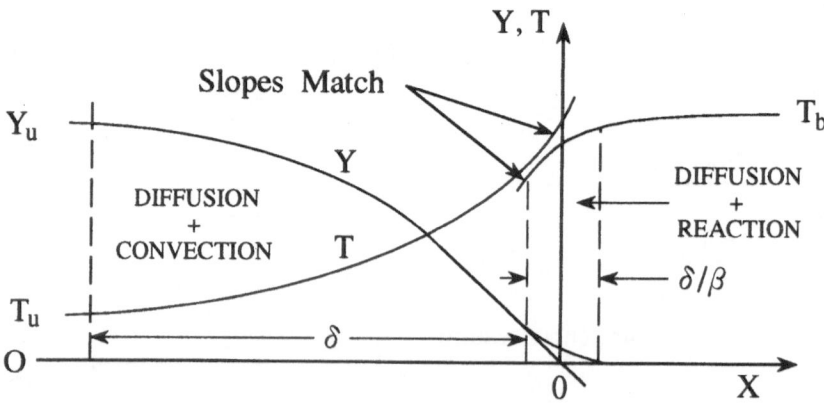

Fig. 2   Asymptotic structure of premixed flames.

$$SDR = D|\nabla Z|^2_{Z=0}\,, \qquad \tau_D = 1/SDR\,. \tag{4}$$

On the other hand, the burning velocity S_b is given as a function of τ_D and this dependence may be calculated for any laminar flames by using detailed kinetics.

$$S_b = S_b(\tau_D)\,. \tag{5}$$

Now the time-dependent behavior of the reaction sheet can be predicted by solving Equations (2) through (5), simultaneously with Navier-Stokes equation and equation of state. This is shown schematically in Fig. 3.

4 Is Burning Velocity a Unique Function of Diffusion Time?

The most important problem with the proposed method is if it is really possible to represent all effects of flow and flame configurations in terms of τ_D alone. We may take two approaches to answer this question. The first one is to make asymptotic analysis, based on one-step kinetics, to study the characteristics of the curved surface of reaction sheet in three-dimensional space. The second is to make a numerical experiment by adopting detailed kinetics and accurate transport properties to ascertain if S_b is a unique function of τ_D in actual flames. In the following the results of the second approach will be described.

The mixture adopted was lean methane air mixture of equivalence ratio of 0.7. The calculations were done for the following four flames of different flow and flame configurations.

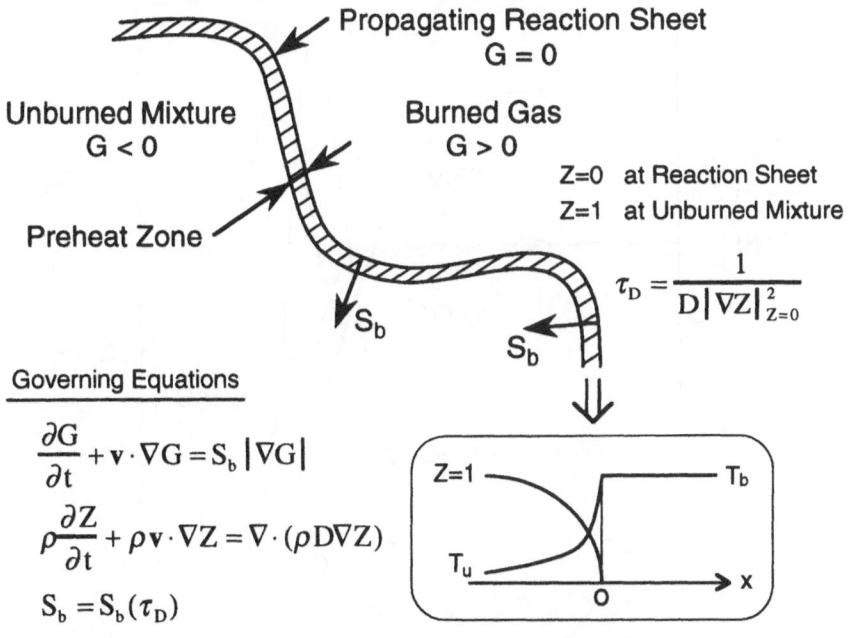

Fig. 3   Time-dependent behavior of propagating reaction sheet.

1) Normal flame (plane one-dimensional flame)
2) Twin flame (axisymmetric plane flame)
3) Tubular flame 1 (flame with curvature convex to unburned mixture)
4) Tubular flame 2 (flame with curvature concave to unburned mixture)

Tubular flame 1 is described in reference [9], whereas tubular flame 2 is in [10]. The adopted reaction scheme to describe combustion reactions in the flames was so called *C2 chemistry* and compiled by Miller and Bowman [11]. The scheme involves 52 species and 235 elementary reactions. The necessary thermochemical and transport properties were obtained from *CHEMKIN* data base [12-14]. The adopted numerical scheme was basically the one developed by Kee et al. for normal flame [15], and was somewhat modified in the calculation of other flames. The central difference formula was used for convective terms, and the adaptive placement of mesh points to form finer meshes was done in such a way that total number of mesh points needed to represent the solution accurately is minimized.

The structure of actual flame is not so simple as is described in Fig. 2. Figure 4 shows an example of the calculated flame structure of normal flame. In the species conservation equation of methane, contributions of convective, diffusive and reactive terms are shown in the figure. The scalar Z is defined as the normalized methane concentration, that is $Z = Y_{CH4} / Y_{CH4,o}$. The distribution of the scalar dissipation rate is shown as well in the figure. In addition, the position of maximum concentration of CH is indicated by an arrow at the top. As is seen in the figure, the

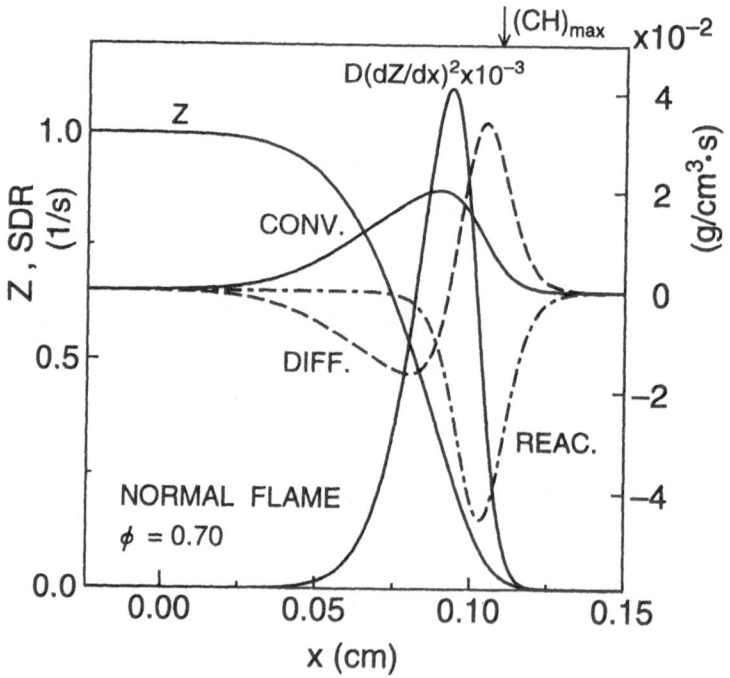

Fig. 4   Flame structure of normal flame.

boundary between the preheat zone and the reaction zone is not so clear. However, we have to define the downstream edge of preheat zone, so that we can deduce the molecular diffusion time in the preheat zone. The edge was defined as the position where the scalar dissipation rate becomes maximum. That is the diffusion time is just the inverse of the peak value of SDR. The concentration gradient at this position will determine the maximum supply rate of reactant to the reaction zone, and we may expect that the diffusion time derived from this rate will govern the whole reactions proceeding in the reaction zone. On the other hand, S_b is defined as the velocity at downstream infinity in normal flame. In case of other flames, the axial velocity u normal to the reaction zone accelerates as the fluid particle enters into the preheat zone and then decelerates to zero at the other boundary, as is shown in Fig. 5. In this case S_b is identified as the peak velocity in the preheat zone.

Figure 6 shows the final results obtained. The burning velocity S_b is plotted as a function of molecular diffusion time τ_D for four flames. In the three flames, which are subject to stretch, the diffusion time decreases as the stretch is increased, and the burning velocity decreases accordingly. At a certain critical diffusion time, the flame is forced to cause the extinction. On the contrary, the burning velocity increases with an increase in diffusion time. It is interesting to note that the point for normal flame is situated just in the extension of three flames. This means that normal flame is the flame with maximum molecular diffusion time and the highest burning velocity. The problem is that the three flames do not give a unique response. The response of

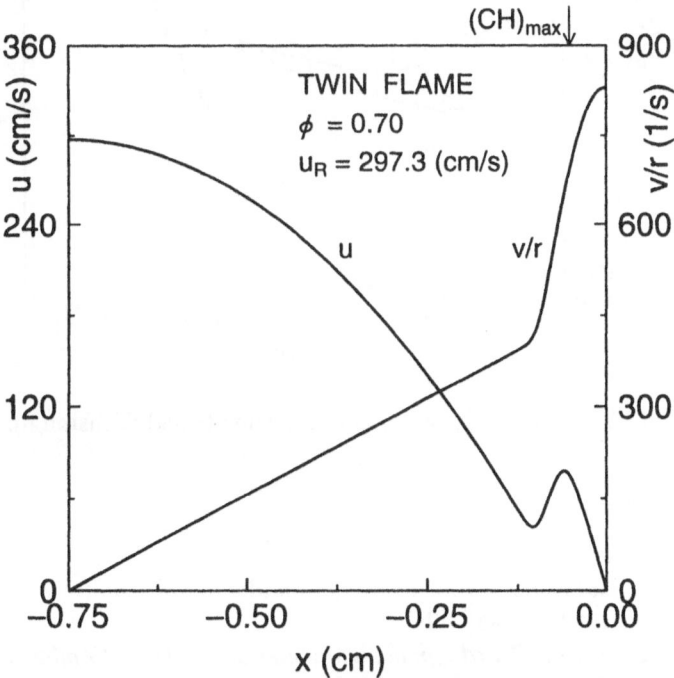

Fig. 5   Flow field of twin flame.

tubular flame 2 is somewhat similar to that of twin flame, whereas that of tubular flame 1 is quite different.

5 Some Concluding Remarks

The above result implies that the diffusion time alone cannot characterize the preheat zone structure. The most important difference among the three flames is the flame configuration. That is the effects of flame curvature. Twin flame is a plane flame, whereas tubular flame 1 and 2 are flames with curvature. Our recent studies on tubular flames [10,16] have shown that the effects of stretch are decelerated in tubular flame 2, whereas they are accelerated in tubular flame 1, indicating the significance of direction of curvature. The curvature, including its sign, is a possible candidate of another parameter required to characterize the preheat zone structure. Further studies are being conducted on this problem.

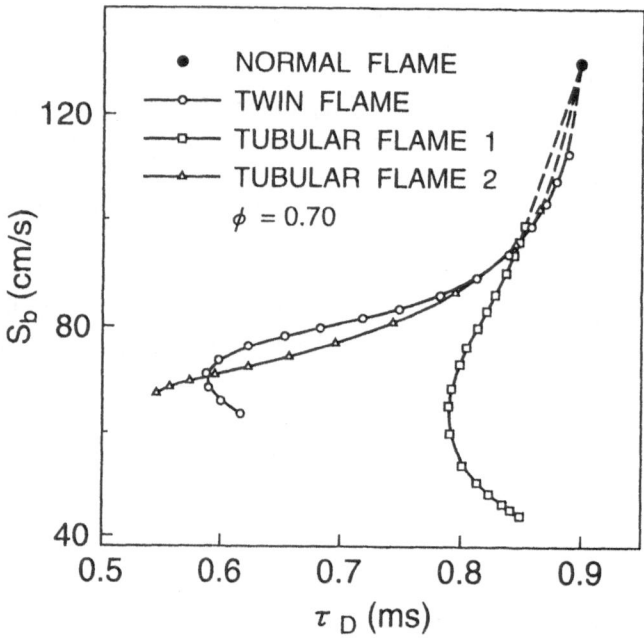

Fig. 6   Relation between burning velocity and diffusion time.

References

1. Oran, E. S. and Boris, J. P., *Numerical Simulation of Reactive Flow*, Elsevier, New York, 1987, pp. 455-476.
2. Pope, S. B., *Twenty-Third Symposium (International) on Combustion*, The Combustion Institute, Pittsburgh, 1990, pp. 591-612.
3. Takeno, T., Nishioka, M. and Yamashita, H., in *Turbulence and Molecular*

Processes in Combustion (T. Takeno Ed.), Elsevier, Amsterdam, 1993, pp. 375-391.

4. Takeno, T., Nishioka, M. and Yamashita, H., Prediction of NO_x emission index of turbulent diffusion flame, *Fifth International Conference on Numerical Combustion*, p. 115, 1993.

5. Takeno, T., Nishioka, M. and Yamashita, H., Effects of local flow field on NO_x emission index of turbulent diffusion flame, *Third Tsukuba International Workshop on Chaos/Turbulence*, p. 83, 1993. also in *The 71st JSME Spring Annual Meeting , Vol. III* (in Japanese), 1994, pp. 618-620.

6. Yamashita, H., Nishioka, M. and Takeno, T., Prediction of NO_x emission index of turbulent diffusion flame, in this proceedings.

7. Williams, F. A.,*Twenty-Fourth Symposium (International) on Combustion*, The Combustion Institute, Pittsburgh, 1992, pp. 1-17.

8. Williams, F. A., *Combustion Theory*, Second Ed., Benjamin/Cummings, Menlo Park, 1985, pp. 154-165.

9. Nishioka, M., Inagaki, K., Ishizuka, S. and Takeno, T., *Combust. Flame* 86 : 90-100 (1991).

10. Takeno, T., Zhu, X. L. and Nishioka, M., Theoretical studies on stability of a tubular flame, (in Japanese), *Combustion Research* 97: 47-59 (1994).

11. Miller, J. A. and Bowman, C. T., *Prog. Ener. Combust. Sci.* 15 : 287-338 (1989).

12. Kee, R. J., Rupley, F. M. and Miller, J. A., Sandia Report SAND 89-8009B, November 1991.

13. Kee, R. J., Rupley, F. M. and Miller, J. A., Sandia Report SAND 87-8215B, March 1990.

14. Kee, R. J., Dixon-Lewis, G., Warnatz, J., Coltrin, M. E. and Miller, J. A., Sandia Report SAND 86-8246, December 1986.

15. Kee, R. J., Grcar, J. F., Smooke, M. D. and Miller, J. A., Sandia Report SAND 85-8240, December 1985.

16. Zhu, X. L., Nishioka, M., Tamura, M., Nakamura, Y. and Takeno, T., Effects of detailed kinetics and heat loss on a tubular flame (in Japanese), *Combustion Research*, in press.

2. Turbulence Non-Premixed Flame

Direct Numerical Simulation of Chemically Reacting Turbulence

Toshio Miyauchi and Mamoru Tanahashi

Department of Mechano-Aerospace Engineering, Tokyo Institute of Technology, 2-12-1 Ookayama, Meguro-ku, Tokyo 152, Japan

Abstract: In this paper, we present two results of direct numerical simulation of chemically reacting flows. One is direct numerical simulation of chemically reacting two-dimensional mixing layer and the other is direct numerical simulation of chemically reacting compressible isotropic turbulence. As for the mixing layer, a low Mach number approximation was used to take into account the variable density effects on the flow fields and to clarify the effects of heat release and density difference of a mean flow. In the case of density difference, expansion and baroclinic torque has a negative contribution to the local vorticity transport in the high density side and a positive contribution in the low density side, which results in an asymmetric vortical structure. The density difference suppresses the growth of mixing layer and causes the overshoot of mean velocity only in the high density side which coincides with an experimental result. Coupling effects of heat release and density difference are also investigated. As for the homogeneous turbulence, fully compressible Navier-Stokes equations are solved to clarify the interaction between turbulence and chemical reaction in turbulent diffusion flame. The chemical reaction is suppressed by the increase of heat release because of the decrease of density and local Reynolds number. However, the decay of enstrophy with heat release is slower than that without heat release because of strong baroclinic torque which is generated near the reaction zone. Also, large amount of heat release causes increase in turbulent energy through the pressure dilatation term. The pressure dilatation term shows the periodic fluctuation which has an acoustic time scale. The fluctuation is enhanced by the heat release and travels in the turbulent field as pressure and dilatation waves.

1 Introduction

Turbulent combustion includes many instabilities such as the fluid dynamical instability, thermal instability and acoustic instability. The nonlinear interactions between these instabilities cause difficulty in analyzing and modeling the turbulent combustion. Especially, the interaction between turbulence and chemical reaction is one of the most difficult problem because the chemically reacting compressible turbulence which includes several chemical species must be treated.

Recent progress in the computer technology provides a new tool for the research of turbulent combustion: a direct numerical simulation (DNS). The DNS approach requires no turbulence models or combustion models but large computer resources. In the past, DNS studies of turbulent reacting flow as reviewed by Givi [1] had assumed the very simple chemical reaction and incompressible flow field. Leonard and Hill [2] calculated chemical reaction in homogeneous turbulence and Riley et al. [3] have conducted DNS of chemically reacting turbulent mixing layers.

Large scale structure in the turbulent shear flows play a significant role in the mass and heat transfer across the shear layer. This vortical structure, which is due to the Kelvin-Helmholtz instability of a parallel flow, is first clarified by the experimental study of Brown and Roshko [4]. In the chemically reacting shear flow such as turbulent diffusion flames, heat release with chemical reaction and density difference of two reactants may affect the development of large scale vortex in the shear layer and the mixing itself.

DNS study for reactive mixing layer was conducted by Riley et al. [3]. They have shown the ability and possibility of DNS for understanding the physical details of reaction in turbulent flows. On the assumptions of the negligible heat release with chemical reaction, concentration fields were treated as passive scalars. In the mixing layer, two familiar vortical structures develop. One is the spanwise large vortex and the other is the streamwise vortical structure called lib. As both structures dominate the mixing and reaction in mixing layers, DNS with the passive scalar assumption has been usefully used to investigate the chemical reaction in turbulent flows [5,6].

However, most of reactive flows observed in the real applications are very complicated. McMurtry et al. [7] have adopted the low Mach number approximation and performed direct simulation to make clear the effect of density changes with heat release on the spanwise and streamwise vortical structures. Large structure developed in the mixing layer shows lower growth rate in the case with heat release. Heat release effects on turbulent statistical properties are also investigated [8,9].

Because very large amount of heat is released in the turbulent flame and causes large density variation, full compressible direct simulations are required to understand and model the turbulent flames. Full compressible simulations of two-dimensional turbulent premixed flames have conducted by Haworth and Poinsot [10]. They have shown the details of turbulent premixed flames; Lewis number effects and strain-curvature relation etc. Recently Baum et al. [11] have attempted to simulate three-dimensional turbulent premixed flames including complex chemistry. Effects of differential diffusion on the flame statistics in turbulent diffusion flames have been investigated by Vervisch et al. [12] on the assumption of a single- and two- step reaction of Arrhenius type. By these fully compressible simulations, the turbulent flame structures are made clear. However, the behaviors of turbulent fields with heat release have not been made clear in detail. To understand the characteristics of fluid motion with heat release is necessary to develop turbulence model such as k-ϵ models and subgrid scale models in large eddy simulations for turbulent combustion.

In addition to the heat release, an important factor affecting chemically reacting flows is the density difference of the mean flows. In section 2, effects of heat release, density difference and these coupling effects on the development of chemically reacting mixing layer are discussed. In section 3, direct numerical simulations of two dimensional chemically reacting compressible turbulence are conducted to clarify the interaction between turbulence and chemical reaction.

2 Effects of Heat Release and Density Difference of a Mean Flow on the Development of Chemically Reacting Mixing Layers

2.1 Numerical Approach

In this study, a low Mach number approximation is adopted to take into account the density variations [7,13]. Governing equations are expanded by γM^2 and higher order terms are assumed to be negligible. Here γ is specific heat ratio and M is a Mach number. As the zeroth-order momentum equation reduces to equations for pressure, the first-order momentum conservation equations are used to describe the velocity field. Details of this approximation can be found in papers of McMurtry et al. [7,13]. The chemical reaction was idealized to be a simple form of $A + B \rightarrow Product + Heat$.

To solve the governing equations, spectral methods are used. Variables are expanded by Fourier series in the streamwise direction and by sine or cosine series in the transverse direction. Computational domain is selected to be $4\lambda \times 4\lambda$ and 128×129 grids are used, where λ is the most unstable wave length for the mean velocity profile. In the streamwise direction, periodic boundary conditions are used and in the transverse direction, free slip boundary conditions are used. Initial mean velocity profile is assumed to be hyperbolic tangent. On the mean velocity profile, solutions of Reyleigh equation are superposed [14]. Initial fluctuation intensities are set equal to 2.8%, 1.4% and 0.7% of mean velocity difference for fundamental, first-subharmonic and second-subharmonic, respectively. Thus amalgamations of spanwise vortical structures occur 2 times in a computational period. Figure 1 shows the initial conditions for present simulations. Initial density of mean flows were assumed to be equal to ρ_1 and ρ_2 for each flow, and density difference is represented by $r = \rho_2/\rho_1$. In this work, non-dimensional heat release is represented by $Ce = \Delta H C_0 / \rho_0 c_v T_0$.

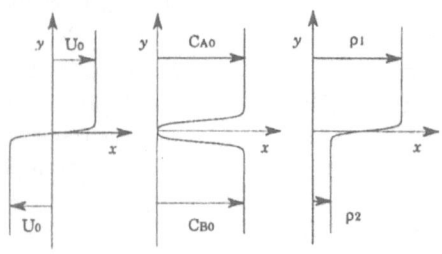

Fig. 1. Initial mean velocity, concentration and density distributions

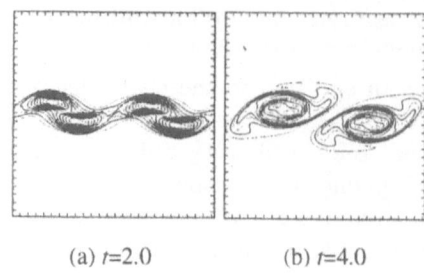

(a) $t=2.0$ (b) $t=4.0$

Fig. 2. Contour plots of vorticity of no heat release case

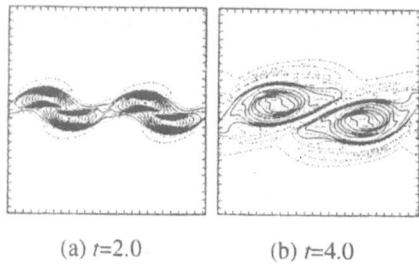

(a) $t=2.0$ (b) $t=4.0$

Fig. 3. Contour plots of vorticity of heat release case

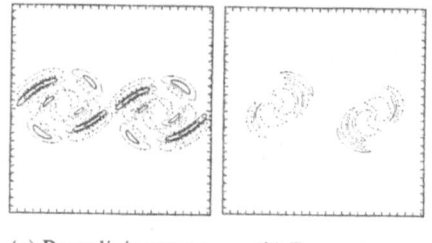

(a) Baroclinic torque   (b) Expansion

Fig. 4. Contour plots of baroclinic torque and expansion terms at $t = 3.0$ ($Ce = 5$, $r = 1.0$)

2.2 Effects of Heat Release

Figures 2 and 3 shows the contour plots of the two-dimensional vortical structure for no heat release and heat release case with no density difference in the mean flow. In these figures, the solid lines represent positive vorticity and dotted lines represent negative vorticity. In both cases, familiar large scale structures develop from the Kelvin-Helmholtz instability. However, in the case with heat release, negative vorticity region appears around the roller. These counter-rotating region are due to baroclinic torque which is caused by the pressure gradient and density gradient [7,13] as shown in Fig. 4. This torque appears in the transport equations only for the variable density flows. Another term appearing in that equation is expansion term. This term, as shown in Fig. 4, slightly affects the development of two-dimensional rollers.

As reported by McMurtry et al. [13], the growth of momentum thickness is suppressed by the density decrease with heat release. Figure 5 shows the development of momentum thickness for three cases ($Ce = 0, 2, 5$). Mean momentum thickness is defined as follow;

$$\delta = \frac{1}{\rho(2U_0)^2} \int <\rho> (U_0- <u>)(<u> -U_0)dy. \tag{1}$$

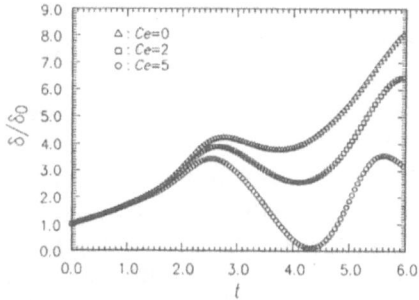

Fig. 5. The development of mean momentum thickness ($r = 1.0$)

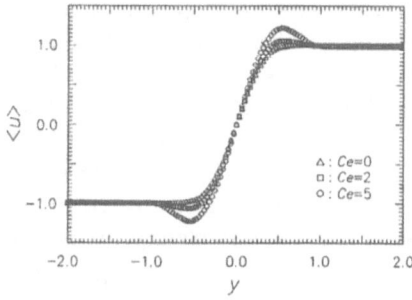

Fig. 6. Mean velocity profile at $t = 4.0$ ($r = 1.0$)

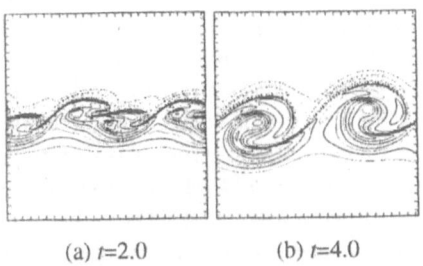

(a) $t=2.0$ (b) $t=4.0$

Fig. 7. Contour plots of vorticity for the case of $Ce = 0$ and $r = 0.2$

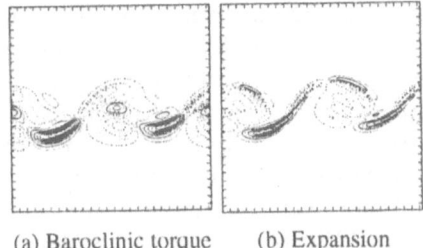

(a) Baroclinic torque   (b) Expansion

Fig. 8. Baroclinic torque and expansion term at $t = 3.0$ ($Ce = 0$, $r = 0.2$)

At the initial stage, the momentum thickness grows with roll-up of the spanwise structure, while this thickness decreases after the roll-up. This is because reactants in each stream, which enter into the central region of mixing layer with roll-up, react in the core region and the density is decreased. As baroclinic torque accelerates and decelerates the velocity field around large scale vortex, the overshoot and undershoot of mean velocity profile is observed as shown in Fig. 6. The total amount of product also decrease with increasing heat release because of the suppression of the growth of the mixing layer with heat release.

2.3 Effect of density difference

Figure 7 shows the contour plots of the density difference case with no heat release ($\rho = 0.2$, $Ce = 0$). The density difference causes asymmetric vortical structure and produces negative vortical region only in the high density side. This phenomenon can be understood by considering the vorticity transport equations in the same way as in the heat release cases. Contour plots of baroclinic torque and expansion terms at $t = 3.0$ are shown in Fig. 8. These contours suggest that, in the braid region, both terms show a negative contribution to the transport of vorticity in the high density side and a positive contribution in the low density side. So the negative vorticity region is produced only in the high density side. In the case with heat release, the contribution of the baroclinic torque to the mixing layer development is more important than that of expansion term, while in the case of density difference, both terms show same order and affect the vorticity transport.

Figure 9, which represents the development of mean momentum thickness, shows that the growth of mixing layers is suppressed by the density difference of the mean flow. As shown above, this suppression is a result of suppressed growth of the spanwise vortex by the baroclinic torque and expansion. Figure 10 shows the mean velocity distribution. The overshoot of a mean velocity profile is observed only in the high density side. The definition of mean momentum thickness is the defects of the momentum from a step-like velocity distribution so that the overshoots of the mean velocity profile results in the decrease of the mean momentum thickness.

Effect of density on the development of the mixing layer has investigated in a familiar experiment of Brown and Roshko [4]. They have conducted their measurement at various density ratios. Minimum density ratio conducted in their experiments is $r = 1/7$ which is smaller than the smallest ratio computed in our present simulations. Present results suggest that the mean velocity exceeds the free stream velocity in the high density side, if density of two streams is different. The experimental results reported by Brown and Roshko [4] shows that this is the case. Namely, velocity overshoot of a few percent can be observed in their results. Figure 11 shows the development of the mean product thickness, which is defined as follows;

$$\delta_P = \int < C_P > dy. \tag{2}$$

Larger density difference suppresses the growth of the product thickness. This is because of the negative vortical region in the high density side.

2.4 Coupling effect

The coexistence of heat release and density difference of a mean flow is also examined. Figure 12 shows the growth of the vorticity field for the case with $r = 0.2$ and $Ce = 5$. In this case, density difference of a mean flow has a dominant effect on the development of the flow field. As mentioned above, density decrease with heat release causes the large amount of baroclinic torque which produces the symmetric counter-rotating regions, while density difference produces asymmetric vortical structure. Figure 13 shows the mean velocity profiles. Equal density case shows overshoot and undershoot of the mean velocity, while the magnitudes of the undershoot decreases with the decrease of the density ratio. For the case with $r = 0.4$, the undershoot disappears. The overshoot of the mean velocity profiles increases its magnitude with the decrease of the density ratio.

The developments of the mean product thickness for several density ratios are shown in Fig. 14. From this figure, it can be concluded that the density difference of mean flow suppresses the development of the product thickness at constant heat release.

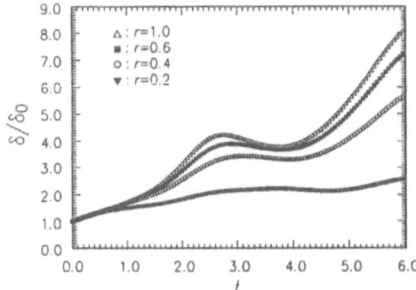

Fig. 9. The development of mean momentum thickness ($Ce = 0$)

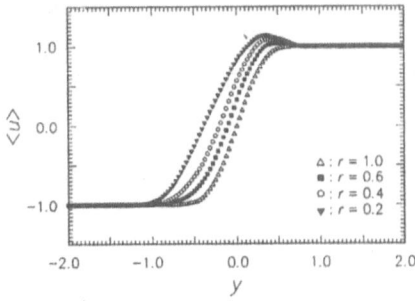

Fig. 10. Mean velocity profile at $t = 4.0$ ($Ce = 0$)

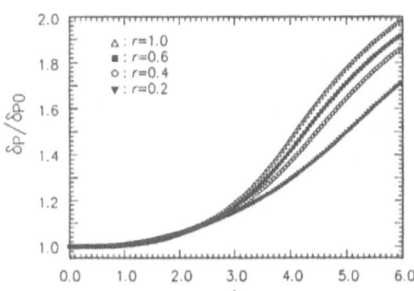

Fig. 11. The development of mean product thickness ($Ce = 0$)

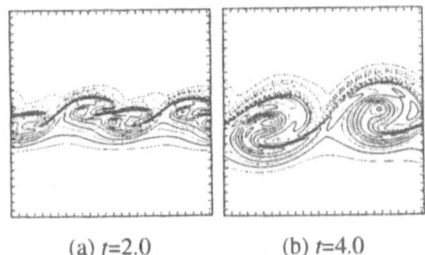

(a) $t=2.0$ (b) $t=4.0$

Fig. 12. Contour plots of vorticity for the case of $Ce = 5$ and $r = 0.2$

3 Pressure and Density Waves in Turbulent Diffusion Flame

3.1 Numerical Approach

In this section, we solved fully compressible governing equations. The governing equations are non-dimensionalized by the initial r. m. s. velocity U_0, the length of a computational domain L_0, initial mean temperature T_0, and density ρ_0. Time is scaled by the initial eddy turn over time. In this section, temperature dependence of viscosity μ is taken into account. However, the temperature dependence of thermal conductivity and diffusivity are neglected.

The computational domain is selected to be $L_0 \times L_0$, and periodic boundary conditions are applied in the x and y directions. A spectral method is used to discritize the governing equations. Aliasing errors from nonlinear terms in the governing equations are fully removed by 3/2 rule. Time integration is conducted by the second-order Adams-Bashforth scheme. The computations are performed in 384×384 grid points which is enough to resolve the smallest scale of turbulence and chemical field. Fully developed compressible homogeneous isotropic turbulent field ($Re_\lambda = 60$, Re_λ is Reynolds number base on Taylor micro scale) is used as initial velocity, temperature and density fields. This initial turbulent field is created by the procedure which has been used to create the initial velocity field for incompressible homogeneous isotropic turbulence [15]. In the present

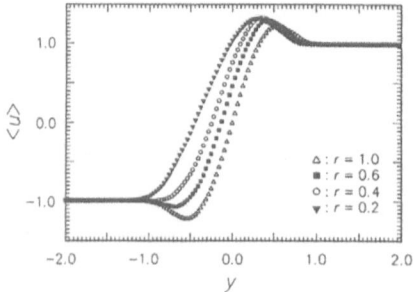

Fig. 13. Mean velocity profile at $t = 4.0$ ($Ce = 5$)

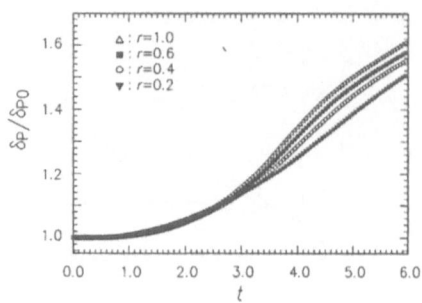

Fig. 14. The development of mean product thickness ($Ce = 5$)

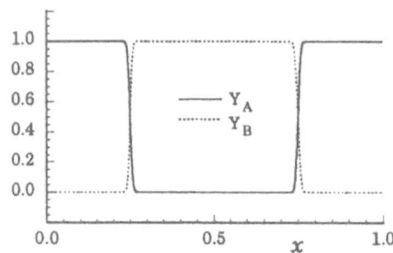

Fig. 15. Initial mass fraction distributions

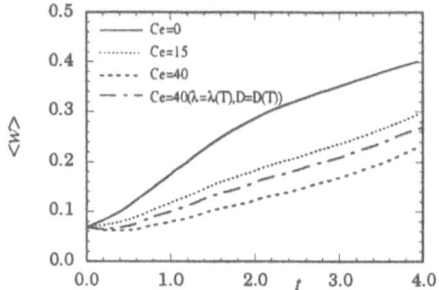

Fig. 16. Temporal development of mean reaction rate

work, calculation of compressible homogeneous turbulence without chemical reaction is conducted for 1 eddy turn over time. Eddy turn over time is defined by l_0/U_0 where U_0 and l_0 is initial r. m. s. velocity and initial integral length, respectively. The period is enough to attain a fully developed compressible turbulence which coincides with a given Mach number. Smoothed step functions as shown in Fig. 15 are used for initial mass fraction distributions. These smoothed functions prevent the Gibbs phenomena [2].

3.2 Global Effects of Heat Release

Figure 16 shows the temporal development of the mean reaction rate for different heat release rates. This figure suggests that the chemical reaction is suppressed by the increase of heat release through density and local Reynolds number decrease in high temperature regions. In Fig 16, the result of variable thermal conductivity and diffusivity case is also shown for comparison. This result shows that the increase of diffusivity with temperature does not enhance the reaction enough to overcome the density and viscosity effects. Present results show that the effects of variable density and transport properties is very important. However, temperature dependencies of thermal conductivity and diffusivity do not affect statistics which is discussed in the rest of this paper.

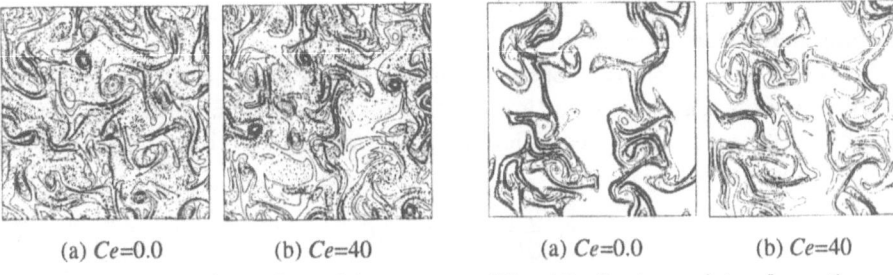

(a) Ce=0.0   (b) Ce=40   (a) Ce=0.0   (b) Ce=40

Fig. 17. Contour plots of vorticity at $t = 4.0$

Fig. 18. Contour plots of reaction rate at $t = 4.0$

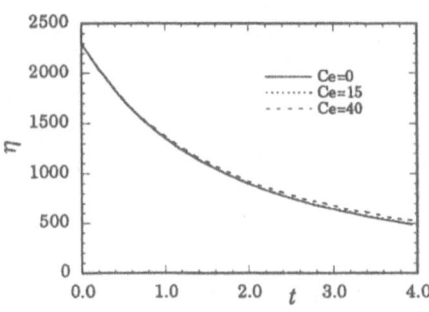

Fig. 19. Development of enstrophy

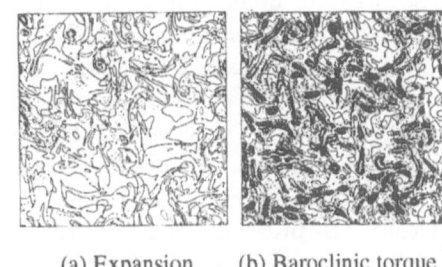

(a) Expansion   (b) Baroclinic torque

Fig. 20. Contour plots of expansion and baroclinic torque at $t = 4.0$

Figures 17 and 18 shows the contour plots of vorticity and instantaneous reaction rate for the cases of no heat release ($Ce = 0$) and heat release ($Ce = 40$) at $t = 4.0$. In Figs. 17 and 18, increments of vorticity and reaction rate contour is 10 and 0.1 respectively. In the case of heat release, temperature increases up to $4T_0$ in the reaction zone. Because of the high temperature and low density, the vorticity in the reaction zone is weakened as shown in Fig. 17 (b). However, vortices near the reaction zone are strengthened. These phenomena suggest the importance of vortex dynamics in variable density flow as discussed in the previous section.

The decays of enstrophy $\eta = \omega^2/2$ are shown in Fig. 19. It seems that the decay of enstrophy becomes slower with the increase of heat release. This phenomenon can be understood by considering the vorticity transport equation. Figure 20 shows the contour plots of the expansion and baroclinic torque terms at $t = 4.0$. Because of the density decrease, strong baroclinic torque appear near reaction zone, which distorted the vorticity field and strengthened the vorticity. The expansion term calculated from this DNS data is very small compared with baroclinic torque. These two figures suggest that the slower decay of enstrophy is caused by the baroclinic torque.

Figure 21 shows the turbulent energy based on Reynolds average and Favre average. In the case of no heat release, the turbulent energy decays monotonously. However, the turbulent energy of the heat release case decays jaggedly. Moreover, growths of the turbulent energy are observed in some phases in spite of the decaying homogeneous isotropic turbulence.

To clarify the mechanisms of the turbulent energy increase, we examined the turbulent energy transport equation. For the compressible homogeneous isotropic turbulence, the transport equation for Favre averaged turbulent energy q can be written as follow;

$$\frac{\partial \bar{\rho} q}{\partial t} = -\epsilon + \overline{p' d'}, \tag{3}$$

where d' is dilatation and d is defined by $d = \nabla \cdot \boldsymbol{u}$. The first term of right hand side is the dissipation term and the second one is the pressure dilatation term. The pressure dilatation term represents energy transfer between the pressure fluctuation and turbulent energy. In the turbulent diffusion flames, as the pressure and density variation is large, the pressure dilatation term has possibility to play an important role in the turbulent flow field. The turbulent dissipation term always results in the decay of the turbulent energy so that the energy increase from the turbulent dissipation term is expected.

The pressure dilatation term is shown in Fig. 22. In the case of no heat release, the pressure dilation term is almost zero. However, with heat release, this term shows a periodic fluctuation which is characterized by an acoustic time scale. These fluctuation causes increase in the turbulent energy for the case of large heat release.

To understand the behavior of the pressure dilatation term, we analyze the linearized equations. Neglecting the viscous and heat conduction terms, one can obtain the following equations for pressure and dilatation fluctuations from the linearized equations [16,17],

$$\frac{\partial p'}{\partial t} = -\gamma \bar{p} d' + \Delta H \dot{w}', \tag{4}$$

where p', d' and Q' are the pressure, dilatation and heat release fluctuations respectively. Combining these two equations, the equation for the pressure fluctuation can be obtained as follow,

$$\frac{\partial^2 p'}{\partial t^2} - \bar{c}^2 \nabla^2 p' = (\gamma - 1)\Delta H \frac{\partial \dot{w}'}{\partial t}, \tag{5}$$

where $\bar{c}$ is the mean speed of sound defined by $\bar{c} = (\gamma R \tilde{T})^{1/2}$. This equation shows that the pressure fluctuation caused by the heat release fluctuates with the acoustic time scale. Figure 23 and 24 show the x-t diagrams of pressure and dilatation on the line at $y = 0$ from $t = 0.0$ to $t = 0.75$. This figure suggests that the pressure and dilatation waves caused by chemical reaction are traveling in the turbulent flow field with exchanging energy between pressure fluctuation and turbulent motion.

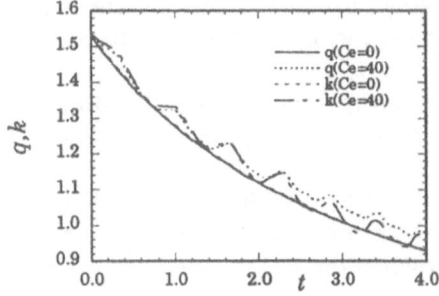

Fig. 21. Development of turbulent energy

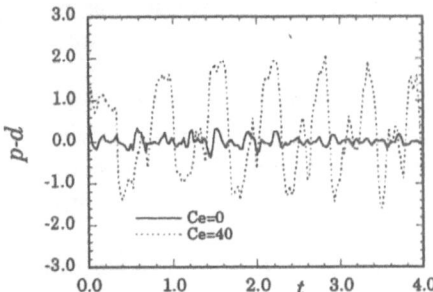

Fig. 22. Development of pressure dilatation term

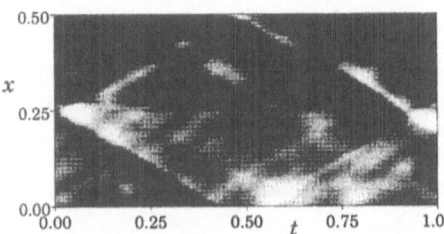

Fig. 23. x-t diagram of pressure fluctuation

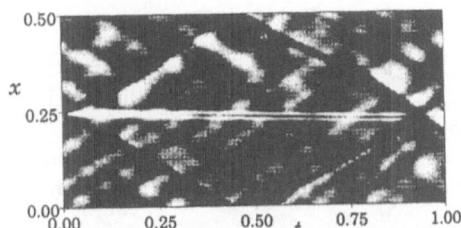

Fig. 24. x-t diafram of dilatation

4 Conclusion

To clarify the effects of heat release and density variation, direct numerical simulations of two-dimensional mixing layers and compressible homogeneous turbulence were conducted. For the mixing layer, following conclusions were obtained by using low Mach number approximation.

(1) Heat release with chemical reaction suppresses the development of the mixing layers and total amount of product.

(2) The density difference of mean flow induces negative vortical regions in the high density side, which is mainly due to the baroclinic torque and expansion terms in the vorticity transport equations, and this negative vortical region suppresses the growth of the mean momentum thickness and the product thickness.

(3) The density difference of mean flow produces the overshoot of a mean velocity profile only in the high density side.

(4) The growth of the mixing layer and product thickness are further suppressed by the coexistence of heat release and density difference.

Direct numerical simulations of chemically reacting compressible turbulence were conducted to clarify the interaction between turbulence and chemical reaction in turbulent diffusion flames and following conclusions were obtained.

(1) Chemical reaction is suppressed with increase of heat release because of the decrease of density and local Reynolds number in the reaction zone.

(2) With heat release, the enstrophy decays slower than that without heat release. This slower decay is caused by strong baroclinic torque near the reaction zone.

(3) In the case of heat release, turbulent energy decays jaggedly. Moreover, large heat release induces growths of the energy in some phases in spite of decaying homogeneous turbulence.

(4) The energy increase with heat release is due to the pressure dilatation term in the energy transport equation. The pressure dilatation term shows periodic fluctuation which is governed by linear equations of pressure and dilatation fluctuation.

References

1. P. Givi: Prog. Energy Combust. Sci. **15**, 1 (1989)
2. A. D. Leonard, J. C. Hill: J. Sci. Comput. **3**, 25 (1988)
3. J. J. Riley, R. W. Metcalfe, S. A. Orszag: Phys. Fluids **29**-2, 406 (1986)
4. G. L. Brown, A. Roshko: J. Fluid Mech. **64**, 775 (1974)
5. C. Lee, W. Metcalfe, F. Hussain: Turbulent Shear Flows **7**, 331 (1991) Springer-Verlag
6. T. Miyauchi, M. Tanahashi: JSME Int. J. **36B**-2, 307 (1993)
7. P. A. McMurtry, W. -H. Jou, J. J. Riley, R. W. Metcalfe: AIAA J. **24**-6, 962 (1986)
8. C. Chen, J. J. Riley, P. A. McMurtry: Combust. Flame **87**, 257 (1991)
9. S. F. Son, P. A. McMurtry, M. Queiroz: Combust. Flame **85**, 51 (1991)
10. D. C. Haworth, T. J. Poinsot: J. Fluid Mech. **244**, 405 (1992)
11. M. Baum, T. J. Poinsot, D. C. Haworth: Proc. 9th Symp. Turbulent Shear Flows, 25-2-1 (1993)
12. L. Vervisch, J. H. Chen, S. Mahalingam, I. K. Puri: Proc. 9th Symp. Turbulent Shear Flows, 25-3-1 (1993)
13. P. A. McMurtry, J. J. Riley, R. W. Metcalfe: J. Fluid Mech. **199**, 297 (1989)
14. A. Michalke: J. Fluid Mech. **109**, 543 (1974)
15. T. Miyauchi, M. Tanahashi, F. Gao, Combust. Sci. Tech. **96**, 135 (1994)
16. G. A. Blaisdell, N. N. Mansour, W. C. Reynolds: J. Fluid Mech. **256**, 443 (1993)
17. O. Zeman: Phys. Fluids **2**, 178(1991)

Prediction of *NOx* Emission Index of Turbulent Diffusion Flame

H. Yamashita, M. Nishioka and T. Takeno
Department of Mechanical Engineering, Nagoya University
Chikusa-ku, Nagoya, 464-01 Japan

Keywords. turbulent diffusion flame, numerical model, *NOx*, emission index,
scalar dissipation rate

1 Introduction

There is an urgent need to develop reliable numerical methods to predict *NOx* emission index in turbulent diffusion flames. The rigorous calculation from first principles will require the detailed kinetics calculation simultaneously with the calculation of time-dependent three-dimensional flow, concentration and temperature fields. The required computational load would be enormous and would be almost impossible to implement even by the most sophisticated super-computers [1,2]. We have to develop some plausible physical models to reduce the computational load so that we can implement the calculation. One possibility is to separate the kinetics calculation from the flow calculation. This will not be always possible, and only in certain limited cases we may make these calculations separately, and then may combine the results later to predict the emission index. Fuel jet diffusion flames developing in coflowing air streams, in laminar flamelet regime [3], should definitely be one of the cases. Figure 1 shows schematically the instantaneous flame structure.

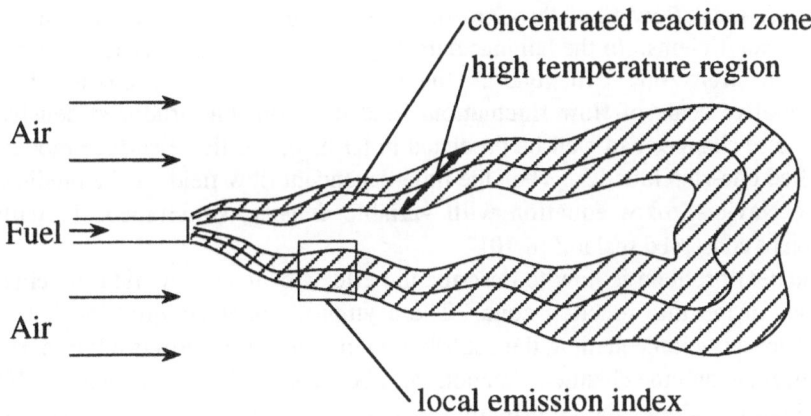

Fig. 1   Instantaneous flame structure of turbulent diffusion flames
developing in a coflowing air stream.

Most chemical reactions are concentrated in rather narrow reaction zone and *NOx* is produced only in this strip-like high temperature region, where Fuel is consumed simultaneously. Then we can define instantaneous local emission index by dividing *NOx* production rate per unit flame surface area by Fuel consumption rate per unit flame surface area. Basically, the local indices must be those of some laminar diffusion flames for which we can calculate the index relatively easily. The index for the whole flame can be deduced by taking an appropriate average of these instantaneous local indices.

$$EI_{NOx} = \left[\frac{NOx \text{ production rate per unit flame surface area}}{Fuel \text{ consumption rate per unit flame surface area}} \right]_{\text{average}} \tag{1}$$

On the basis of these considerations, we have developed a new method to predict the emission index of turbulent diffusion flames in laminar flamelet regime [4-6]. In the following we shall describe the basic idea of the method, and then describe how to calculate the local indices and how to take the average. In addition, the results obtained so far will be described briefly.

2 Interaction between Flow Field and Chemical Reactions

In the beginning, we shall discuss the problem how to describe the interaction between flow field and chemical reactions in the turbulent diffusion flame [7]. First, we shall consider how to describe the effects of chemical reactions on flow field. The respective elementary reactions themselves, involved in the combustion reaction, are not important for the local flow field, and what is important is the resultant heat release and the density change produced through these reactions. That is, the reactions will affect local flow field through the reduction in density, and the increase in transport coefficients. In the laminar flamelet regime, the heat is released only in the narrow reaction zone. In general, the width of this zone is smaller than the representative scale of flow fluctuations, and therefore the produced density and temperature changes can well be predicted in terms of the flame surface model with simplified one-step kinetics. Then the time-dependent flow field can be predicted by solving Navier-Stokes equation with variable density, simultaneously with the equation of conserved scalar Z [8-10].

The next problem is how to describe the effects of local flow field on chemical reactions in the thin reaction zone. The asymptotic analysis predicts that in the species conservation equation, the contribution of convective term is not important in this thin zone, and the chemical reaction must be balanced by the molecular diffusion [11]. Figure 2 shows the contribution of the respective terms in the thin reaction zone, compared to contributions in outer diffusion layer. Then the local flow field itself cannot affect the reactions in the reaction zone. In the outer diffusion layer, on the other hand, there are no chemical reactions and the molecular diffusion is balanced by convection. Hence the local flow field can affect the concentration distribution in

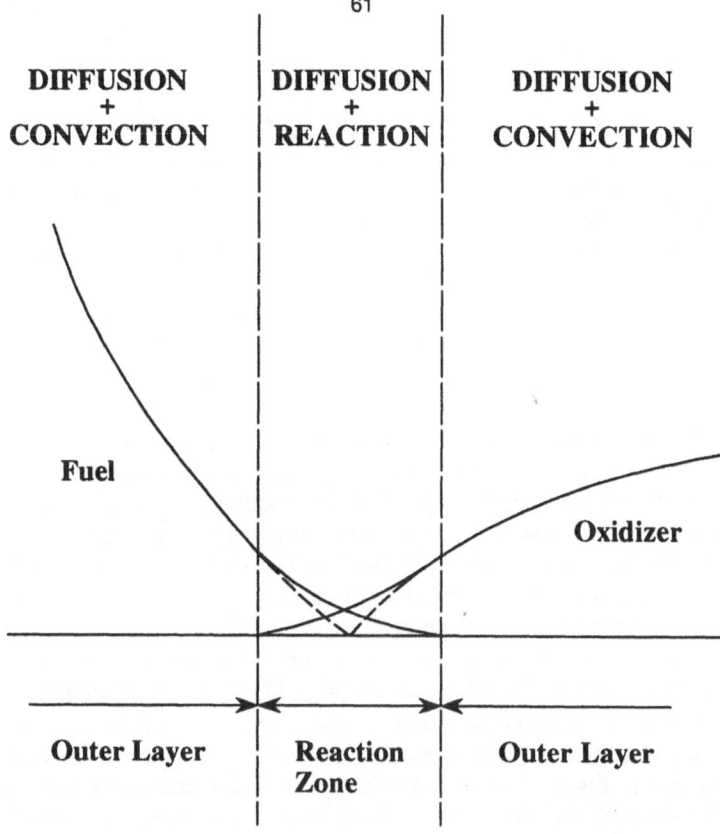

Fig. 2   Contributions of convective, diffusive and reactive terms
in thin reaction zone and outer diffusion layer.

the layer. Since the concentration must be continuous at the boundary, the molecular diffusion and hence the reaction in the reaction zone will be modified accordingly. In this way, the local flow field affects the reaction in the reaction zone through the molecular diffusion in the outer diffusion layer. The effects can be evaluated in terms of the concentration gradient, or the scalar dissipation rate, at the boundary. The inverse of this quantity gives the characteristic molecular diffusion time in the diffusion layer. When we adopt the flame surface model, the reaction zone thickness becomes infinitesimally small and the characteristic diffusion time can be evaluated by the scalar dissipation rate at the flame surface position. After all, we can evaluate the effects of local flow field on chemical reactions in terms of this quantity. On the other hand, the relation between molecular diffusion and reaction in the thin reaction zone is independent of flow field.

3 Method to Predict Emission Index

The procedure for the actual calculation is as follows. First, we calculate the emission index of a counterflow laminar diffusion flame as a function of velocity gradient by using detailed kinetics of large mechanism [12]. The diffusion times in the outer diffusion layer can be defined for the flame structures calculated for the respective velocity gradients to give the relation between the emission index and the diffusion time. It should be remembered that this relation is independent of flow field. Therefore, we can derive the relation in any flow field and we may adopt the counterflow diffusion flame, for which we can make use of similarity solutions to reduce the governing equations into ordinary differential equations to make the calculation relatively easy.

Second step of the calculation is to make the jet flame calculation in terms of the flame surface model. In the calculation we may adopt any turbulence models provided the minimum turbulence scale is larger than the reaction zone thickness. Or, we may make direct numerical calculations by adopting very fine mesh sizes in finite-difference methods. In most cases of laminar flamelet regime, however, mixing process in the jet is governed by relatively large scale fluctuations. The small scale three-dimensional turbulence may modify their evolution process, but will not alter the essential characteristics. In such a case we may adopt the moderate mesh size and the computational load will not become so large [7]. The calculation will yield the scalar dissipation rate distribution at any instant, and we can evaluate the local rate at the flame surface position. The inverse of the rate gives the local diffusion time in the diffusion layer. Figure 3 shows an example of the instantaneous distribution with the instantaneous flame surface superimposed. The time ensemble of these distributions make it possible to derive probability density function (*PDF*) of the rate at flame surface position. The combination of this *PDF* and the relation between the emission index and the diffusion time makes it possible to obtain the average value of the emission index of the whole flame.

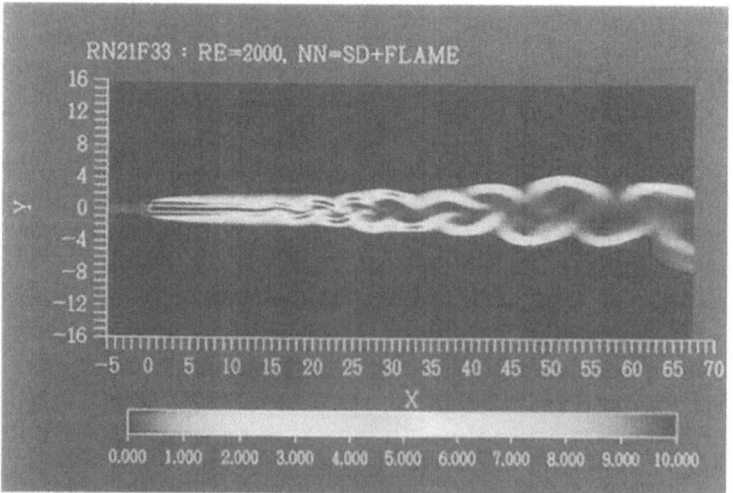

Fig. 3   Instantaneous distribution of scalar dissipation rate with instantaneous flame surface for methane flame with Reynolds number of 2000.

4   Emission Index as a Function of Diffusion Time

The details of the jet flame calculation have been described elsewhere [4,8], and hence in the following we shall describe the counterflow diffusion flame calculation to obtain the emission index as a function of the diffusion time. The Fuel adopted was methane. The theoretical model of the flame, as well as the adopted reaction scheme and numerical scheme, is described in reference [12,13] and hence will be described very briefly here. The Fuel flows upwards through a lower nozzle and meets the air flowing downward through an upper nozzle. The one-dimensional diffusion flame is established in this opposing stagnation flow. The cylindrical coordinate is used with the origin on the center of exit plane of the lower nozzle. The flow and flame are axisymmetric and x represents the axial distance from the lower nozzle exit. The fuel and air injection velocities, u_f and u_o respectively, and the nozzle distance L are variable. The conserved scalar Z is defined in terms of N_2 concentration by

$$Z = \frac{Y_{N_2,O} - Y_{N_2}}{Y_{N_2,O} - Y_{N_2,f}} \qquad (2)$$

where $Y_{N_2,O}$ and $Y_{N_2,f}$ are mass fractions at the upper and lower nozzle exits, respectively.

Figure 4 shows an example of the calculated flame structure for $L = 1.5$ cm.. The

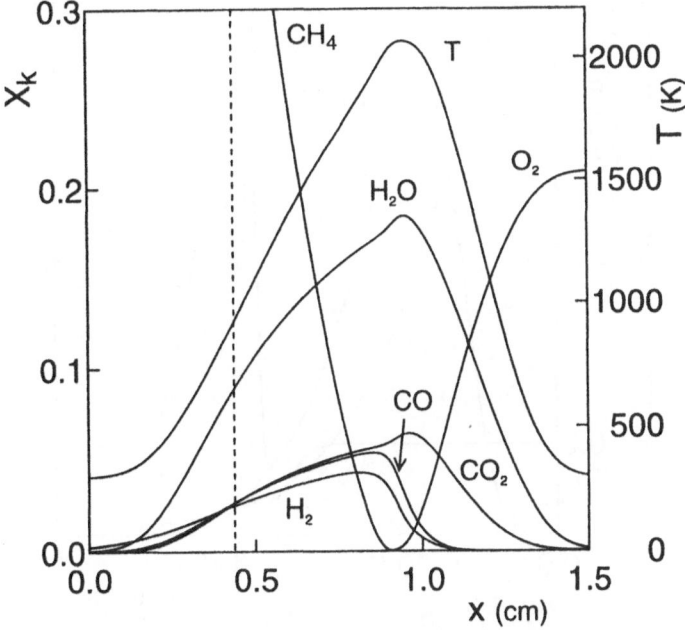

Fig. 4   Calculated flame structure for $u_O = u_f = 5$ cm/s and $L = 1.5$ cm.

injection velocities are equal and are as small as 5.0 cm/s. The temperature T and mole fraction X_k of main species k are shown. The dotted vertical line indicates position of the stagnation plane. The main reaction zone is located in the air side of the plane. The corresponding distributions of the conserved scalar Z and the scalar dissipation rate (SDR) are shown in Fig. 5 with the heat release rate profile. The heat is released in the air side of the flame with two peaks, and there is an endothermic region in the Fuel side next to a sharp highest peak. The latter is presumably due to decomposition of the fuel. We may take the position of the sharp peak in heat release rate profile as the apparent flame surface position x_q. The corresponding values of Z_q and $(SDR)_q$ can be defined accordingly. The inverse of $(SDR)_q$ gives the diffusion time τ_D. On the other hand, the aerodynamic time may be defined by $\tau_a = L/(u_o + u_f)$. The emission index EI_{NOx} can be calculated as explained in reference [14]. In this way, EI_{NOx} was calculated for different values of L, u_o and u_f.

Figure 6 shows Z_q and τ_D as a function of τ_a. For different combinations of L, u_o and u_f, Z_q remains almost constant at values close to 0.055, which is the one for the flame surface model of one-step kinetics. Although most of the data are for the axisymmetric plane flame, some results for the rectangular plane flame are shown as well. In the latter flame, the divergence of stagnation flow in one direction is prohibited. The results for the rectangular flame remain almost constant at values close to 0.055, in the same way as for the axisymmetric flame. The diffusion time is almost proportional to the aerodynamic time, and the data are on the same line for

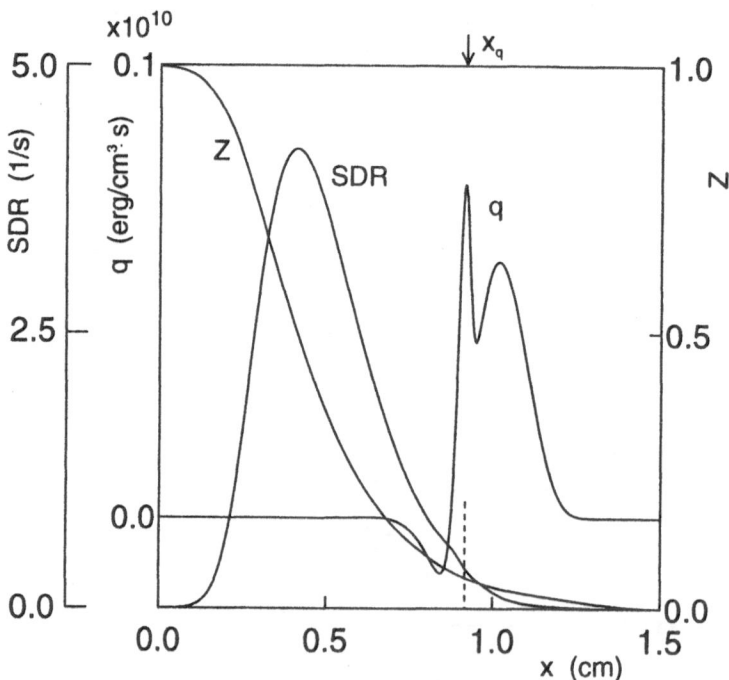

Fig. 5   Distributions of heat release rate, conserved scalar, and
scalar dissipation rate.

different combinations of L, u_o and u_f, and for different flow configurations. The final relation between the diffusion time and the emission index is shown in Fig. 7. As is seen in the figure, the data for different combinations of L, u_o and u_f, and for different flow configurations lie on one curve. That is, the emission index is a unique function of the diffusion time, and increases monotonically with the diffusion time.

5   Some Concluding Remarks

The present numerical study has revealed that in the diffusion flame the emission index is a unique function of the diffusion time of the diffusion layer, being independent of the flow configuration and of flow parameters. Although the flow configurations studied are restricted to the stagnation flows where we can adopt the similarity solutions, the implication seems quite plausible and universally important. Therefore, the proposed method to predict NOx emission index of turbulent diffusion flames in the laminar flamelet regime appears very promising. Further studies will be conducted in this direction.

Acknowledgments

The authors would like to extend their sincere thanks to Mr. Y. Takemoto for his help in doing numerical calculation. The financial support by Toho Gas Co. and Rinnai Co. are greatly acknowledged.

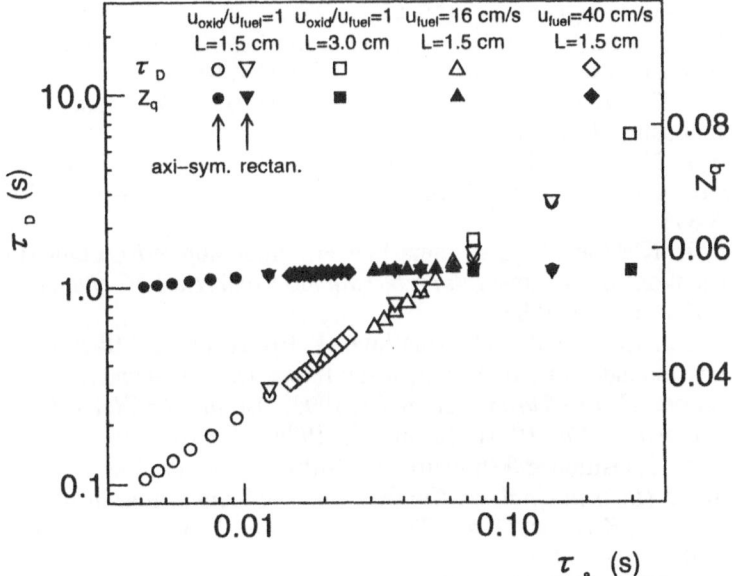

Fig. 6   Dependence on aerodynamic time of diffusion time and value of conserved scalar at apparent flame surface position.

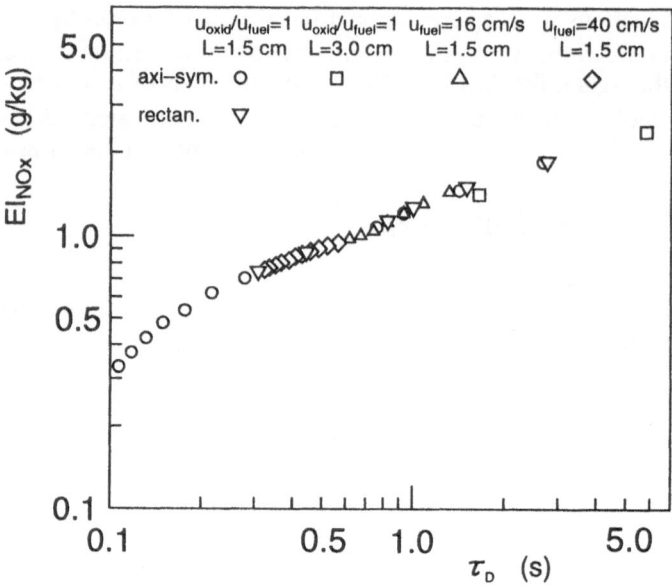

Fig. 7   Dependence of emission index on diffusion time.

References

1. Oran, E. S. and Boris, J. P., *Numerical Simulation of Reactive Flow*, Elsevier,New York, 1987, pp. 455-476.
2. Pope, S. B., *Twenty-Third Symposium (International) on Combustion*, The Combustion Institute, Pittsburgh, 1990, pp. 591-612.
3. Peters, N., *Twenty-First Symposium (International) on Combustion*, The Combustion Institute, Pittsburgh, 1986, pp. 1231-1250.
4. Takeno, T., Nishioka, M. and Yamashita, H., in *Turbulence and Molecular Processes in Combustion* (T. Takeno Ed.), Elsevier, Amsterdam, 1993, pp. 375-391.
5. Takeno, T., Nishioka, M. and Yamashita, H., Prediction of *NOx* emission index of turbulent diffusion flame, *Fifth International Conference on Numerical Combustion*, p. 115, 1993.
6. Takeno, T., Nishioka, M. and Yamashita, H., Effects of local flow field on *NOx* emission index of turbulent diffusion flame, *Third Tsukuba International Workshop on Chaos/ Turbulence*, p. 83, 1993. also in *The 71st JSME Spring Annual Meeting , Vol. III* (in Japanese), 1994, pp. 618-620.
7. Takeno, T., Transition and structure of jet diffusion flames, *Twenty-Fifth Symposium (International) on Combustion*, The Combustion Institute,in press.
8. Yamashita, H., Kushida, G. and Takeno, T., *Proc. R. Soc. Lond. A* **431**: 301 314 (1990).
9. Yamashita, H., Kushida, G. and Takeno, T., *Progress in Astronautics and Aeronautics*, Vol. 131, AIAA, 1991, pp. 193-219.

10. Yamashita, H., Kushida, G. and Takeno, T., *Twenty-Fourth Symposium (International) on Combustion*, The Combustion Institute, Pittsburgh, 1992, pp. 311-316.
11. Williams, F. A., *Combustion Theory*, Second Ed., Benjamin/Cummings, Menlo Park, 1985, pp. 76-80.
12. Nishioka, M., Nakagawa, S., Ishikawa, Y. and Takeno, T., *Combust. Flame* 98: 127-138 (1994).
13. Nishioka, M., Nakagawa, S., Ishikawa, Y. and Takeno, T., *Progress in Astronautics and Aeronautics*, Vol. 151, AIAA, 1993, pp. 141-162.
14. Takeno, T. and Nishioka, M., *Combust. Flame* 92: 465-468 (1993).

Solitary Wave Solution
of Turbulent Mixing Layer
by the Method of Pseudo-Compressibility

Burtsitsig Bai[1], Shunichi Tsuge[1] and Yuichi Matsuo [2]

[1] Institute of Engineering Mechanics, University of Tsukuba, Tsukuba, 305 Japan
[2] National Aerospace Laboratory, Chofu, Tokyo, 182 Japan

Abstract. Incompressible turbulent mixing layers are investigated by solving governing equations proposed by one of the authors using the method of pseudo-compressibility. The finite difference method with Newton iteration and with the upwind technique is employed in the computation, and a solitary wave solution of turbulent fluctuation correlations are obtained in the physical-plus-eddy space. Reynolds stress and turbulent intensities are calculated through a simple integration over the eddy space of the solitary wave solution. Those predicted quantities that are free from any empirical parameters are compared with existing experiments and satisfactory agreement is observed.

Keywords. Mixing layer, turbulent transports, solitary wave

1   Introduction

The eventual goal of turbulence research, namely, the accurate prediction of turbulent transports, i.e., Reynolds stress, turbulent heat flux density, turbulent diffusion flux density and turbulent chemical reaction rate, hinges crucially on the proper description of small eddies. However, all the existing computational methods suffer from the 'small-eddy' difficulties. The origin of the difficulty for direct numerical simulation is the fact that turbulence is fractal in nature, namely, that fluid variables are self-similarly irregular, so are not differentiable. Therefore, if we are to replace differentiations in the Navier-Stokes equation by finite differences, the step-size should be taken smaller than the lower limit of the fractal region that is of the order of Kolmogorov wave length or even smaller[1]. The spectral method, an alternative method which replaces space derivatives with multiplications by the wave number, so is free from the limitation discussed here, faces with another difficulty that unrealistically large number of terms of $O(R^{9/4})$(R; the Reynolds number) in Fourier-series need to be considered.

To alleviate these difficulties, many other computational approaches modeling the small eddies have been proposed. The problem here is the fact that none of these models are founded on first principles, so they are subject to empirical parameters for a best fit with data. Thus they can only meet requirements of engineering needs in somewhat ad hoc manner.

A 'neoclassical' statistical theory[2] has been proposed to by-pass both of these problems. It deals with a set of smooth, averaged variables

$$\{\langle u_j^{'}\hat{u}_{\ell}^{'}\rangle, \langle u_j^{'}\hat{u}_{\ell}^{'}\tilde{u}_r^{'}\rangle, \cdots\}, \quad (u_j^{'} = u_j - \langle u_j\rangle) \tag{1.1}$$

as the equivalent of a single fractal turbulent field quantity $u_j = u_j(x, t)$ The method is applied to turbulent Benard convection [3] and turbulent premixed flames[4], both showing satisfactory agreements with measurements.

In these papers separation of variables are effected in the frequency (scalar) space that limits its applicability to one-dimensional flows. It has turned out[5] that if the variable separation is effected in the wavenumber(vector)space the method can be applicable to more general flow geometry. The actual calculation has been carried out for incompressible mixing layers using marker-and-cell(MAC) method and on the assumption based on the flow visualization that turbulent vortices are nondispersive waves, namely that they propagate in the medium without changing shape. It is shown that the solution has the form of a solitary wave which builds up stationary with elapse of time. Local turbulence intensities such as Reynolds'stress are calculated using the solitary-wave solution, showing a good agreement with experiments[6][7].

The objective of this paper is to check how quantitative result will be changed if we replace the nondispersion assumption of the previous paper by the (local) Taylor hypothesis, namely, that vortices are carried with the mean flow. In fact this assumption seems to hold better for small eddies that are more important for the Reynolds stress.

2 Derivation of Governing Equation in Physical-Plus-Eddy Space

The method of separation of variables as sketched in Sec.1 is a method to decompose multi-point fluctuation-correlations that are 6D,9D,12D,... in the physical space into variables in the 3D(physical)plus 3D(wave number)space. Their actual forms are:

$$\left.\begin{array}{l} \langle u_j^{'}\hat{u}_{\ell}^{'}\rangle = RP\int_{-\infty}^{\infty} q_j\hat{q}_{\ell}\,\delta(k+\hat{k})dkd\hat{k} \\ \langle u_j^{'}\hat{u}_{\ell}^{'}\tilde{u}_r^{'}\rangle = RP\int_{-\infty}^{\infty} q_j^+\hat{q}_{\ell}^+\tilde{q}_r^+\delta(k+\hat{k}+\tilde{k})dkd\hat{k}d\tilde{k} \end{array}\right\} \tag{2.1}$$

where k is the nondimensional wave number and where $q = q(x, k, t)$, $\hat{q} = q(\hat{x}, \hat{k}, t)$ etc. are the separated variables that are complex quantities subject to a supplementary condition

$$q_j(-k) = q_j^*(k) \tag{2.2}$$

This is the condition to warrant the physical property that Reynolds stress tensor $\langle u_j' \hat{u}_\ell' \rangle$ be symmetric.

To truncate the infinite chain of equations governing the set of variables (2.1) at the lowest (two-particel) level of the hierarchy, closure condition

$$q_j^+ = q_j \tag{2.3}$$

has been proposed[2]. From experimenters point of view this assumption may be intuitively acceptable because in measurements double and triple correlations are formed using the *same* hot wire outputs. Theoretically, however, the situation is not so obvious. Precise examination of the three-particle equation shows that the closure assumption corresponds to neglecting some part (roughly 1/3) of four-point correlations.

It should be mentioned that the validity of this assumption can be tested only through comparison with experiments. So far the fact that the application of the same assumption to two different types of turbulence Ref.[3] and Ref.[4], both results in satisfactory agreement with measurements seem to support its soundness.

With closure condition (2.3), equations governing q_j are written as

$$\left.\begin{aligned}
&\frac{\partial q_r}{\partial x_r} = 0 \\
&(-i\omega + \frac{\partial}{\partial t} + u_\ell \frac{\partial}{\partial x_\ell} - \nu \frac{\partial^2}{\partial x_\ell^2})q_j \\
&+ q_\ell \frac{\partial u_j}{\partial x_\ell} + \frac{\partial}{\partial x_\ell} \int q_j(k - \tilde{k})q_\ell(\tilde{k})d\tilde{k} + \frac{1}{\rho}\frac{\partial q_4}{\partial x_j} = 0
\end{aligned}\right\} \tag{2.4}$$

where ω is the separation constant related to the phase velocity C_j as

$$\left.\begin{aligned}
\omega &= K \cdot C \\
K_j &= K_j/\delta \quad (\delta: \text{characteristic length})
\end{aligned}\right\} \tag{2.5}$$

q_4 is the pressure fluctuation and ν is the kinematic viscosity. Boundary conditions for q_j, q_4 are

$$q_j = 0, \quad \text{grad}\, q_4 = 0 \tag{2.6}$$

at solid boundaries and wherever turbulence vanishes. Since both governing equations and boundary conditions are homogeneous, the expected solution must be of the solitary wave type.

In view of the form of nonlinear terms that are convolutional integral, we see that periodic part in q_j

$$q_\alpha(\boldsymbol{x}, \boldsymbol{k}, t) = e^{i\boldsymbol{k}\cdot\boldsymbol{x}/\delta}Q_\alpha(\boldsymbol{x}, \boldsymbol{k}, t) \tag{2.7}$$

may be separated out from the equations. Furthermore, if Q_j is Fourier-transformed as

$$Q_\alpha(\boldsymbol{x}, \boldsymbol{k}, t) = \frac{1}{(2\pi)^3} \int_{-\infty}^{\infty} e^{i\boldsymbol{k}\cdot\boldsymbol{s}}F_\alpha(\boldsymbol{x}, \boldsymbol{s}, t)\, d\boldsymbol{s} \tag{2.8}$$

then Eqs.(2.4) transform to

$$\left.\begin{array}{l} \partial_j F_j = 0 \\ (\dfrac{\partial}{\partial t} - C_\ell \dfrac{\partial}{\partial S_\ell} + u_\ell \partial_\ell - \nu \partial_\ell^2)F_j + \dfrac{1}{\rho}\partial_j F_4 + \dfrac{\partial u_j}{\partial x_\ell}F_\ell + \partial_\ell F_j F_\ell = 0 \end{array}\right\} \tag{2.9}$$

with $S_j = \delta s_j$ and

$$\partial_j \equiv \frac{\partial}{\partial x_j} + \frac{\partial}{\partial S_j}. \tag{2.10}$$

The boundary condition for F_α are

$$F_\alpha \to 0 \quad (\alpha; j, 4) \quad as \quad |\boldsymbol{S}| \to \infty \tag{2.11a}$$

and

$$F_j \to 0, \quad \mathrm{grad}\, F_4 = 0 \tag{2.11b}$$

at a solid boundaries or wherever turbulence vanishes.

Once the solution has been obtained, the fluctuation correlations are calculated through (2.1) which are shown to have the very simple form in the S-space;

$$\left.\begin{array}{l} \langle u_j' \hat{u}_\ell' \rangle = \dfrac{1}{(2\pi\delta)^3}\displaystyle\int_{-\infty}^{\infty} F_j(\boldsymbol{x}, S)F_\ell(\hat{\boldsymbol{x}}, S + \hat{\boldsymbol{x}} - \boldsymbol{x})dS \\ \langle u_j' \hat{u}_\ell' \tilde{u}_r' \rangle = \dfrac{1}{(2\pi\delta)^3}\displaystyle\int_{-\infty}^{\infty} F_j(\boldsymbol{x}, S)F_\ell(\hat{\boldsymbol{x}}, S + \hat{\boldsymbol{x}} - \boldsymbol{x})F_r(\boldsymbol{x}, S)dS \end{array}\right\} \tag{2.12}$$

The Reynolds stress tensor is then obtained from (2.12) by putting $\hat{\boldsymbol{x}} = \boldsymbol{x}$.

3 Equations Governing a Turbulent Mixing Layer

A mixing layer is generated by two parallel flows with different velocities $U_{\pm\infty}$, separated by a semi-infinite spilitter plate, starting to merge at its end. It is observed experimentally that the flow becomes self-similar asymptotically in the well-developed region, where the mixing layer grows as

$$\delta = \alpha x_1 \tag{3.1}$$

This experimental evidence indicates that Eqs.(2.9) are governed by reduced number of self-similar variables $\eta = x_2/\delta, s = S/\delta$. This observation together with the group-theoretical consideration of self-similarity[8] require that the terms with viscosity be negligibly small, reflecting the fact that the viscous stress is overwhelmed by the Reynolds stress everywhere in the flow. Under this condition Eqs.(2.9) reduce to the following set of 'inviscid' equations:

$$\left.\begin{aligned}
\partial_1 f_1 + \partial f_2 + \frac{\partial f_3}{\partial s_3} &= 0 \\[2mm]
NL f_1 + \partial_1 f_4 - \alpha\eta\frac{\partial u}{\partial \eta}f_1 + \frac{\partial u}{\partial \eta}f_2 &= 0 \\[2mm]
NL f_2 + \partial_2 f_4 - \alpha\eta\frac{\partial v}{\partial \eta}f_1 + \frac{\partial v}{\partial \eta}f_2 &= 0 \\[2mm]
NL f_3 + \frac{\partial f_4}{\partial s_3} &= 0
\end{aligned}\right\} \tag{3.2}$$

with

$$\left.\begin{aligned}
\partial_1 &\equiv (1 - \alpha s_1)\frac{\partial}{\partial s_1} - \alpha(\eta\frac{\partial}{\partial \eta} + s_2\frac{\partial}{\partial s_2} + s_3\frac{\partial}{\partial s_3}) \\[2mm]
\partial_2 &\equiv \frac{\partial}{\partial \eta} + \frac{\partial}{\partial s_2} \\[2mm]
NL &\equiv \frac{\partial}{\partial t} - c\frac{\partial}{\partial s_1}(u + f_1)\partial_1 + (v + f_2)\partial_2 + f_3\partial_3
\end{aligned}\right\} \tag{3.3}$$

where $(u(\eta), v(\eta), 0)$ is the mean velocity made nondimensional using the velocity difference $U = U_\infty - U_{-\infty}$ and f_α and c are the nondimensional fluctuations and the phase velocity defined similarly;

$$\left.\begin{aligned}
f_j &= F_j/U \quad f_4 = F_4/\rho U^2 \\[2mm]
c &= C/U
\end{aligned}\right\} \tag{3.4}$$

In what follows we will neglect terms of $\partial/\partial s_2$, since the flow is not periodic in x_2-direction. To facilitate the analysis, we adopt a generalized curvilinear coordinate system,

$$\xi = \xi(x, y, z)$$
$$\eta = \eta(x, y, z) \tag{3.5}$$
$$\zeta = \zeta(x, y, z)$$

to reduce Eqs.(3.2) to an (modified) conservation form;

$$\partial_t \hat{Q} + \partial_\xi \hat{E} + \partial_\eta \hat{F} + \partial_\zeta \hat{G} = J^{-1}H \tag{3.6}$$

where

$$\hat{Q} = J^{-1}\begin{bmatrix} 0 \\ f_1 \\ f_2 \\ f_3 \end{bmatrix} \quad \hat{E} = \begin{bmatrix} F_1 \\ Uf_1 + f_1F_1 + \hat{\xi}_x f_4 \\ Uf_2 + f_2F_1 + \hat{\xi}_y f_4 \\ Uf_3 + f_3F_1 + \hat{\xi}_z f_4 \end{bmatrix} \quad \hat{G} = \begin{bmatrix} F_3 \\ f_1F_3 + \hat{\zeta}_x f_4 \\ f_2F_3 + \hat{\zeta}_y f_4 \\ f_3F_3 + \hat{\zeta}_z f_4 \end{bmatrix}$$

$$\hat{F} = \begin{bmatrix} F_2 \\ Vf_1 + f_1F_2 + \hat{\eta}_x f_4 \\ Vf_2 + f_2F_2 + \hat{\eta}_y f_4 \\ Vf_3 + f_3F_2 + \hat{\eta}_z f_4 \end{bmatrix} \quad H = \begin{bmatrix} \tilde{\nabla}f_1 \\ \tilde{\nabla}((u + f_1)f_1 + f_4) + \alpha f_1 y u_y f_2 u_y \\ \tilde{\nabla}((u + f_1)f_2) + \alpha f_1 y v_y f_2 v_y \\ \tilde{\nabla}((u + f_1)f_3) \end{bmatrix}$$

with

$$\left.\begin{array}{l} \tilde{\nabla} = \alpha(x\partial_x + y\partial_y + z\partial_z) \\ FF = \hat{\kappa}_x f_1 + \hat{\kappa}_y f_2 + \hat{\kappa}_z f_3 \quad (FF;\ \ F_1, F_2, F_3) \\ UU = \hat{\kappa}_x u + \hat{\kappa}_y v \quad (UU;\ \ U, V) \\ \hat{\kappa}_x = J^{-1}\partial \kappa/\partial x \ \ \text{etc.} \quad (\kappa;\ \ \xi, \eta, \zeta) \end{array}\right\} \tag{3.7}$$

In Eqs.(3.7) J is the Metric Jacobian and $\{s_1, \eta, s_3\}$ are written as $\{x, y, z\}$ here until the end of Sec.4 for simplicity in expression.

We have yet to prescribe the actual form of phase velocity c defined by Eq.(2.5) to complete the formulation. Here we invoke the (local) Taylor hypothesis

$$c = u \tag{3.8}$$

namely, that vortices flow with the local mean velocity, and will see how the result compares with that of the preceding paper[5] based on the assumption.

$$c = \frac{U_\infty + U_{-\infty}}{2} \tag{3.9}$$

Eqs.(3.2) through (3.8) with boundary conditions(2.11) make a well-posed problem once the mean flow profile $u(y)$ and $v(y)$ have been prescribed. Here $u(y)$ will be taken from existing experiment[6] and $v(y)$ from the equation of continuity as

$$v = \int_0^y y \frac{du}{dy} dy. \tag{3.10}$$

4 The Method of Solution

The solitary wave solution now in sought will be reached as an asymptotic steady-state limit of time-dependent equations (3.2). For this purpose the method of pseudo-compressibility is employed. This method assumes a fictitious term, namely, time derivative of pressure fluctuation in the continuity equation;

$$\frac{1}{\beta}\frac{\partial f_4}{\partial t} + \partial_j f_j = 0 \tag{4.1}$$

where β is the pseudo-compressibility constant. The Euler implicit scheme is used in calculating the time derivative. After the local time linearization, we have equations of the following form

$$[I + h(\partial_\xi A + \partial_\eta B + \partial_\zeta C)]\Delta \hat{Q}^n = -h(R^n - H) \tag{4.2}$$

with

$$R^n = \partial_\xi \hat{E}^n + \partial_\eta \hat{F}^n + \partial_\zeta \hat{G}^n \tag{4.3}$$

here $A = \partial \hat{E}/\partial \hat{Q}$, $B = \partial \hat{F}/\partial \hat{Q}$ and $C = \partial \hat{G}/\partial \hat{Q}$ are the flux Jacobian and $\Delta Q = Q^{n+1} - Q^n$ and $h = \Delta t$

Then the left-hand-side of Eqs.(4.2) are factorized approximately, thereby diagonal matrices in each direction are obtained[9]. This procedure saves both time and memory load. Updating variables from n to $n+1$ time steps are carried out through the following procedures[10][11]

$$\left. \begin{array}{l} [I + h\partial_\xi \Lambda_\xi]\Delta Q^{(1)} = -hX_\xi^{-1}R \\ [I + h\partial_\eta \Lambda_\eta]\Delta Q^{(2)} = X_\eta^{-1}X_\xi \Delta Q^{(1)} \\ [I + h\partial_\zeta \Lambda_\zeta]\Delta Q^{(3)} = X_\zeta^{-1}X_\eta \Delta Q^{(2)} \end{array} \right\} \tag{4.4}$$

$$Q^{n+1} = Q^n + JX_\xi \Delta Q^{(3)} \tag{4.5}$$

where $X_\kappa A X_\kappa^{-1} = \Lambda_\kappa$, Λ_κ are diagonal matrices whose elements are given by $(\Theta, \Theta, \Theta + \sqrt{\Theta^2 - \beta \nabla \xi^2}, \Theta - \sqrt{\Theta^2 - \beta \nabla \xi^2})$ for $\Theta = (\boldsymbol{F} + \boldsymbol{U}/2) \cdot \nabla \kappa$.

For spatial discretization, an upwind TVD scheme is applied for the right hand side of convection terms and a first order flux split upwind scheme is used for the left hand side of convection terms.

Initial values for f_α are chosen somewhat arbitrarily as far as the amplitudes are not too small.

5 Results and Discussions

The computation is run until convergence criterion for ΔQ is satisfied to the level of 10^{-7} to reach the steady-state solution. The flow conditions chosen are $U_\infty/U_{-\infty} = 5/3$ that are the same as Wygnanski[6] and Bell-Mehta[7] experiments. Computational domains for the solitary wave are $-10 < s_1 < 10$, $-20 < \eta < 20$, $-15 < s_3 < 15$ with $61 * 61 * 61$ meshes taken, respectively.

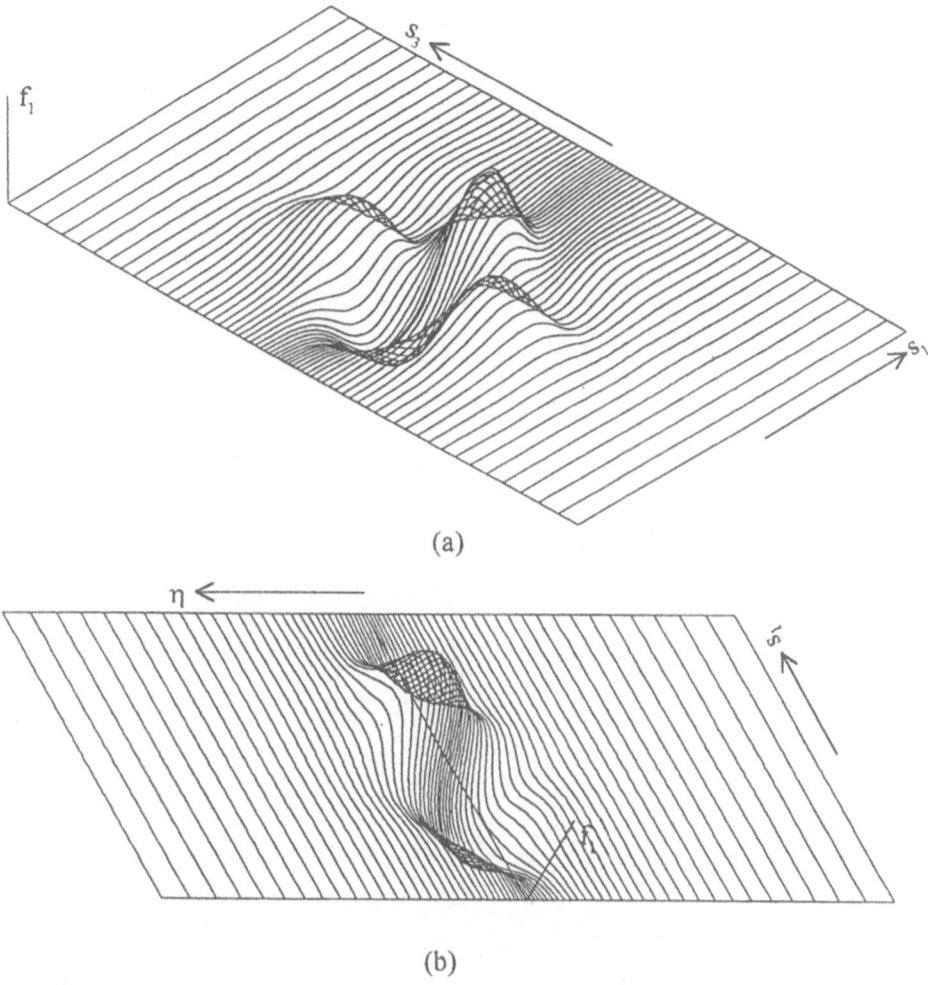

(a)

(b)

Figure 5.1: Solitary wave solution of f_1 on the plane (a)$\eta = 0$, (b)$s_3 = 0$.

Figs.5.1 show the existence of a solitary wave solution for f_1 on the plane $\eta = 0$ (Fig.5.1a) and $s_3 = 0$(Fig.5.1b), respectively. Physical quantities such as turbulent fluctuation correlations are calculated using this solution

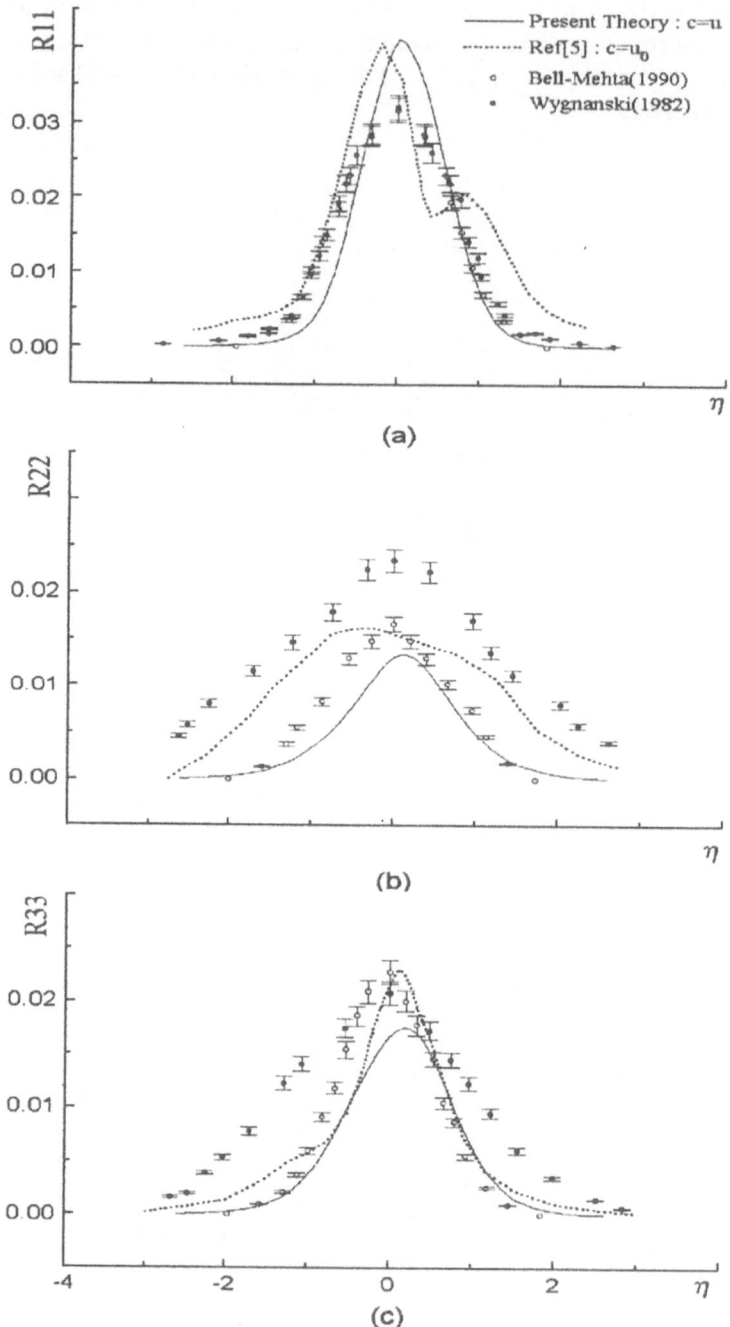

Figure 5.2: Turbulence intensities

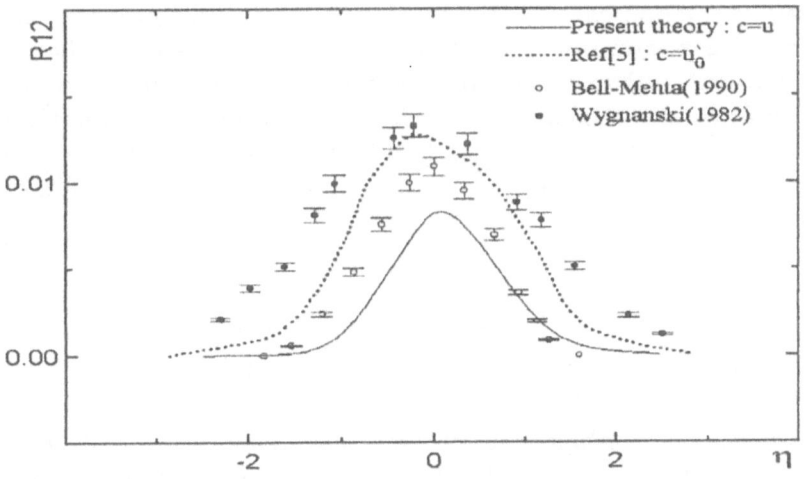

Figure 5.3: Reynolds Stress Profile

through the following formula, a 2D and self-similar version of Eqs.(2.12);

$$\langle u'_j \hat{u}'_\ell \rangle = \frac{1}{(2\pi)^2} \iint_{-\infty}^{\infty} f_1(\eta, s_1, s_3) f_1(\hat{\eta}, s_1 + \frac{\hat{x}_1 - x_1}{\delta}, s_3 + \frac{\hat{x}_3 - x_3}{\delta}) ds_1 ds_3 \quad (5.1)$$

The Reynolds stress tensor is obtained by putting $\hat{x}_1 = x_1, \hat{x}_3 = x_3$ in this expression, and the distribution of the main elements across the mixing layer are plotted and compared with existing experiments [7][6] and with the previous result[5] based on condition (3.9)(see Figs.5.2 and Figs.5.3). Roughly speaking, the computed result shows better agreement with experiments by Bell and Mehta[7] rather than those by Wygnanski[6]. It is observed that the current result using the local Taylor hypothesis(3.8) are qualitatively similar to the preceding one based on the nondispersion relationship(3.9). It is hard to tell which of the assumptions fits better to the reality at this point. It is also observed that the current result has a tendency of slight under estimation of reality for turbulence intensities. The agreement, however, seems to be more than reasonable considering that no empirical parameters are involved in the entire process of computations.

6   Concluding Remark

Reynolds stress and other turbulence intensities of incompressible turbulent mixing layer are calculated based on a 'neoclassical' statistical theory that are free from empirical parameters. Those quantities which are computed from a solitary-wave solution obtained in the physical-plus-eddy space are shown to agree with existing experimental data more than reasonably. Also the current result based on the local Taylor hypothesis is qualitatively similar to that of the preceding work based on the nondispersion assumption. This seems to be an indication of the robust nature of the solitary–wave solutions.

References

[1] K.R.Sreenivasan and C.Meneveau, "The Fractal Facets of Turbulence," *J.Fluid Mech.* vol. 173, pp.357, 1986.

[2] S.Tsuge, "Separability into coherent and chaotic time dependencies of turbulent fluctuations," *Phys.Fluids*, vol. 27, pp.1370–1376, Jun. 1984.

[3] K.Ishibashi, "Solitary Wave Solution of Turbulent Benard Convection," *Ph.D Thesis,Ins.Eng.Mech.Univ.Tsukuba Japan, 1991.*

[4] S.Tsuge and S.Ogawa, "Molecular and turbulent transports competing in premixed flames," *Turbulent and Molecular Processes in Combustion(Edited by T.Takeno, Elsevier,1993)* pp. 35–50.

[5] S.Tsuge and S.Ogawa, "Turbulence as a Solitary Wave in Physical-plus-eddy Space," *Trans. Jpn Soc. Aero. Space Sci., to appear.*

[6] D.Oster and I.Wygnanski, "The Forced Mixing Layer Between Parallel Streams," *J.Fluid Mech.* vol. 123, pp.91–130, 1982.

[7] James H. Bell and Rabindra D. Mehta, "Development of a Two-Stream Mixing Layer from Tripped and Untripped Boundary Layer," *AIAA JOURNAL*, vol. 28, pp.2034–2042, 1990.

[8] G.Birkhoff, "Hydrodynamics—A study in logic, fact and similitude," *Pub. Taft Fund, Univ. Cincinnati, 1993* p.116.

[9] Pulliam, T.H. and Chaussee, D.S., "A Diagonal Form of an Implicit Approximate Factorization Algorithm," *Journal of Computational Physics*, vol. 39, pp.347–363, 1981.

[10] S.Obayashi, "Numerical Simulation of Underexpended Plumes Using Upwind Algorithm," *AIAA Paper*, 88-4360-CP, 1988.

[11] Y.Matsuo, C. Arakawa, S.Saito and H.Kobayashi, "Navier-Stocks Computations for Flowfield of an Advanced Turboprop," *AIAA Paper*, 88-3094, 1988.

3. Modeling

SIMPLIFIED TRANSPORT AND REDUCED CHEMISTRY MODELS OF PREMIXED AND NONPREMIXED COMBUSTION

Mitchell D. Smooke[1] and V. Giovangigli[2]

[1]Department of Mechanical Engineering, Yale University
New Haven, Connecticut
[2]Ecole Polytechnique and CNRS Centre de Mathématiques Appliquées
91128, Palaiseau cedex, France

Abstract. In this paper we investigate the application of simplified transport models and reduced chemistry approximations to a set of premixed and nonpremixed methane-air flames. The two models we consider are the laminar premixed and the laminar counterflow diffusion flame in a Tsuji configuration. We examine 1) modifications of the governing conservation equations 2) transport model simplifcations and 3) reduced kinetic mechanisms that reduce substantially the CPU costs of these problems without compromising the predictive capabilities of the models. Numerical results focus of flame speeds, extinction strain rates and temperature and species profiles.

Keywords. Combustion, reduced chemistry, reduced transport, flames

1. Introduction

Combustion models that are used in the simulation of pollutant formation, ignition phenomena and in the study of chemically controlled extinction limits often combine detailed chemical kinetics with complicated transport phenomena. As the number of chemical species and the geometric complexity of the computational domain increases, the modeling of such systems becomes computationally prohibitive on even the largest supercomputer. While parallel architectures and algorithmic improvements have the potential of enhancing the level of problems one can solve, the modeling of three-dimensional time-dependent systems with detailed transport and finite rate chemistry will remain beyond the reach of combustion researchers for several years to come. The situation is even less promising if one wants to consider direct numerical simulation of turbulence with finite rate chemistry. While some applications can be studied effectively by lowering the dimensionality of the computational domain, there are many systems in which this is neither feasible nor scientifically sound.

The difficulty centers primarily on the number of chemical species and on the size of the different length scales in the problem. While local mesh refinement will ultimately solve the length scale problem, the size of the chemical system depends in large part upon the fuel one considers. For matrix based solution methods such as Newton's method, the Jacobian matrix formation dominates the cost of a flame computation. For detailed transport computations, in the limit of large species number, each Jacobian function evaluation scales with the square of the number of species. If one doubles the number of species, then the cost of the function evaluation quadruples. The actual function evaluations used in forming the Jacobian are divided between transport and chemistry computations. While reducing the cost of the transport evaluation can increase the overall efficiency of the computation, the largest CPU reduction is obtained when the transport is simplified and the number of species is reduced. It is clear from these arguments that, if one can simplify the transport evaluation and reduce the number of species in a reaction network while still retaining the predictive capabilities of the models, then potentially larger problems could be solved with existing technology. In addition, current large scale problems could be moved to smaller workstation computers.

In this paper we investigate the application of simplified transport models [1] and reduced chemistry approximations [2-5] to a set of premixed and nonpremixed methane-air flames. The two models we will consider are the laminar premixed and the laminar counterflow diffusion flame. In the next section we formulate each problem as a coupled nonlinear two-point boundary value problem with separated boundary conditions. We then consider 1) modifications of the governing equations 2) transport model simplifications and 3) reduced kinetic mechanisms that lower substantially the CPU costs of these problems while maintaining the accuracy of more complex models. Numerical results are presented in Section 4.

2. Problem Formulations

Premixed Flame Problem

The formulation of the premixed flame problem we consider closely follows the one originally proposed by Hirschfelder and Curtiss [6]. Our goal is to predict theoretically the mass flow rate (adiabatic flame speed), the mass fractions of the species and the temperature as functions of the independent coordinate x. Upon neglecting viscous effects, body forces, radiative heat transfer and the diffusion of heat due to concentration gradients, the equations governing the structure of a steady, one-dimensional, freely propagating, isobaric flame (with expansion angle $\alpha = 0$) are

$$\frac{d\dot{M}}{dx} = 0, \tag{2.1}$$

$$\dot{M}\frac{dY_k}{dx} + \frac{d}{dx}\left(\rho Y_k V_k\right) - \dot{w}_k W_k = 0, \qquad k = 1, 2, \ldots, K, \tag{2.2}$$

$$c_p \dot{M}\frac{dT}{dx} - \frac{d}{dx}\left(\lambda \frac{dT}{dx}\right) + \sum_{k=1}^{K} \rho Y_k V_k c_{p_k}\frac{dT}{dx} + \sum_{k=1}^{K} \dot{w}_k h_k W_k = 0, \tag{2.3}$$

$$\rho = \frac{p\overline{W}}{RT}. \tag{2.4}$$

In these equations x denotes the independent spatial coordinate; $\dot{M} = \rho u$ the mass flow rate with u the velocity of the fluid mixture and ρ the mass density; T, the temperature; Y_k, the mass fraction of the kth species; p, the pressure; W_k, the molecular weight of the kth species; $\overline{W}$, the mean molecular weight of the mixture; R, the universal gas constant; λ, the thermal conductivity of the mixture; c_p, the constant pressure heat capacity of the mixture; c_{pk}, the constant pressure heat capacity of the kth species; $\dot{w}_k$, the molar rate of production of the kth species per unit volume; h_k, the specific enthalpy of the kth species; and V_k, the diffusion velocity of the kth species. The form of the diffusion velocities, the transport coefficients and the chemical production rates is described in detail in [7].

The boundary conditions for the system in (2.1-2.4) are given by (see also [8]).

$$T(0) = T_i, \tag{2.5}$$

$$\dot{M}Y_k(0) + \rho(0)Y_k(0)V_k(0) = \dot{M}\epsilon_k(\phi), \quad k = 1, 2, \ldots, K, \tag{2.6}$$

$$\lambda(0)\frac{dT}{dx}(0) = \sum_{k=1}^{K} \epsilon_k(\phi)[h_k(T(0)) - h_k(T_u)]. \tag{2.7}$$

and as $x \to \infty$ by

$$\frac{dT}{dx} = 0, \tag{2.8}$$

$$\frac{dY_k}{dx} = 0, \quad k = 1, \ldots, K, \tag{2.9}$$

where T_u is the unburnt temperature and $\epsilon_k(\phi)$ is the known incoming mass flux fraction of the kth species which depends on the reactant stream equivalence ratio ϕ. The temperature T_i is used to remove the translational invariance in the problem. The premixed flame model then consists of equations (2.1-2.4) together with (2.5-2.9).

Counterflow Diffusion Flame Problem

We consider counterflow diffusion flames in a Tsuji configuration [9-12]. The one-dimensional governing equations can be derived by considering a boundary layer model. If we introduce the free stream (tangential) velocity at the edge of the boundary layer $u_\infty = ax$ where a is the strain rate together with the notation

$$U = \frac{u}{u_\infty}, \tag{2.10}$$

$$V = \rho v, \tag{2.11}$$

where U is related to the derivative of a modified stream function (see e.g., [13]), then the boundary layer equations can be transformed into a system of ordinary differential equations valid along the stagnation-point streamline $x = 0$. We have

$$\frac{dV}{dy} + a\rho U = 0, \tag{2.12}$$

$$V\frac{dU}{dy} - \frac{d}{dy}\left(\mu\frac{dU}{dy}\right) - a(\rho_\infty - \rho U^2) = 0, \tag{2.13}$$

$$V\frac{dY_k}{dy} + \frac{d}{dy}(\rho Y_k V_{ky}) - \dot{w}_k W_k = 0, \quad k = 1, 2, \ldots, K, \tag{2.14}$$

$$c_p V\frac{dT}{dy} - \frac{d}{dy}\left(\lambda\frac{dT}{dy}\right) + \sum_{k=1}^{K} \rho Y_k V_{ky} c_{pk}\frac{dT}{dy} + \sum_{k=1}^{K} \dot{w}_k W_k h_k = 0. \tag{2.15}$$

The system is closed with the ideal gas law. In addition to the quantities already defined, x and y denote independent spatial coordinates in the tangential and transverse directions, respectively; u and v the tangential and the transverse components of the velocity, respectively; μ the viscosity of the mixture; and V_{ky} is the diffusion velocity of the kth species in the y direction.

The boundary conditions for the Tsuji configuration are given by

$$V(0) = V_w, \tag{2.16}$$

$$U(0) = 0, \tag{2.17}$$

$$V_w Y_k(0) + \rho Y_k(0) V_k = V_w \epsilon_k, \quad k = 1, 2, \ldots, K, \tag{2.18}$$

$$T(0) = T_w, \tag{2.19}$$

at the cylinder wall ($y = 0$) and

$$U = 1, \tag{2.20}$$

$$Y_k = Y_{k_\infty}, \quad k = 1, 2, \ldots, K, \tag{2.21}$$

$$T = T_\infty, \tag{2.22}$$

as $y \to \infty$ The mass flux, temperature and the incoming mass flux fractions (V_w, T_w and ϵ_k) at the wall are specified, as are the mass fractions of the species and the temperature (Y_{k_∞} and T_∞) at the edge of the boundary layer.

3. Model Simplifications

One of the goals of this paper is to be able to reduce the overall CPU time for premixed and nonpremixed flames without compromising the accuracy of the models by simplifying the form of the governing conservation equations. We will systematically examine simplifications to the energy equation and the transport and chemistry submodels.

Enthalpy Flux Terms

Both the premixed and nonpremixed energy equations contain a term of the form

$$H = \sum_{k=1}^{K} c_{p_k} \rho Y_k V_k \frac{dT}{dx}. \tag{3.1}$$

We point out that if all the species heat capacities are equal, i.e., $c_{p_k} = c$ =constant, then

$$H = c\rho \frac{dT}{dx} \sum_{k=1}^{K} Y_k V_k, \tag{3.2}$$

which is identically equal to zero since

$$\sum_{k=1}^{K} Y_k V_k = 0. \tag{3.3}$$

In practice, we do not expect all of the heat capacities to be equal but we anticipate that their variation will be small so that $|H|$ will be small compared to the other three terms in the energy equation. To investigate these ideas, we have plotted in Figure 1 the convective term, the conduction term, the chemistry term and the enthalpy flux term in the energy equation as a function of the independent spatial coordinate for a one atmosphere, stoichiometric, premixed methane-air flame. The computations were performed with the premixed flame model described in the previous section. The transport coefficients were evaluated as in [7] and the reaction mechanism was the skeletal mechanism listed in Table 3.2. We note that the enthalpy flux term is negligible in all regions of the flame. Similar results hold, for example, for a one atmosphere premixed flame with $\phi = 0.6$ and for a stoichiometric flame at a pressure of 30.0 atmospheres. The situation is almost identical for a counterflow flame at a strain rate of 100 sec^{-1}. Based upon these results, we will eliminate the enthalpy flux term from our premixed and nonpremixed models.

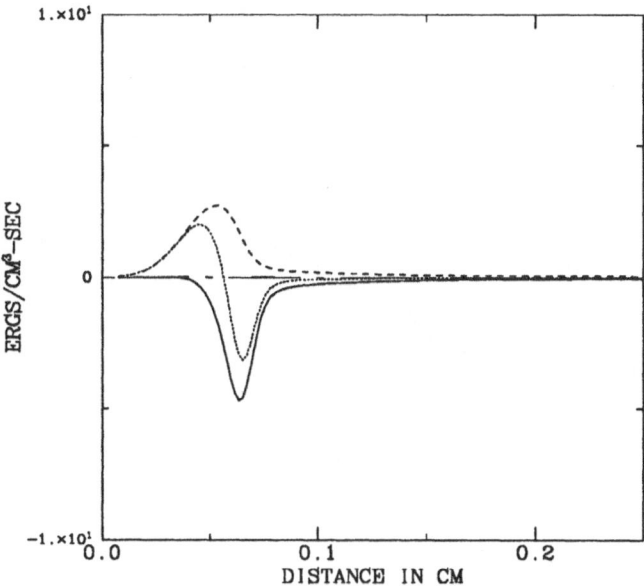

Figure 1. Comparison of the size of the diffusive (dot), convective (dash), enthalpy flux (solid-dash) and chemistry terms (solid) in the energy equation for an atmospheric pressure, stoichiometric, premixed methane-air flame with detailed transport and skeletal chemistry.

Thermal Diffusivity Approximation

In detailed flame models the heat capacity of the mixture is often approximated by a polynomial fit in the temperature to the JANNAF data. The thermal conductivity is approximated by a complex relation involving translational, vibrational and rotational factors. While each approximation by itself can be fairly complicated, the ratio of the two quantities can often be approximated by a simple expression. To investigate this possibility, we postulate the relation

$$\frac{\lambda}{c_p} = A \left(\frac{T}{T_o}\right)^r ,$$ (3.4)

for a constant A, a reference temperature $T_o = 298K$ and an exponent r. Using the detailed transport-finite rate skeletal chemistry model employed in the enthalpy flux evaluation, we generated vaues of λ/c_p as a function of the independent spatial coordinate for a stoichiometric, one atmosphere, premixed methane-air flame. We then determined the parameters A and r with a nonlinear least squares fitting procedure. Specifically, we found

$$A = 2.58 \times 10^{-4} \text{ g/cm-sec} \text{ and } r = 0.7.$$ (3.5)

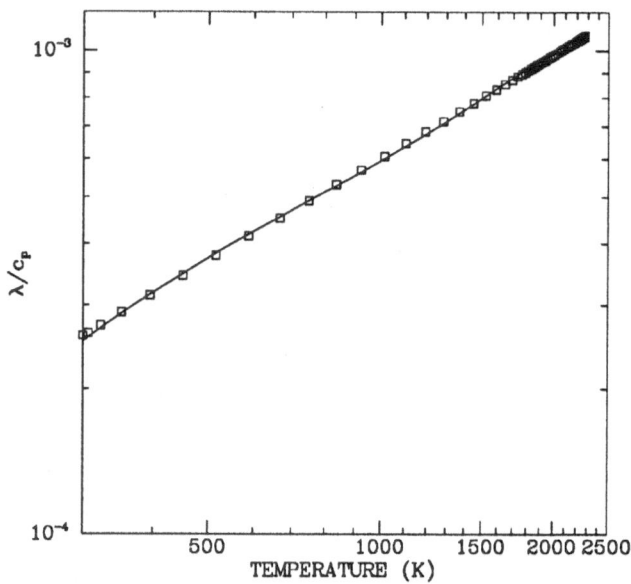

Figure 2. A comparison between the detailed transport-skeletal chemistry value of λ/c_p (⊓) and the least squares fit (solid) as a function of the temperature for an atmospheric pressure, stoichiometric, premixed methane-air flame.

In Figure 2 we compare the results of these calculations. Similar comparisons for a one atmosphere $\phi = 0.6$ flame and a stoichiometric 30 atmosphere flame can also be obtained. In all cases we find excellent agreement over the entire temperature range. Results for the counterflow diffusion flame with $a = 100 \text{ sec}^{-1}$ are contained in Figure 3. We note, however, that the simple power law is only able to approximate the detailed transport form of λ/c_p on the fuel lean side. A two zone model could be

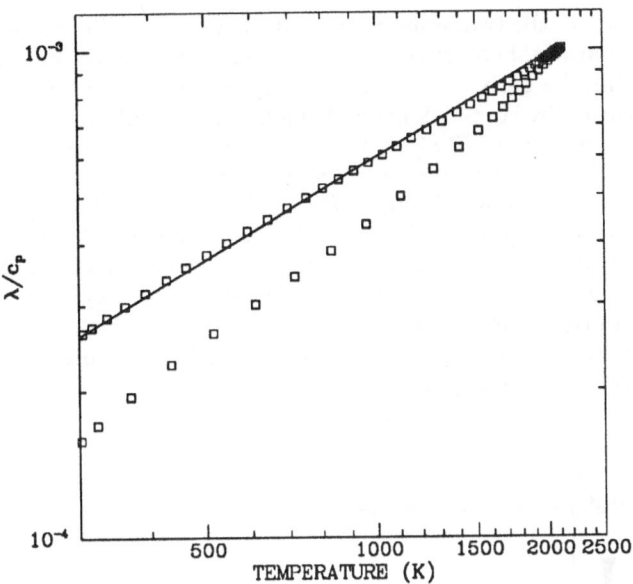

Figure 3. A comparison between the detailed transport-skeletal chemistry value of λ/c_p ($\sqcap$) and the least squares fit (solid) as a function of the temperature for a methane-air, counterflow diffusion flame with $a = 100$ sec^{-1}.

developed in an analagous fashion but, for simplicity and consistency, we will only use the relations in (3.4) and (3.5) in the test problems in Section 4.

Heat Conduction Term

The premixed and nonpremixed energy equation contain terms of the form

$$C = \frac{1}{c_p} \frac{d}{dx} \left(\lambda \frac{dT}{dx} \right). \tag{3.6}$$

Since we have combined the thermal conductivity and the heat capacity of the mixture into a single ratio, it would be reasonable to rewrite the heat conduction term using this expression, i.e., we would like to write

$$C = \frac{d}{dx} \left(\frac{\lambda}{c_p} \frac{dT}{dx} \right). \tag{3.7}$$

However, we note that

$$\frac{1}{c_p} \frac{d}{dx} \left(\lambda \frac{dT}{dx} \right) = \frac{d}{dx} \left(\frac{\lambda}{c_p} \frac{dT}{dx} \right) + \frac{\lambda}{c_p^2} \frac{dc_p}{dx} \frac{dT}{dx}. \tag{3.8}$$

We observe that an additional term in the energy equation is generated by this reformulation. To investigate whether this term can be neglected, we compare in

Figure 4 the size of the two terms on the right-hand side of (3.8) as a function of distance for a 30 atmosphere, stoichiometric, premixed methane-air flame. While the term containing the derivative of the heat capacity of the mixture is not dominant throughout the entire spatial domain, it can be comparable in size to the other term on the right-hand side of (3.8). If the term containing the derivatives of the heat capacity is neglected, the burning velocity for the one atmosphere, stoichiometic methane-air flame becomes 27.07 cm/sec and the counterflow diffusion flame extinction strain rate becomes $a = 212$ sec^{-1}. These numbers differ significantly from the experimentally measured flame speed of 37-40 cm/sec and the extinction strain rate of $a = 350$ sec^{-1}. As we will see in Section 4, by keeping the heat capacity derivative term the computed flame speed becomes 37.67 cm/sec and the extinction strain rate becomes $a = 353$ sec^{-1}. Based upon these results, the heat capacity derivative term in the energy equation will be retained.

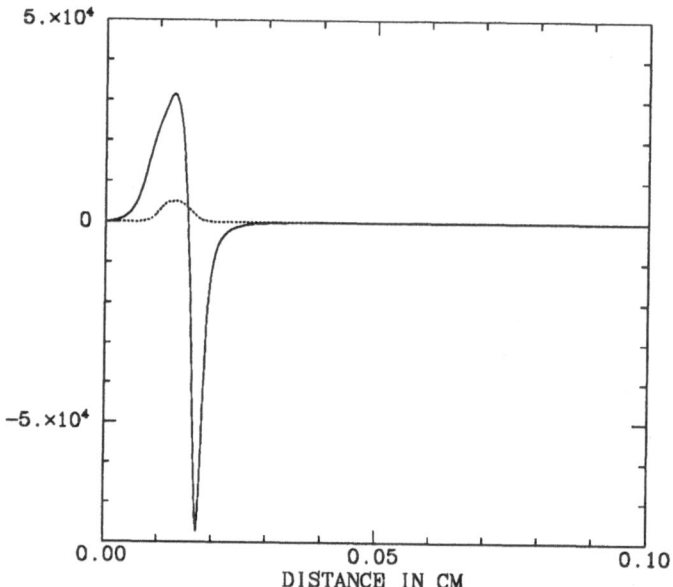

Figure 4. A comparison between the size of the heat conduction term in Eqn. (3.7) (solid) and the second term on the right-hand side of Eqn. (3.8) (dot) for a 30 atmosphere, stoichiometric, premixed methane-air flame.

Lewis Number Approximations

We define the Lewis number as the ratio of the thermal diffusivity to the mass diffusivity, i.e.,

$$\text{Le}_k = \frac{\lambda}{\rho D_k c_p}. \tag{3.9}$$

Our transport model can be simplified dramatically if the Lewis numbers of the various species in our kinetic model can be taken to be approximately constant. Using the detailed transport-finite rate skeletal chemistry model for a one atmosphere, stoichiometric, premixed methane-air flame, we generated the spatially dependent Lewis numbers for each of the 16 chemical species. This data was then fit to a constant. These constant values are plotted in Figures 5-8 versus the detailed transport-skeletal chemistry Lewis numbers for a counterflow methane-air diffusion flame with $a = 100$ sec^{-1}. We note that the fits are remarkably good. Similar results hold for the premixed flames. Table 3.1 lists the Lewis numbers we computed for each of the 16 species.

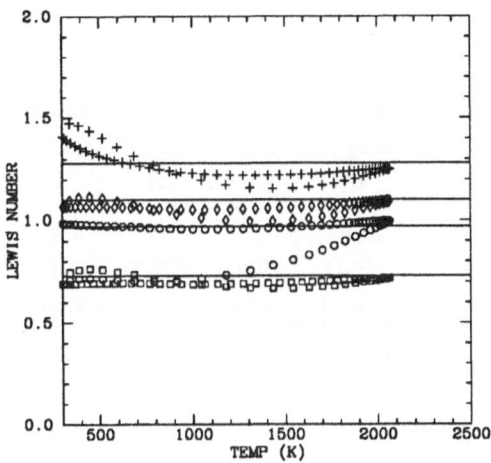

Figure 5. A comparison between the detailed transport-skeletal chemistry Lewis numbers and the premixed flame least squares fits (solid) as a function of the temperature for $CH_2O(+)$, $HO_2(\diamond)$, $CH_4(\circ)$ and $OH(\sqcap)$ for a counterflow methane-air diffusion flame with $a = 100$ sec^{-1}.

Diffusion Velocity Approximations

The species balance equations contain terms of the form

$$D = \frac{d}{dx}\left(\rho Y_k V_k\right), \tag{3.10}$$

where V_k is the diffusion velocity of the kth species. To relate the diffusion velocity to the mass diffsion coefficients, we employ the approximation

$$Y_k V_k \approx -D_k \frac{dY_k}{dx}, \tag{3.11}$$

where Y_k is the mass fraction of the kth species. Hence, we can write

$$\frac{d}{dx}\left(\rho Y_k V_k\right) \approx -\frac{d}{dx}\left(\rho D_k \frac{dY_k}{dx}\right). \tag{3.12}$$

To conserve mass we will solve for only $K - 1$ chemical species with the mass fraction of N_2 determined from mass conservation, i.e.,

$$Y_{N_2} = 1 - \sum_{k=1}^{K-1} Y_k. \tag{3.13}$$

With this formulation, the mass diffusivity can be related to the transport model in (3.4-3.5) via the Lewis numbers, i.e.,

$$\rho D_k = \frac{1}{Le_k} \left(\frac{\lambda}{c_p} \right) = \left(\frac{2.58 \times 10^{-4}}{Le_k} \right) \left(\frac{T}{298} \right)^{0.7}. \tag{3.14}$$

TABLE 3.1

Simplified Transport Model Lewis Numbers

Species	CH_4	O_2	H_2O	CO_2	H	O	OH	HO_2
Value	0.97	1.11	0.83	1.39	0.18	0.70	0.73	1.10
Species	H_2	CO	H_2O_2	HCO	CH_2O	CH_3	CH_3O	N_2
Value	0.30	1.10	1.12	1.27	1.28	1.00	1.30	1.00

Viscosity Model

The momentum equation in the counterflow model contains a term involving the viscosity of the mixture. From the definition of the Prandtl number we have

$$Pr = \frac{\mu c_p}{\lambda}. \tag{3.15}$$

By selecting a constant Prandtl number equal to 0.75 we can write

$$\frac{d}{dy} \left(\mu \frac{dU}{dy} \right) = Pr \frac{d}{dy} \left(\frac{\lambda}{c_p} \frac{dU}{dy} \right). \tag{3.16}$$

Chemistry Model

The skeletal reaction mechanism is listed in Table 3.2. It contains 10 reversible and 15 irreversible reactions. The choice of this mechanism is somewhat arbitrary. It was chosen as a compromise between large reaction networks [14-15] and reduced mechanisms [2-5]. When used in conjunction with a detailed transport model, flame speeds, species profiles and extinction strain rates compare well with available experimental data. It also serves as the starting point for the development of the reduced mechanism discussed below. Of particular importance, however, is the reaction

$$CH_4 + (M) \rightleftharpoons CH_3 + H + (M) \tag{3.17}$$

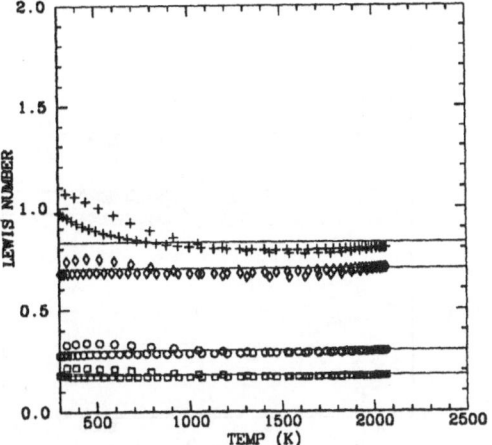

Figure 6. A comparison between the detailed transport-skeletal chemistry Lewis numbers and the premixed flame least squares fits (solid) as a function of the temperature for $H_2O(+)$, $O(\diamond)$, $H_2(\circ)$ and $H(\sqcap)$ for a counterflow methane-air diffusion flame with $a = 100$ sec^{-1}.

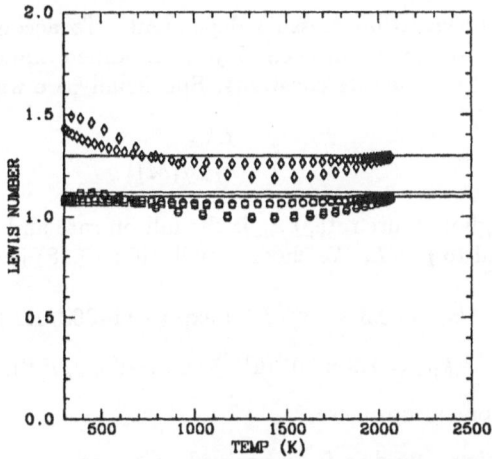

Figure 7. A comparison between the detailed transport-skeletal chemistry Lewis numbers and the premixed flame least squares fits (solid) as a function of the temperature for $CH_3O(\diamond)$, $O_2(\circ)$ and $H_2O_2(\sqcap)$ a counterflow methane-air diffusion flame with $a = 100$ sec^{-1}.

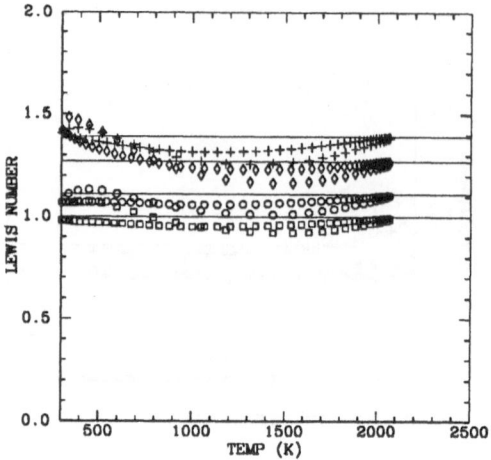

Figure 8. A comparison between the detailed transport-skeletal chemistry Lewis numbers and the premixed flame least squares fits (solid) as a function of the temperature for $CO_2(+)$, $HCO(\diamond)$, $CO(\circ)$ and $CH_3(\sqcap)$ for a counterflow methane-air diffusion flame with $a = 100$ sec^{-1}.

This reaction is known to be pressure dependent. To account for the functional dependence on the pressure we have employed a modified Lindemann approximation for the forward and reverse rate constants. Specficially, we write

$$k = \frac{\alpha k_\infty}{1 + k_{fall}/[M]}, \tag{3.18}$$

where k_∞ is the high pressure rate, k_{fall} is the fall-off rate and $[M]$ is the third body concentration equal to p/RT. We chose α such that (3.18) is equal to the Warnatz rates [16]

$$k_{10_f} = 2.3 \times 10^{38}(T^{-7}) \exp(-114360/RT), \tag{3.19}$$

$$k_{10_b} = 1.9 \times 10^{36}(T^{-7}) \exp(-9050/RT), \tag{3.20}$$

for reaction 10 at one atmosphere.

Reduced chemistry models for premixed and nonpremixed flames have been studied, for example, by Peters [2], Peters and Kee [3], Peters and Williams [4] and Bilger and Kee [5]. These models employ a combination of truncated steady-state and partial equilibrium assumptions to produce a reaction network with a smaller number of species and reaction steps than the original mechanism from which the reduction process originated. Of critical importance to this process is the form of the chemical production rates for the non steady-state species. By assuming $OH, O, HO_2, H_2O_2, CH_3, CH_3O, HCO$ and CH_2O are in steady-state, Peters has developed the following four-step reduced mechanism

$$CH_4 + 2H + H_2O \rightleftharpoons CO + 4H_2 \tag{I}$$

$$CO + H_2O \rightleftharpoons CO_2 + H_2 \qquad (II)$$

$$2H + M \rightleftharpoons H_2 + M \qquad (III)$$

$$O_2 + 3H_2 \rightleftharpoons 2H + 2H_2O \qquad (IV)$$

where the global reaction rates of the four steps (I-IV) are given by

$$w_I = w_{10} + w_{11} + w_{12}, \qquad (3.21)$$

$$w_{II} = w_9, \qquad (3.22)$$

$$w_{III} = w_5 - w_{10} + w_{16} - w_{18} + w_{19} - w_{22} + w_{24} + w_{25}, \qquad (3.23)$$

$$w_{IV} = w_1 + w_6 + w_{18} + w_{22}, \qquad (3.24)$$

with $w_i, i = 1, \ldots 25$, the rate of progress of elementary reaction i.

The concentration of the steady-state species are given by

$$[OH] = \frac{w_{3_b}}{[H_2]k_{3_f}}, \qquad (3.25)$$

$$[O] = \frac{-BB + \sqrt{(BB^2 - 4AA \cdot CC)}}{2AA}, \qquad (3.26)$$

where

$$AA = k_{13}B, \qquad (3.27a)$$

$$BB = BD + k_{13}(C - A) \qquad (3.27b)$$

$$CC = -AD, \qquad (3.27c)$$

and

$$A = w_{1_f} + w_{2_b} + w_{4_f}, \qquad (3.28a)$$

$$B = k_{1_b}[OH] + k_{2_f}[H_2] + k_{4_b}[H_2O], \qquad (3.28b)$$

$$C = w_{11_f} + w_{12_f}, \qquad (3.28c)$$

$$D = k_{10_b}[H] + k_{11_b}[H_2] + k_{12_b}[H_2O]. \qquad (3.28d)$$

The remaining steady-state species concentrations are given by

$$[HO_2] = \frac{w_5}{k_6[H] + k_7[H] + k_8[OH]}, \qquad (3.29)$$

$$[H_2O_2] = \frac{w_{21} + w_{22_b} + w_{23_b}}{k_{22_f}[M] + k_{23_f}[OH]}, \qquad (3.30)$$

$$[CH_3] = \frac{w_{11_f} + w_{12_f}}{k_{10_b}[H] + k_{11_b}[H_2] + k_{12_b}[H_2O] + k_{13}[O]}, \qquad (3.31)$$

$$[CH_3O] = \frac{w_{18}}{k_{19}[H] + k_{20}[M]}, \qquad (3.32)$$

$$[HCO] = \frac{w_{13} + w_{19} + w_{20}}{k_{16}[H] + k_{17}[M]}. \qquad (3.33)$$

In (3.25-3.33) [·] denotes species concentration and k_{i_f} and k_{i_b} denote the forward and reverse rate constants, respectively, for reaction i.

TABLE 3.2

Skeletal Methane-Air Reaction Mechanism
Rate Coefficients in the Form $k_f = AT^\beta \exp(-E/RT)$.
Units are moles, cubic centimeters, seconds, Kelvins and calories/mole.

	REACTION	A	β	E
1f.	$H + O_2 \rightarrow OH + O$	2.000E+14	0.000	16800.
1b.	$OH + O \rightarrow H + O_2$	1.575E+13	0.000	690.
2f.	$O + H_2 \rightarrow OH + H$	1.800E+10	1.000	8826.
2b.	$OH + H \rightarrow O + H_2$	8.000E+09	1.000	6760.
3f.	$H_2 + OH \rightarrow H_2O + H$	1.170E+09	1.300	3626.
3b.	$H_2O + H \rightarrow H_2 + OH$	5.090E+09	1.300	18588.
4f.	$OH + OH \rightarrow O + H_2O$	6.000E+08	1.300	0.
4b.	$O + H_2O \rightarrow OH + OH$	5.900E+09	1.300	17029.
5.	$H + O_2 + M \rightarrow HO_2 + M^a$	2.300E+18	-0.800	0.
6.	$H + HO_2 \rightarrow OH + OH$	1.500E+14	0.000	1004.
7.	$H + HO_2 \rightarrow H_2 + O_2$	2.500E+13	0.000	700.
8.	$OH + HO_2 \rightarrow H_2O + O_2$	2.000E+13	0.000	1000.
9f.	$CO + OH \rightarrow CO_2 + H$	1.510E+07	1.300	-758.
9b.	$CO_2 + H \rightarrow CO + OH$	1.570E+09	1.300	22337.
10f.	$CH_4 + (M) \rightarrow CH_3 + H + (M)^b$	6.300E+14	0.000	104000.
10b.	$CH_3 + H + (M) \rightarrow CH_4 + (M)^b$	5.200E+12	0.000	-1310.
11f.	$CH_4 + H \rightarrow CH_3 + H_2$	2.200E+04	3.000	8750.
11b.	$CH_3 + H_2 \rightarrow CH_4 + H$	9.570E+02	3.000	8750.
12f.	$CH_4 + OH \rightarrow CH_3 + H_2O$	1.600E+06	2.100	2460.
12b.	$CH_3 + H_2O \rightarrow CH_4 + OH$	3.020E+05	2.100	17422.
13.	$CH_3 + O \rightarrow CH_2O + H$	6.800E+13	0.000	0.
14.	$CH_2O + H \rightarrow HCO + H_2$	2.500E+13	0.000	3991.
15.	$CH_2O + OH \rightarrow HCO + H_2O$	3.000E+13	0.000	1195.
16.	$HCO + H \rightarrow CO + H_2$	4.000E+13	0.000	0.
17.	$HCO + M \rightarrow CO + H + M$	1.600E+14	0.000	14700.
18.	$CH_3 + O_2 \rightarrow CH_3O + O$	7.000E+12	0.000	25652.
19.	$CH_3O + H \rightarrow CH_2O + H_2$	2.000E+13	0.000	0.
20.	$CH_3O + M \rightarrow CH_2O + H + M$	2.400E+13	0.000	28812.
21.	$HO_2 + HO_2 \rightarrow H_2O_2 + O_2$	2.000E+12	0.000	0.
22f.	$H_2O_2 + M \rightarrow OH + OH + M$	1.300E+17	0.000	45500.
22b.	$OH + OH + M \rightarrow H_2O_2 + M$	9.860E+14	0.000	-5070.
23f.	$H_2O_2 + OH \rightarrow H_2O + HO_2$	1.000E+13	0.000	1800.
23b.	$H_2O + HO_2 \rightarrow H_2O_2 + OH$	2.860E+13	0.000	32790.
24.	$OH + H + M \rightarrow H_2O + M^a$	2.200E+22	-2.000	0.
25.	$H + H + M \rightarrow H_2 + M^a$	1.800E+18	-1.000	0.

[a] Third body efficiencies: $CH_4 = 6.5, H_2O = 6.5, CO_2 = 1.5, H_2 = 1.0, CO = 0.75, O_2 = 0.4, N_2 = 0.4$ All other species $= 1.0$
[b] Modified Lindemann form, $k = \alpha k_\infty / (1 + k_{fall}/[M])$ where $k_{fall} = .0063 \exp(-18000/RT)$.

4. Numerical Results

In this section we present the results of applying the model simplifications and the reduced chemistry discussed in the previous section to a sequence of premixed and nonpremixed flames. We focus our results on adiabatic flame speeds, extinction strain rates, and temperature and species profiles. Corresponding results for full transport calculations are often provided as a means of verifying the simplified transport model. Results are reported in the independent spatial coordinate, a normalized spatial coordinate and the mixture fraction.

Solution Method

In both the premixed and nonpremixed computations we are interested in following the solution as a system parameter is varied. For example, in the premixed problems we are interested in allowing the equivalence ratio to change as the pressure is held fixed. Similarly, we are also interested in varying the pressure for a fixed equivalence ratio. For the diffusion flames, the strain rate is the parameter of interest. While we could compute a single flame with specified values of these parameters and then use this computed solution as a starting estimate for a new problem with different parameter values, this is extremely inefficient. Instead, we apply an arclength continuation method such that the grid and the solution smoothly change as the parameter is varied. Specifically, the solution algorithm we implement proceeds with a phase-space, pseudo-arclength continuation method with Newton-like iterations and global adaptive gridding [17-18]. After we replace the continuous spatial derivatives by finite difference expressions, the premixed model and the diffusion flame model reduce to a system of the form

$$\mathcal{F}(\mathcal{X}, \gamma) = 0, \tag{4.1}$$

where $\mathcal{X}$ is the solution vector and γ is a system parameter (such as the equivalence ratio, the pressure or the strain rate). The solutions $(\mathcal{X}, \gamma)$ in (4.1) form a one-dimensional manifold which, as a result of the presence of turning points, cannot be parameterized in the form $(\mathcal{X}(\gamma), \gamma)$. The upper part of the manifold denotes the stable solutions and the lower part the unstable ones assuming there are no Hopf bifurcations.

To generate this solution set, $(\mathcal{X}, \gamma)$ is reparameterized into $(\mathcal{X}(s), \gamma(s))$ where s is a new independent parameter and γ becomes an eigenvalue. The system in (4.1) can now be written

$$\mathcal{F}(\mathcal{X}(s), \gamma(s)) = 0, \tag{4.2}$$

and the dependence of s on the augmented solution vector $(\mathcal{X}, \gamma)$ is specified by an extra scalar equation

$$\mathcal{N}(\mathcal{X}(s), \gamma(s), s) = 0, \tag{4.3}$$

which is chosen such that s approximates the arclength of the solution branch in a given phase space (see [19]). The system in (4.2-4.3) is solved by combining a first-order Euler predictor and a corrector step involving Newton-like iterations and adaptive gridding. To resolve the high activity regions of the dependent solution components, the mesh must be refined in the continuation calculations.

Premixed Flames

The first set of calculations we consider focuses on the variation of flame structure as a function of the equivalence ratio ϕ. We are interested in using the simplified transport model along with the skeletal and reduced reaction mechanisms reported in the previous section to compute flame structure for stoichiometric to lean methane-air flames at one atmosphere. Specifically, flames with equivalence ratios of $\phi = 1.0, 0.9, 0.8, 0.7$, and 0.6 are to be computed. By reparameterizing the premixed flame problem so that the equivalence ratio is the free parameter, we can apply the arclength continuation method discussed above to investigate premixed flame structure as the equivalence ratio changes. In this way we can follow accurately the movement of the flame while adaptively refining the flame front. This is in distinction to simply computing a flame for a given value of ϕ and then using this solution as the starting estimate for another computation with a different ϕ. Ordinarily, this approach will result in a mesh which could include the union of the two grids for the two flames. If this process is carried out over the entire equivalence ratio range, a very inefficient computation could result.

Three sets of equivalence ratio calculations were performed. One with the simple transport and the skeletal chemistry (denoted hereafter as "simple transport") described in the previous section, another with kinetic theory transport and skeletal chemistry (denoted hereafter as "full transport") and another with simple transport and the reduced four-step chemistry. In Figure 9 we illustrate the flame speed variation as ϕ is changed. The solid line corresponds to the simple transport calculations, the dashed line to the full transport calculations and the dotted line to the reduced chemistry computations. We note exceptional agreement over the entire range of ϕ considered. The flame speeds rarely differ by more than a couple of cm/sec.

The arclength continuation procedure outlined above can be modified to allow the pressure to become the free parameter in question. In this way we can investigate flame structure for variable pressure. This is particularly useful for flame studies in which the thermodynamic pressure varies by several atmospheres. The results in Figure 10 illustrate the variation of the adiabatic flame speed as a function of the pressure. Computations are reported for the 1-30 atmosphere regime for the skeletal chemistry cases employing both simple (solid) and full transport (dash) approximations. No steady-state reduced chemistry solutions (dot) were obtainable above 10.5 atmospheres. As was the case for the equivalence ratio computations, the flame speed results are in excellent agreement across the full range of comparable pressures.

Premixed flame structure for an atmospheric pressure, stoichiometric flame with simple transport and both skeletal and reduced chemistry are illustrated in Figures 11-13. The profiles are reported in terms of a normalized distance $\hat{x}$ through the flame. We define

$$\hat{x} = \dot{M}(x - x^\circ)c_p/\lambda, \tag{4.4}$$

where $\dot{M}$ is the mass flux through the flame, x° is the location of the maximum fuel consumption and λ/c_p is evaluated from Eqns. (3.4-3.5) of Section 3. We observe excellent agreement in terms of peak heights and the general shape of the profiles as a function of the normalized distance coordinate. Specific tabulations of premixed flame structure for three of the flames with simple transport and both skeletal and

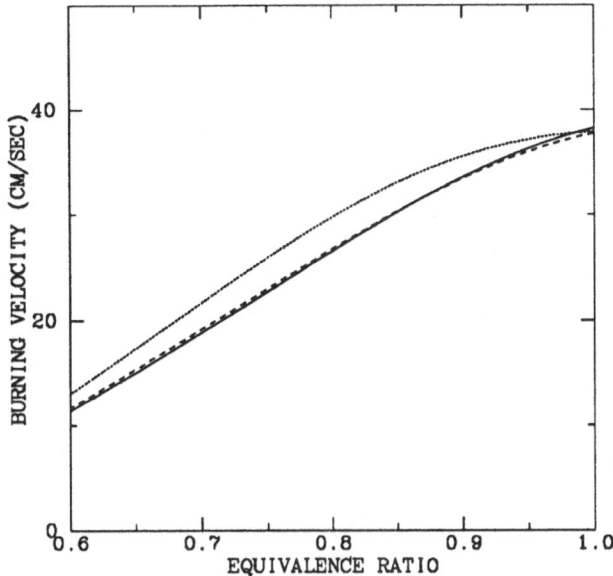

Figure 9. Variation of the burning velocity as a function of the equivalence ratio for simple transport-skeletal chemistry (solid), full transport-skeletal chemistry (dash) and simple transport-reduced chemistry (dot) models of premixed methane-air flames.

reduced chemistry are given in Tables 4.1 and 4.2, respectively. T° represents the temperature at the point of maximum fuel consumption.

Counterflow Diffusion Flames

Counterflow flames in the Tsuji configuration were studied from low strain rates until extinction using arclength continuation methods. As in the premixed case, computations were carried out for simple and full transport approximations with skeletal chemistry and for a simple transport-reduced chemistry model. In Figure 14 we illustrate the "C-shaped" extinction curves for the three flame models. The solid line is for simple transport and skeletal chemistry, the dashed line is for full transport and skeletal chemistry and the dotted line is for simple transport and reduced chemistry. The two skeletal chemistry models produced almost identical results except for the region near extinction. Extinction for the simple transport flame occurred at $a = 353$ sec^{-1} and at $a = 361$ sec^{-1} for the full transport model. The upper portion of the reduced chemistry curve compares quite favorably with the results of the other two models. In particular, extinction occurs at $a = 323$ sec^{-1}. Significant variations exist in the lower unphysical branch compared with the skeletal chemistry solutions. Temperature and major species profiles for a counterflow flame ($a = 100$ sec^{-1}) with simple transport and both skeletal and reduced chemistry are

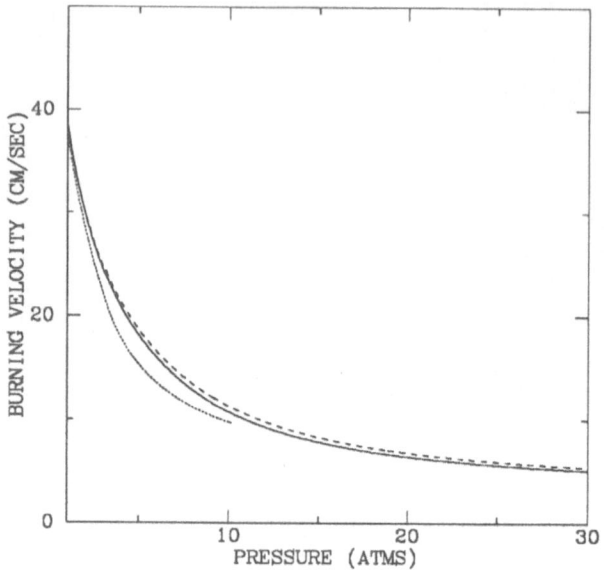

Figure 10. Variation of the burning velocity as a function of the pressure for simple transport-skeletal chemistry (solid), full transport-skeletal chemistry (dash) and simple transport-reduced chemistry (dot) models of premixed methane-air flames. No steady-state reduced chemistry solutions were obtained above 10.5 atmospheres.

reported in Figures 15-17 in terms of the mixture fraction ξ where

$$\xi = \frac{2Z_C/W_C + .5Z_H/W_H + (Z_{O,o} - Z_O)/W_O}{2Z_{C,F}/W_F + .5Z_{H,F}/W_H + Z_{O,o}/W_O}, \tag{4.1}$$

which is defined in terms of the corresponding element mass fractions. Here $Z_{j,F}$ is the mass fraction at any location of element j contained in the fuel stream and $Z_{j,o}$ is the mass fraction at any location of element j contained in the oxidizer stream. The subscripts C H and O denote, respectively, the elements carbon, hydrogen and oxygen. We can write

$$Z_j = \sum_{i=1}^{K}(a_{ij}W_jY_i)/W_i, \tag{4.2}$$

where a_{ij} is a stoichiometric coefficient denoting the number of atoms of element j in molecule i. We notice exceptional agreement between the two profiles of the temperature, $CH_4, O_2, N_2, H_2O, CO_2$ and H. The major differences between the two chemistry models appear in the peak heights of CO and H_2. The reduced chemistry model predicts somewhat higher values of these species compared to the skeletal chemistry model. Finally, in Tables 4.3 and 4.4 we summarize some of the results of the counterflow computations for simple transport-skeletal chemistry and simple

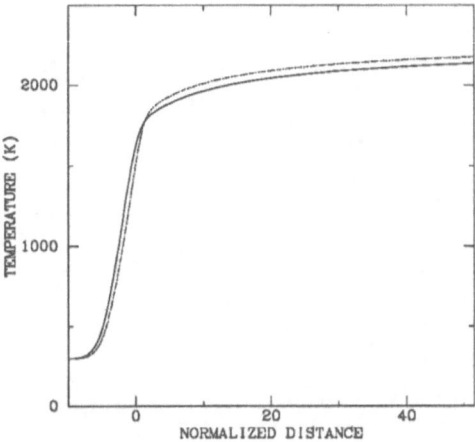

Figure 11. Computed temperature profiles as a function of the normalized distance for an atmospheric pressure, stoichiometric, methane-air flame employing a simple transport-skeletal chemistry (solid) and a simple transport-reduced chemistry (dot) model.

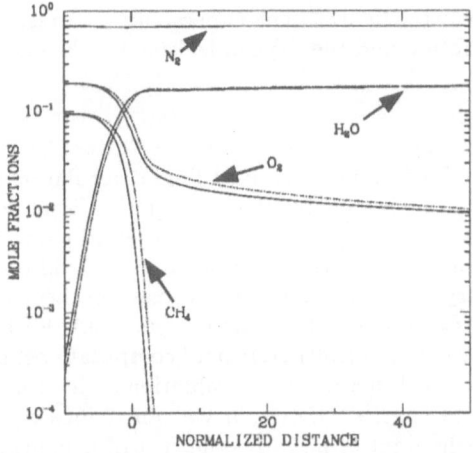

Figure 12. Computed major species profiles as a function of the normalized distance for an atmospheric pressure, stoichiometric, methane-air flame employing a simple transport-skeletal chemistry (solid) and a simple transport-reduced chemistry (dot) model.

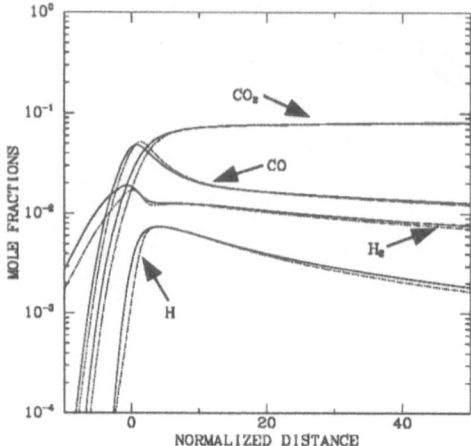

Figure 13. Computed major species profiles as a function of the normalized distance for an.atmospheric pressure, stoichiometric, methane-air flame employing a simple transport-skeletal chemistry (solid) and a simple transport-reduced chemistry (dot) model.

transport-reduced chemistry flames. The superscript ° corresponds to the point of maximum fuel consumption and the oxygen leakage is taken at $\xi = 0.1$

CPU Implications

The cost of a premixed or nonpremixed flame calculation in which Newton's method is employed is dominated by the cost of the formation of the Jacobian matrix. One of the goals of this paper has been the reduction of the overall CPU time needed to solve problems of this type with detailed kinetic theory transport and finite rate chemistry. The Jacobian matrices used in the solution algorithm are block tridiagonal [16]. Each matrix formation requires $3N + 1$ function evaluations where $N = K + 2$. For detailed (full) transport computations, in the limit of large K (species number), the cost of a function evaluation scales approximately with the square of the number of species. Hence, if we can reduce the number of species by a factor β, then in the limit of large K, the cost of a single function evaluation is reduced by a factor β^2 and the overall Jacobian cost is reduced by β^3. If in addition to the lower CPU costs due to a smaller number of species, we have a cost reduction α due to transport simplifications, then the overall cost reduction for a one-dimensional flame calculation will be $\alpha\beta^3$.

We have found that the CPU time required for the formation of the Jacobian matrix for a one atmosphere, stoichiometric, premixed methane-air flame with simple transport and skeletal chemistry is a factor $\alpha = 1.8$ lower than the formation of the corresponding full transport and skeletal chemistry Jacobian. Similarly, we have found the CPU time for the formation of the Jacobian matrix for a one atmo-

TABLE 4.1

Numerical Solutions for Premixed Flames
Simple Transport-Skeletal Chemistry
Pressure in Atmospheres

Flame Parameters	$\phi = 1.0, p = 1.0$	$\phi = 0.6, p = 1.0$	$\phi = 1.0, p = 10.0$
Flame Speed (cm/sec)	37.67	11.59	11.29
Peak H (mole)	7.38×10^{-3}	7.14×10^{-4}	1.04×10^{-3}
T° (K)	1621	1360	1921
T_{max}(K)	2272	1668	2329

TABLE 4.2

Numerical Solutions for Premixed Flames
Simple Transport-Reduced Chemistry
Pressure in Atmospheres

Flame Parameters	$\phi = 1.0, p = 1.0$	$\phi = 0.6, p = 1.0$	$\phi = 1.0, p = 10.0$
Flame Speed (cm/sec)	37.90	13.02	9.75
Peak H (mole)	7.39×10^{-3}	8.81×10^{-4}	9.68×10^{-4}
T° (K)	1577	1195	2037
T_{max}(K)	2285	1668	2314

sphere, stoichiometric, premixed methane-air flame with full transport and reduced chemistry to be a factor $\beta^3 = 4.8$ lower than the formation of the corresponding matrix with full transport and skeletal chemistry. If both the simplified transport and reduced chemistry are put together, we find the Jacobian CPU time to be reduced by a factor of $\alpha\beta^3 = 8.64$. In this problem the number of unknowns is reduced from 18 to 10 implying a $\beta^3 = 5.8$ and an overall reduction factor of $\alpha\beta^3 = 10.4$. These results indicate that we are obtaining 83% of the maximum CPU reduction for a problem with only 16 species. These results are extremely encouraging for multidimensional flame studies. In particular, as the number of species increases, we should be able to approach the theoretical results more closely. Moreover, larger multidimensional problems can now be performed on existing machines and some multidimensional problems can be moved to smaller high speed workstations.

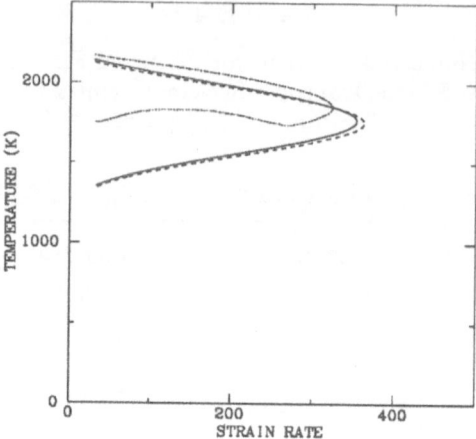

Figure 14. C-shaped extinction curves for the simple transport-skeletal chemistry (solid), the full transport-skeletal chemistry (dash) and the simple transport-reduced chemistry (dot) model of a counterflow, methane-air, diffusion flame. The upper branch corresponds to physical solutions and the lower branch to unphysical solutions.

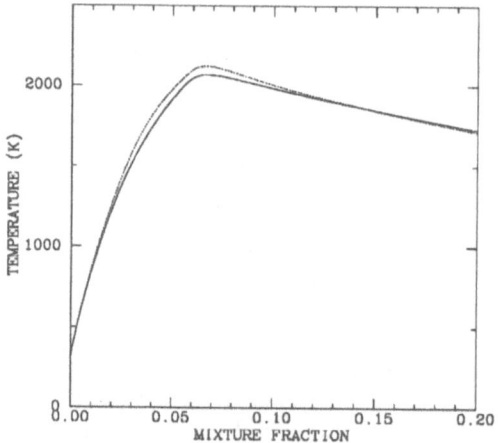

Figure 15. Computed temperature profiles as a function of the mixture fraction for a counterflow, methane-air, diffusion flame ($a = 100$ sec^{-1}) employing a simple transport-skeletal chemistry (solid) and a simple transport-reduced chemistry model (dot).

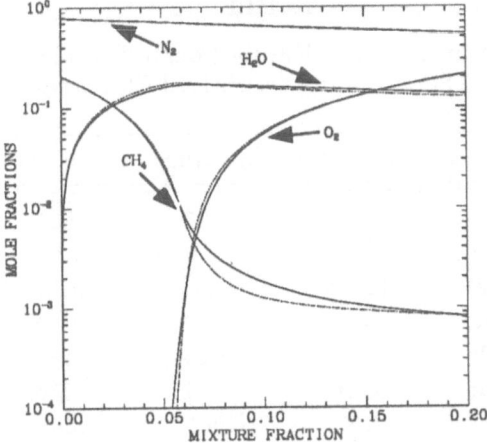

Figure 16. Computed major species profiles as a function of the mixture fraction for a counterflow, methane-air, diffusion flame ($a = 100$ sec^{-1}) employing a simple transport-skeletal chemistry (solid) and a simple transport-reduced chemistry model (dot).

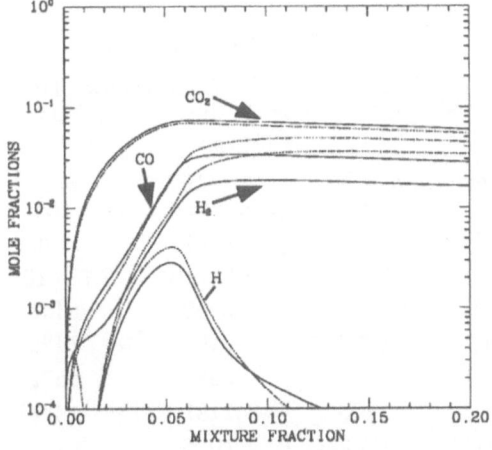

Figure 17. Computed major species profiles as a function of the mixture fraction for a counterflow, methane-air, diffusion flame ($a = 100$ sec^{-1}) employing a simple transport-skeletal chemistry (solid) and a simple transport-reduced chemistry model (dot).

TABLE 4.3

Numerical Solutions for Counterflow Flames
Simple Transport-Skeletal Chemistry

Flame Parameters	$a = 30$	$a = 100$	$a = 300$	$a_{ext} = 353$
Peak T (K)	2137	2067	1887	1766
Peak H (mole)	2.50×10^{-3}	2.87×10^{-3}	2.70×10^{-3}	7.88×10^{-5}
Peak H_2 (mole)	.02124	.0186	.0157	1.04×10^{-3}
Peak CO (mole)	.0325	.0334	.0414	.0470
Peak CO_2 (mole)	.0779	.0734	.0581	.0785
Peak H_2O (mole)	.1736	.1711	.1636	.1040
T°(K)	2137	2065	1883	1760
O°_2 (mole)	2.17×10^{-3}	5.48×10^{-3}	.0204	.0354
H°_2 (mole)	.0169	.0158	.0136	.0115
CO° (mole)	.0293	.0316	.0395	.0424
O_2 Leakage	1.53×10^{-4}	1.89×10^{-3}	.0141	.0280

TABLE 4.4

Numerical Solutions for Counterflow Flames
Simple Transport-Reduced Chemistry

Flame Parameters	$a = 30$	$a = 100$	$a = 300$	$a_{ext} = 323$
Peak T (K)	2171	2120	1936	1847
Peak H (mole)	4.07×10^{-3}	4.127×10^{-3}	2.78×10^{-3}	2.22×10^{-3}
Peak H_2 (mole)	.0544	.0358	.0138	.0102
Peak CO (mole)	.0639	.0491	.0425	.0432
Peak CO_2 (mole)	.0675	.0689	.0579	.0510
. Peak H_2O (mole)	.1878	.1781	.1631	.1560
T°(K)	2171	2108	1886	1799
O°_2 (mole)	1.81×10^{-3}	6.74×10^{-3}	.0303	.0409
H°_2 (mole)	.0253	.0192	.0096	7.72×10^{-3}
CO° (mole)	.0669	.0331	.0355	.0375
O_2 Leakage	9.30×10^{-5}	1.25×10^{-3}	.0141	.0234

References

[1] Smooke, M. D. and Giovangigli, V., "Premixed and Nonpremixed Test Problem Formulation," in Reduced Chemistry and Asymptotic Approximations of Methane-Air Flames, M. D. Smooke, Ed., Springer-Verlag, (1991).

[2] Peters, N., "Numerical and Asymptotic Analysis of Systematically Reduced Reaction Schemes for Hydrocarbon Flames," in Numerical Simulation of Combustion Phenomena, R. Glowinski et al., Eds., Lecture Notes in Physics, Springer-Verlag, (1985), p. 90.

[3] Peters, N. and Kee, R. J., "The Computation of Stretched Laminar Methane-Air Diffusion Flames Using a Reduced Four-Step Mechanism," *Comb. and Flame*, **68**, (1987), p. 17.

[4] Peters, N. and Williams, F. A., "The Asymptotic Structure of Stoichiometric Methane Air Flames," *Comb. and Flame*, **68**, (1987), p. 185.

[5] Bilger, R. W., Starner, S. H. and Kee, R. J., "On Reduced Mechanisms for Methane-Air Combustion in Non-Premixed Flames," *Comb. and Flame*, **80**, (1990), p. 135.

[6] Curtiss, C. F. and Hirschfelder, J. O., "Transport Properties of Multicomponent Gas Mixtures," *J. Chem. Phys.*, **17**, (1949), p. 550.

[7] Giovangigli, V. and Darabiha, N., "Vector Computers and Complex Chemistry Combustion," in : Proceedings of the Conference on Mathematical Modeling in Combustion, Lyon, France, NATO ASI Series, (1987).

[8] Giovangigli, V. and Smooke, M. D., "Application of Continuation Methods to Plane Premixed Laminar Flames," *Comb. Sci. and Tech.*, **87**, (1992), p. 241.

[9] Tsuji, H. and Yamaoka, I., "The Counterflow Diffusion Flame in the Forward Stagnation Region of a Porous Cylinder," Eleventh Symposium (International) on Combustion, Reinhold, New York, (1967), p. 979.

[10] Tsuji, H. and Yamaoka, I., "The Structure of Counterflow Diffusion Flames in the Forward Stagnation Region of a Porous Cylinder," Twelfth Symposium (International) on Combustion, Reinhold, New York, (1969), p. 997.

[11] Tsuji, H. and Yamaoka, I., "Structure Analysis of Counterflow Diffusion Flames in the Forward Stagnation Region of a Porous Cylinder," Thirteenth Symposium (International) on Combustion, Reinhold, New York, (1971), p. 723.

[12] Tsuji, H., "Counterflow Diffusion Flames," *Progress in Energy and Comb.*, **8**, (1982), p. 93.

[13] Dixon-Lewis, G., David, T., Haskell, P. H., Fukutani, S., Jinno, H., Miller, J. A., Kee, R. J., Smooke, M. D., Peters, N., Effelsberg, E., Warnatz, J. and Behrendt, F., "Calculation of the Structure and Extinction Limit of a Methane-Air Counterflow Diffusion Flame in the Forward Stagnation Region of a Porous Cylinder," Twentieth Symposium (International) on Combustion, Reinhold, New York, (1985), p. 1893.

[14] Miller, J. A., Mitchell, R. E., Smooke, M. D., and Kee, R. J., "Toward a

Comprehensive Chemical Kinetic Mechanism for the Oxidation of Acetylene: Comparison of Model Predictions with Results from Flame and Shock Tube Experiments," Nineteenth Symposium (International) on Combustion, Reinhold, New York, (1982), p. 181.

[15] Miller, J. A., Smooke, M. D., Green, R. M. and Kee, R. J., "Kinetic Modeling of the Oxidation of Ammonia in Flames," *Comb. Sci. and Tech.*, **34**, (1983), p. 149.

[16] Warnatz, J., "The Mechanism of High Temperature Combustion of Propane and Butane," *Comb. Sci. and Tech*, **34**, (1983), p. 177.

[17] Giovangigli, V. and Smooke, M. D., "Extinction of Strained Premixed Laminar Flames with Complex Chemistry," *Comb. Sci. and Tech.*, **53**, (1987), p. 23.

[18] Giovangigli, V. and Smooke, M. D., "Adaptive Continuation Algorithms with Application to Combustion Problems,", *App. Num. Math.*, **5**, (1989), p. 305.

[19] Keller, H. B., "Numerical Solution of Bifurcation and Nonlinear Eigenvalue Problems," in Applications of Bifurcation Theory, P. Rabinowitz, Ed., Academic Press, New York, (1977), p. 359.

Preferential Diffusion Effects in Diffusion Flames

Toshimi Takagi(1), Zhe Xu(2) and Masaharu Komiyama(1)

(1)Department of Mechanical Engineering, Osaka University
   Suita, Osaka, Japan
(2)Combustion Division, Central Research Institute of Electric Power Industry
   Yokosuka, Kanagawa, Japan

Abstract. Numerical computations were made of axisymmetric laminar jet diffusion flames taking into account detailed chemical kinetics and multicomponent diffusion. It was shown that preferential diffusion of heat and species causes significant amount of excess and deficit of enthalpy, and increase and decrease of mole fraction in the flame. These effects result in higher flame temperature than the maximum adiabatic temperature for the original fuel and/or significantly lower flame temperature than that estimated from the adiabatic temperature without preferential diffusion. The processes of the preferential diffusion effects were analyzed to reveal the factors inducing such higher and lower temperature. The present studies also emphasize the importance of the preferential diffusion effects in diffusion flames of high ambient pressure and different flow configurations.

Keywords. laminar diffusion flame, preferential diffusion, numerical simulation

1. Introduction

Reacting flows such as diffusion flames involve combined processes of momentum, heat and species transfer with chemical reactions of extensive heat release. The diffusion flame has been extensively studied because it is a typical flame configuration of general use and the investigation of the fundamental processes of the diffusion flame is useful for understanding the structure of various kinds of similar flames and microscopic structure of the turbulent diffusion flames. Early pioneering studies of diffusion flames employed flame sheet model without taking account of preferential diffusion. Although the effects of the preferential diffusion of heat and species in diffusion flames are pointed out by experiments (1), the detailed processes and the significance of the preferential diffusion effects have not yet been well analyzed so far by taking into account detailed chemical kinetics and multicomponent transport processes. Recent numerical computations of diffusion flames take into account chemical kinetics and multicomponent transport processes (2),(3),(4). Attention has been paid to the importance of preferential diffusion (5).

In the present study, computations are made of axisymmetric laminar jet diffusion flames taking into account detailed chemical kinetics and multicomponent diffusion. Significant effects are pointed out of the preferential diffusion of heat and species on the flame temperature generated in the diffusion flames. The study also reveals that the effects are much more stimulated in the flames at higher ambient pressure and of an inverted flame configuration. The processes of the preferential diffusion are elucidated in terms of the excess and deficit of enthalpy and H2 component.

2. Formulation

Formulations are made for the computations of axisymmetric laminar jet diffusion flames where gaseous fuel jet from a round tube nozzle is surrounded by a coaxial air. By considering flames in which the axial convective fluxes are much larger than diffusive fluxes, streamwise diffusion can be neglected to allow use of boundary layer type conservation equations of mass, axial momentum, species and energy :

Mass

$$\frac{\partial}{\partial x}(\rho u) + \frac{1}{r}\frac{\partial}{\partial r}(r\rho v) = 0 \qquad (1)$$

Momentum

$$\rho\left(u\frac{\partial u}{\partial x} + v\frac{\partial u}{\partial r}\right) = -\frac{\partial p}{\partial x} + \frac{1}{r}\frac{\partial}{\partial r}\left(r\mu\frac{\partial u}{\partial r}\right) + \rho g_x \quad (2)$$

Species

$$\rho\left(u\frac{\partial m_j}{\partial x} + v\frac{\partial m_j}{\partial r}\right) = \frac{1}{r}\frac{\partial}{\partial r}\left(r\rho D_{jm}\frac{\partial m_j}{\partial r}\right) + R_j \qquad (3)$$

Energy

$$\rho\left(u\frac{\partial h}{\partial x} + v\frac{\partial h}{\partial r}\right) = \frac{1}{r}\frac{\partial}{\partial r}\left(r\frac{\lambda}{C_{pm}}\frac{\partial h}{\partial r}\right)$$

$$+\frac{1}{r}\frac{\partial}{\partial r}\left\{r\sum_{j=1}^{n}\left(\rho D_{jm} - \frac{\lambda}{C_{pm}}\right)h_j\frac{\partial m_j}{\partial r}\right\} \qquad (4)$$

In these equations, x and r represent the axial and radial coordinates, respectively; u and v, the axial and radial fluid velocities; p, pressure; T, temperature; ϱ, density; h, enthalpy; mj, Cp,j and hj, mass fraction, specific heat and enthalpy of the j-th species; Rj, species production by chemical reactions; μ, viscosity; λ, thermal conductivity; and Djm, the effective diffusivity of the j-th species in the mixture. Radiant heat transfer, thermal diffusion and viscous dissipation are neglected in this analysis. Density is obtained from the ideal gas equation. Temperature in the flame is related to enthalpy and species concentrations. The thermodynamic properties of specific heat Cpj, Cpm and enthalpy are from JANAF Table (6) and transport properties of viscosity, heat conductivity and diffusivity are from the reference (7),(8).

3. Reaction Scheme and Rate Constant

We take into consideration 10 species of H2, O2, H2O, N2, O, H, OH, HO2, N, and

Table 1 Reaction and rate constant

$$k_f = AT^n exp(-E/R_0 T) \quad \text{(unit: mol, J, s, cm, K)}$$

Reaction		A	n	E
(1) $H + O_2$	$= OH + O$	2.24E14	0.0	70300
(2) $O + H_2$	$= OH + H$	1.74E13	0.0	39600
(3) $H_2 + OH$	$= H + H_2O$	2.19E13	0.0	21600
(4) $OH + OH$	$= H_2O + O$	5.75E12	0.0	3270
(5)a) $H + H + H_2$	$= H_2 + H_2$	9.20E16	-0.6	0
b) $H + H + N_2$	$= H_2 + N_2$	1.00E18	-1.0	0
c) $H + H + O_2$	$= H_2 + O_2$	1.00E18	-1.0	0
d) $H + H + H_2O$	$= H_2 + H_2O$	6.00E19	-1.25	0
(6) $O + O + M_1$	$= O_2 + M_1$	2.62E16	-0.84	0
(7) $OH + H + M_2$	$= H_2O + M_2$	1.17E17	0.0	0
(8) $H + O_2 + M_3$	$= HO_2 + M_3$	2.70E18	-0.86	0
(9) $H + HO_2$	$= OH + OH$	2.50E14	0.0	7950
(10) $H + HO_2$	$= O_2 + H_2$	2.50E13	0.0	2910
(11) $H + HO_2$	$= H_2O + O$	5.00E13	0.0	4190
(12) $O + HO_2$	$= OH + O_2$	4.80E13	0.0	4190
(13) $OH + HO_2$	$= H_2O + O_2$	5.00E13	0.0	4190
(14) $N_2 + O$	$= NO + N$	1.40E14	0.0	315700
(15) $N + O_2$	$= NO + O$	6.40E09	1.0	6250
(16) $N + OH$	$= NO + H$	4.20E13	0.0	0

Reaction	Third Body and Factor of Reaction Rate
(6)	$M_1 = N_2$
(7)	$M_2 = H_2O + 0.25H_2 + 0.25O_2 + 0.2N_2$
(8)	$M_3 = H_2 + 0.44N_2 + 0.35O_2 + 6.5H_2O$

NO. The reaction scheme and the forward reaction rate constants of H-O system including 13 pairs of elementary reactions are listed in Table 1 (3). The rate constant kf of each forward reaction used in the computations are also listed. R0 is the universal gas constant. Backward reaction rate constants are given in terms of the forward reaction rate constant and the equilibrium constants. Reaction rate Rj per unit volume and time in eq.(3) is evaluated by the following equation.

$$R_j = M_j \sum_{i=1}^{m} (\nu''_{j,i} - \nu'_{j,i}) k_i \prod_{l=1}^{s} [C_l]^{\nu'_{l,i}} \qquad (5)$$

where, $\nu'_{j,i}$ or $\nu''_{j,i}$ is stoichiometric coefficient of reactant or product of j species in reaction i; k_i, rate constant of reaction i; $[C_l]$, molar concentration of l species. Forward and backward reactions are considered separately and so 32 reactions are included in Table 1. The value of m in eq.(5) is 32 and s is 10.

4. Numerical Procedure and Conditions

The differential equations are discretized by a control volume method. For the flame at atmospheric pressure, 103 meshes are located in radial direction of the flame with nonuniform interval. The minimum mesh size in radial direction near the reaction zone is 0.025 mm. The mesh size in axial direction is 0.002 mm. At higher pressure, smaller mesh size in radial direction is used. A marching integration is performed. The numerical integration method is basically that of the reference (9). Incorporation of an implicit computation of the reaction source terms is essential to stable integration. The upstream computational boundary is set at the plane of the fuel nozzle tip. The fuel or air of 25°C flows at the inner or outer parts of the nozzle rim. Axial velocity profile at the exit of the fuel nozzle and outside the nozzle is given. Radial velocity component is given to be zero. Adiabatic equilibrium concentration of species and the temperature of a stoichiometric mixture of fuel and air are given at the fuel nozzle rim of 0.3 mm thickness to initiate the reactions. It was checked that the upstream boundary conditions given at the rim make little difference to the computational results at 5 mm downstream. On the central axis, a zero gradient species concentration, temperature and velocity are given. At the outer free boundary, values of all variables are given.

5. Results and Discussion

5.1 Comparison of Computation and Experiments

The measured radial profiles of species concentration of H2, O2, N2, and OH concentration and temperature (3),(10) at the cross section of 30 mm from the fuel nozzle exit are compared with computations in a flame under the following conditions: the fuel is the mixture of H2 and N2 with mole fraction of 0.3 : 0.7; the average velocity at the fuel nozzle exit and the surrounding average air velocity is 5.2 and 1.7 m/s, respectively; the inner diameter of the fuel nozzle is 4.2 mm. In the experiments, H2, O2 and N2 concentration was analyzed by a gas chromatograph after gas sampling from the flame, OH radical concentration which plays important roles in the flame reactions was obtained by a laser induced fluorescence method and temperature was detected by a laser Rayleigh scattering method (10). H2 concentration is large near the axis and O2 concentration is large in the outer region. H2 and O2 concentration decreases at r=3 mm where stoichiometric condition prevails. The maximum temperature and OH concentration are observed at the radial distance near the stoichiometric condition. The computation predicts well the measured temperature, major species and OH concentration(3) which confirms that the present computation can describe the diffusion flame of H2/N2.

5.2 Flame Properties of a H2/N2 Flame

Figures 1(a),(b),(c) show computed radial profiles of major species concentration of H2, O2, H2O and temperature at various cross sections in the flame. The flame tip is at x=140 mm where H2 disappears and the maximum temperature is located on the central axis. The radial temperature profiles have maximum at off-axis region in the upstream up to the flame tip. The profiles of temperature are similar to those of H2O concentration. Maximum temperature at each cross section (so called flame temperature) changes significantly al
ong with x direction. The highest flame temperature is 1760 K located at about x=20 mm and the lowest is 1294 K at the flame tip. The highest and lowest flame temperature is higher and lower than the adiabatic maximum equilibrium temperature 1660 K of the original fuel and air. The difference between the highest and lowest flame temperature is considerable amount of 466 K which suggests that the preferential diffusion of heat and species plays important roles because the flame temperature should be nearly constant if there were no preferential diffusion as shown in Fig.1(d).

In order to analyze the effects of the preferential diffusion between heat and species, profiles of enthalpy at several cross sections in the flame are shown in Fig.2. The enthalpy is the sum of sensible heat and the heat of formation of each species. The enthalpy at the inlet cross section is uniformly zero. It is noted that the excess enthalpy is observed near or outside the maximum temperature region, while an enthalpy deficit occurs in the outer and inner part of the flame. The excess and deficit of enthalpy are induced by the second term on the right hand side of the energy conservation equation (4). This term (so called term 2) comes from the preferential

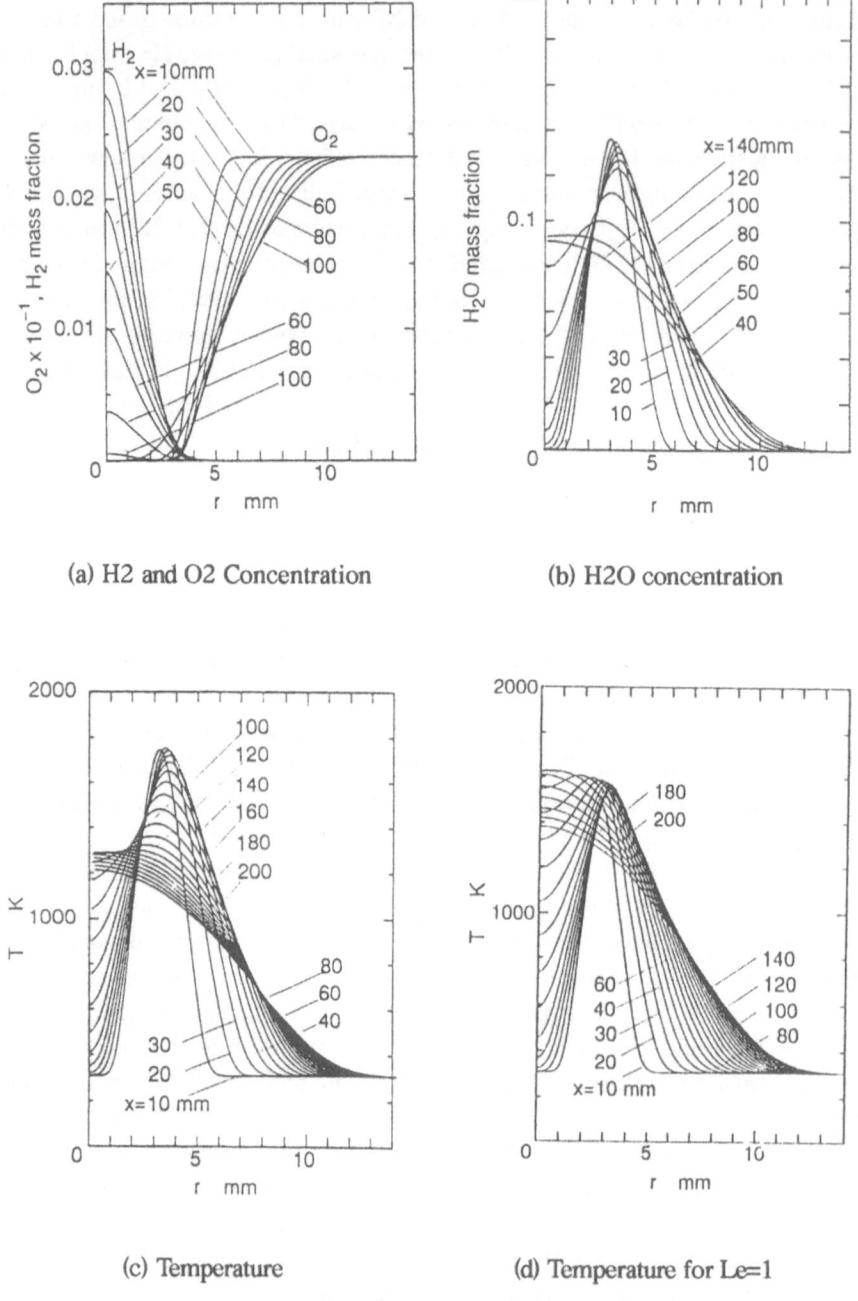

(a) H2 and O2 Concentration

(b) H2O concentration

(c) Temperature

(d) Temperature for Le=1

Fig.1 Radial profiles of concentration and Temperature in the flame

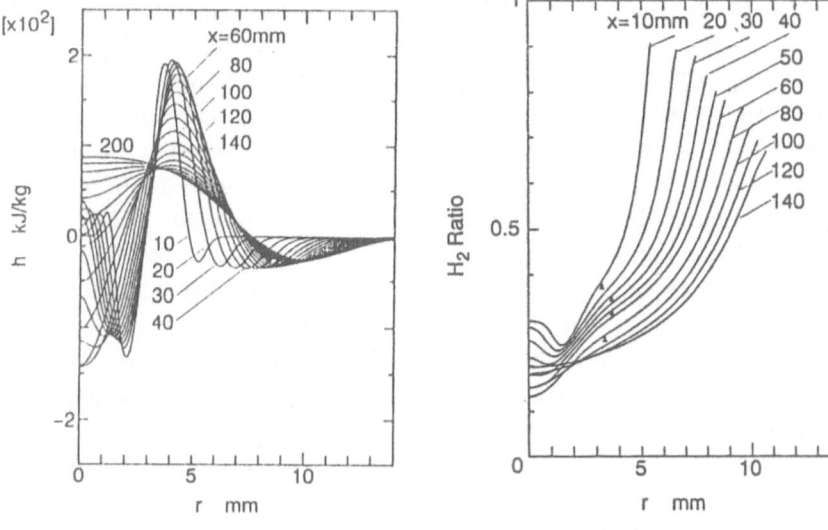

Fig.2 Radial profiles of enthalpy

Fig.3 Radial profiles of H2 Ratio
(Arrows indicates the maximum
temperature of the cross section)

diffusion of heat and species. If there were no such preferential diffusion, that is, Lewis number $Djm/(\lambda/\varrho \; Cpm)$ were unity for all species, the enthalpy profiles would be kept uniform because the term 2 disappears and the first term on the right hand side of equation (4) (so called term 1) is a diffusion term which just drives the enthalpy uniform. The term 2 consists of a sum of each species contribution which is evaluated by computation. It revealed that the excess of enthalpy near the stoichiometric region is dominated by the species of H, H2 and H2O, all of which have Lewis number larger than unity. In the downstream region where little H and H2 exist, H2O is the dominant species governing the excess and deficit of enthalpy. The excess or deficit of enthalpy induced as noted above is one of the reasons responsible for a' higher or lower flame temperature than the adiabatic temperature of the original fuel.

In order to investigate the effect of preferential diffusion among each species, the quantity, so called H2 Ratio, is introduced. It is defined by the mole fraction of H2 in the fuel which should be before combustion. If there were no preferential diffusion, H2 Ratio is 0.3 in the above flame which corresponds to the predetermined mole fraction of H2 in the original fuel. But, it changes from the mole fraction in the original fuel due to the preferential diffusion. It is the measure indicating how much H2 in the original fuel is condensed or diluted by the preferential diffusion in the flame. H2 Ratio is evaluated from the constraints of conservation of elements of H, C, N and O during combustion and species concentrations computed based on the equations taking account of preferential diffusion as described above.

Figure 3 shows radial profiles of H2 Ratio in the same flame of Fig.1. In the upstream region up to x=50 mm, H2 Ratio near the position of maximum temperature at about r=3 mm becomes larger than the H2 Ratio of 0.3 of the original fuel. This increase of H2 Ratio is caused by the diffusion of highly diffusive H and H2 from the central part to the reaction zone. In the downstream, the H2 Ratio near the maximum temperature region at about r=3 mm decreases below 0.3 of the original fuel. This decrease comes from the fact that H2O formed from H2 at the reaction zone diffuses fast as compared with N2 in the original fuel. So, N2 tends to remain at the reaction zone to dilute H2 at the reaction zone and H2 Ratio increases in the outer part of the flame.

The flame temperature in the upstream region exceeds the maximum adiabatic flame temperature of the original fuel which results from the excess enthalpy and the concentrating effect of H2 Ratio and its decrease in the downstream results from the decrease of the excess enthalpy and the decrease of H2 Ratio.

5.3 Effects of Pressure

Effects of pressure were investigated by computations, because fundamental

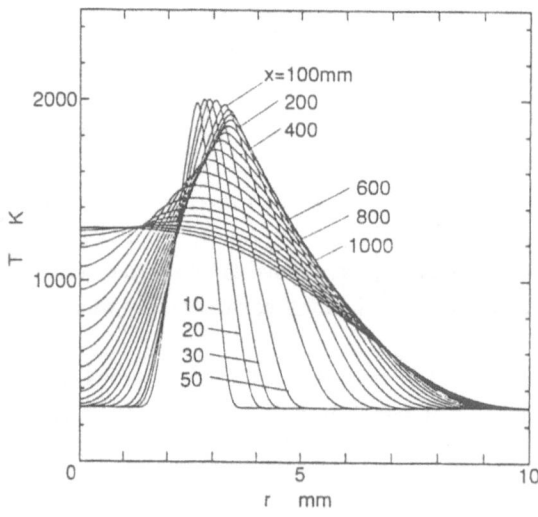

Fig.4 Radial profiles of temperature
(Ambient pressure is 1Mpa)

understanding of combustion at high pressure is useful for high pressure combustion in furnaces, gas turbine combustors and internal combustion engines. Ambient pressure is changed from 0.1 to 4.4 Mpa. The average velocity at the fuel nozzle exit is 3.5 m/s. Other conditions are the same with those of section 5.1.

Figure 4 is radial profiles of temperature in the flame at 1 Mpa. It is noted that the profiles become steeper and the maximum flame temperature (2005K) is much higher and the flame tip temperature (1280K) is much lower than the adiabatic flame temperature (1661K) of the original fuel. This tendency is more noticeable at higher pressure even though the maximum adiabatic flame temperature for the original fuel is not so influenced by the pressure. So, the much higher maximum flame temperature and much lower flame tip temperature than the equilibrium temperature of the original fuel should result from the preferential diffusion effects.

It was confirmed that the more excessive enthalpy and H2 Ratio at higher pressure induce higher maximum flame temperature and the decrease of enthalpy and lower H2 Ratio near the flame tip region results in the lower temperature than that of the atmospheric flame.

5.4 Effects of Flow Configuration

Two flames of different flow configurations are computed and compared with each

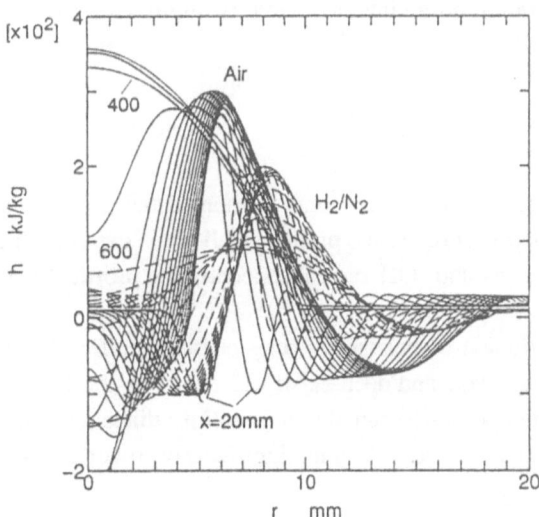

Fig.5 Radial profiles of temperature. Broken lines: usual flame, profiles are at 20mm interval for x=20~200mm, 50mm interval for x=200~600mm.   Solid lines: inverted flame, profiles are at 20mm interval.

other. One is a usual diffusion flame where gaseous fuel comes from a round tube nozzle surrounded by a coaxial air flow. Another is a flame where air instead of the fuel comes from the round nozzle surrounded by a coaxial fuel flow. Here we call the former flame a usual flame and the latter an inverted flame.

Figure 5 shows radial temperature profiles in two flames. Broken lines are for the usual flame and solid lines for the inverted flame where the fuel is the same with that of Fig.1. The inner diameter is 10 mm and uniform velocity of fuel and air for both flames at the inlet cross section is 5 m/s. From Fig. 5, it is pointed out that (1)in the usual flame, the flame temperature defined by the peak temperature at each cross section decreases in the x direction and it is about 1270 K at the flame tip of x=600 mm, (2)on the contrary, the flame temperature in the inverted flame increases along with flow until it reaches 2430 K at the flame tip of x=360 mm which is extraordinarily higher than that of the usual flame.

In order to investigate the reasons for the unusually high temperature generated in the inverted flame, enthalpy and H2 Ratio are evaluated in the flame. It was confirmed that (1)higher excess enthalpy is induced by the preferential diffusion of heat and species in the reaction zone near the peak temperature and near the flame tip of the inverted flame, (2)H2 Ratio near the reaction zone and flame tip region in the inverted flame reaches 0.5~0.7 which is much larger than the H2 Ratio of 0.3 of the original fuel even though the H2 Ratio is about 0.2 near the flame tip of the usual flame. The increase of the H2 Ratio in the inverted flame comes from the accumulation of highly diffusive H2O and H2 in the central axis. It can be safely said that the reason of the high temperature generation in the inverted flame is highly excessive enthalpy and H2 Ratio.

6. Summary

(1) Numerical computations by taking into account detailed chemical kinetics and multicomponent transport processes predict well the experimental profiles of temperature, major species and OH radical concentration profiles in H2/N2 laminar jet diffusion flames.

(2) The preferential diffusion of heat and species causes significant excess and deficit of enthalpy, and also increase and decrease of H2 Ratio in the flame. These effects induce higher flame temperature than the maximum adiabatic temperature for the original fuel in the upstream region, and significantly lower temperature near the flame tip region.

(3) For higher pressure, the effects of excess and deficit of enthalpy and H2 Ratio due to the preferential diffusion become larger which induce significantly high flame temperature in the upstream region and low temperature near the flame tip.

(4) In the inverted flame where air instead of fuel comes from central nozzle surrounded by a coaxial fuel flow, extraordinarily high temperature is generated in the

reaction zone and the flame tip region. This high temperature results from the highly excessive enthalpy and accumulation of H2 component induced from preferential diffusion of heat and species.

References

(1) Ishizuka,S. and Sakai,Y., Structure and Tip-opening on Laminar Diffusion Flames, Proc. 21st Symposium (International) on Combustion, (1986), pp.1821-1828, The Combustion Institute.

(2) Smooke, M.D., Lin, P., Lam, J.K. and Long, M.B., Computational and Experimental Study of a Laminar Axisymmetric Methane-Air Diffusion Flame, 23rd Symposium (International) on Combustion, (1990), pp.575-582, The Combustion Institute.

(3) Takagi, T., Tada, K. and Komiyama, M., 1990, Numerical Simulation of the Hydrogen Diffusion Flame, Bulletin of JSME, 56, 527, pp.2109-2114.

(4) Fukutani, S., Kuniyoshi, N. and Jinno, H., Flame Structure of an Axisymmetric Hydrogen-Air Diffusion Flame, 23rd Symposium (International) on Combustion, (1990), pp.567-573, The Combustion Institute.

(5) Takagi, T. and Zhe Xu, Numerical Analysys of Laminar Diffusion Flames (Effects of Preferential Diffusion of Heat and Species, Combustion and Flame, 96, 1/2, (1994), pp.50-59.

(6) Stull,D.R. and Prophet,H., JANAF Thermochemical Tables, 2nd ed., (1971),U.S. Dept. of Commerce, Washington.

(7) Bird,R.B., Stewart,W.E. and Lightfoot,E.N., Transport Phenomena, Chap.2, (1960), John Wiley Sons.

(8) Kee,R.J., Dixon-Lewis,G., Warnats,J., Coltrin,M.E., and Miller,J.A., A Fortran Computer Code Package for the Evaluation of Gas-Phase Multicomponent Transport Properties, (1986), SAND86-8246.

(9) Spalding,D.B., HMT Genmix-A General Computer Program for Two-Dimension Parabolic Phenomena, (1977), Pergamon Press.

(10) Komiyama,M., Kema,T. and Takagi,T., Measurements of OH Concentration and Temperature in Diffusion Flame by Excimer Laser-Induced Fluorescence and Rayleigh Scattering, Bulletin of JSME, 56, 523, (1990), pp.810-816.

Modeling Low Reynolds Number Microgravity Combustion Problems

Howard R. Baum

National Institute of Standards and Technology, Gaithersburg, MD 20899, USA

Abstract. The limit of low Reynolds number and low gravity allows a novel approximation to be developed for the equations of combustion theory. When the velocity field is separated into solenoidal and irrotational parts it is possible to show that the former is negligible to the lowest order in Reynolds number. Hence, the computation of the velocity field from the Navier Stokes equations may be replaced by the simpler task of solving Poissons equation for the irrotational field. This irrotational field is then substituted into the convective terms in energy and species equations, enabling solutions to be found for scalar quantities like the temperature and species mass fractions. The procedure is illustrated by examining two problems; the flow of air past a sphere blowing a light gas like helium, and combustion of volatiles blown from a porous sphere in an oxidizing crossflow.

Keywords. asymptotics, blowing, combustion, diffusion flame, microgravity

1   Introduction

In this article a rational method is examined for solving the equations of combustion theory under the simultaneous limits of low Reynolds number and zero gravity. The method consists of the following steps: First, the velocity field is separated into its irrotational and solenoidal parts, where the former contains the expansion due to heat or mass addition and the latter the vorticity. Second, a simple Poisson equation is derived and solved for the irrotational part of the flow. Finally, the solenoidal part of the flow is demonstrated to be of higher order in Reynolds number than the potential flow uniformly in space.

Rather than attempt to demonstrate this formally in a general way, two prototype problems that are simple enough to be solved analytically are chosen. The first problem, the flow past a small sphere blowing a light gas of a density similar to that achieved in combustion processes, contains the essence of the ideas behind this approach. Both the density and velocity fields are obtained in enough detail to justify the method. The analogous combustion problem is then considered, where the gas issuing from the sphere reacts with the oxidizing stream. Here, the purpose is to demonstrate that the combustion processes themselves modify the details of the calculation, but do not change the overall scheme. Finally, it is noted that the potential equation can be found quite generally in terms of the temperature and heat release.

The objective is not primarily to obtain solutions to the problems chosen, which have been studied in some detail previously. The combustion problem has been analyzed by Fendell et. al. [1] and Gogos et. al. [2], while the vaporizing sphere has also been treated by Fendell [3]. However, none of these investigations, nor any of the other studies of low Reynolds number that followed from the pioneering works of Kaplun and Lagerstrom [4] and Proudman and Pearson [5] noticed that the flow fields used to approximate the convective transport were irrotational at leading order in Reynolds number. The consequences of this observation extend well beyond the class of problems that are amenable to analytical treatment.

The practical value of this solution method arises when the problem of interest is too complicated to be solved analytically. Then, the computation of solutions to the Navier-Stokes equations can be replaced by the much simpler task of solving the Poisson equation for the velocity potential. This approach is especially useful for problems in three dimensions and/or problems with complex geometries, where solutions to the Navier-Stokes equations, even at low Reynolds numbers, become formidible undertakings. Even transient axially symmetric or two dimensional problems with simple geometries in which complex physical or chemical mechanisms must be represented in detail can benefit from this procedure. Computational resources not spent on solving the Navier-Stokes equations can be devoted to studying the mechanisms of direct interest. Indeed, the method was first developed to study just such a problem; the heat and mass transfer phenomena associated with the thermal degradation and combustion of complex materials in a microgravity environment [6], [7], [8].

The physical interpertation of the mathematical analysis developed below is that in regions of the order of a few sphere radii from the surface, the convective terms in the species, energy, and momentum equations are dominated by the blowing from the surface. In the absence of blowing, the convective terms can be ignored altogether, and the classical Stokes flow equations apply. However, the blowing effects can be accomodated by a potential flow since they are determined by the component of the velocity normal to the surface. Far from the body, the convective velocity is dominated by the uniform free stream flow, so that the Oseen approximation is valid. However, a uniform flow is also a potential flow. The irrotational component of the flow satisfies a Poisson equation as a consequence of mass conservation. Hence, a solution to this equation which is uniform far from the body and yields the required blowing velocity at the surface meets all the requirements of a low Reynolds number flow everywhere.

The combustion literature is populated with examples where the fluid mechanics is simplified for the sake of a deeper analysis of the thermal and mass transfer aspects of the problem under consideration. Very few of these simplifications are rationally derived; most are made arbitrarily. For example, de Ris [9] used the Oseen approximation in his landmark study of opposed flow flame spread. Wichman et. al. [10] investigated the effects of assuming different forms of the velocity profile on this problem. Clearly such simplifications, when carefully used, can enhance our understanding of many complex combustion processes. However, there are many cases when one does not want to simplify the fluid mechanics in such an arbitrary manner. The approach studied below, when used in a parameter range for which the approximations can be justified, offers a way around these difficulties.

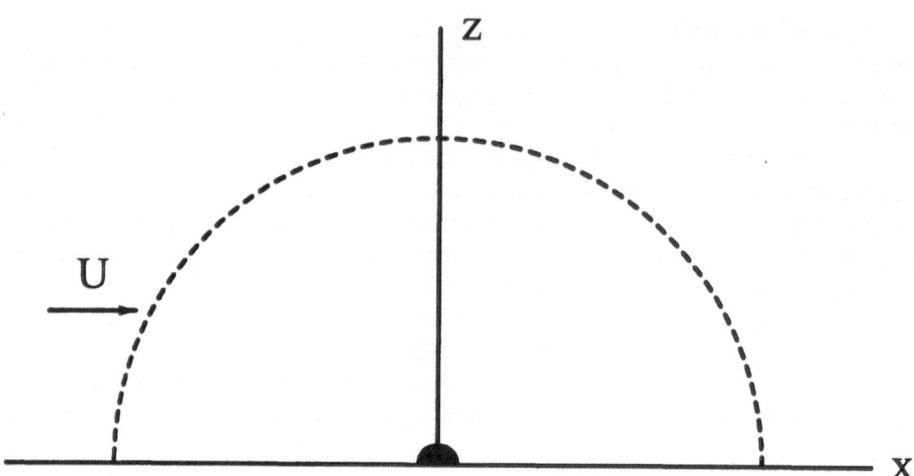

Fig. 1: Schematic showing division of flow domain past porous sphere. The inner region, $r = O(1)$ in units of sphere radius, is dominated by blowing from the surface. The outer region, $r = O(1/Re)$, is dominated by the uniform flow U.

2 Isothermal Blowing of a Light Gas

First consider the uniform blowing of a light gas (e.g. helium) from the surface of a sphere immersed in a stream of heavy gas (e.g. air). In this case there is no chemical reaction, the temperature is uniform, and the pressure and diffusivity are effectively nearly constant. The only unknowns are the local density, velocity, and helium mass fraction. The air stream of speed U and density ρ_∞ blows past the sphere of radius a. The geometry is shown in Figure 1. Let $\vec{u}$ be the local velocity normalized with respect to U, ρ the density made dimensionless with respect to ρ_∞, and Y, the helium mass fraction. Under these assumptions, the mass and species conservation equations, together with the equation of state take the form

$$\frac{D\rho}{Dt} + \rho(\nabla \cdot \vec{u}) = 0 \qquad (1)$$

$$\rho\frac{DY}{Dt} = \frac{1}{ReSc}\nabla \cdot (\rho\nabla Y) \qquad (2)$$

$$\rho = \frac{1}{1 + (\alpha - 1)Y} \qquad (3)$$

Here, the spatial coordinate vector $\vec{r}$ with origin at the sphere center has been normalized with respect to a, the Reynolds number $Re = Ua/\nu$, the Schmidt number $Sc = \nu/D$, and the kinematic viscosity ν and diffusivity D are taken as constant. The parameter $\alpha = W_{air}/W_{helium}$ is the molecular weight ratio of the gases.

The velocity field $\vec{u}$ is decomposed into a solenoidal part $\vec{v}$ and an irrotational flow $\nabla\phi$ as follows:

$$\vec{u} = \vec{v} + \nabla\phi \qquad (4)$$

$$\nabla \cdot \vec{v} = 0 \tag{5}$$

This decomposition is crucial to the analysis, since the key step is the observation that $\vec{v}$ is of higher order in Re than $\nabla\phi$. Without approximation, a linear combination of equations (1) and (2) yields a simple relation between ϕ and Y:

$$\nabla\phi = (\frac{\alpha - 1}{ReSc})\rho\nabla Y + \nabla\Phi \tag{6}$$

$$\nabla^2\Phi = 0 \tag{7}$$

Similarly, the same equations can be combined to obtain an evolution equation for a new scalar ξ defined by the relation:

$$\xi = \alpha\rho Y \tag{8}$$

Then:

$$(\nabla\Phi + \vec{v}) \cdot \nabla\xi = \frac{1}{ReSc}\nabla^2\xi \tag{9}$$

Thus, equations (7) and (9) can be regarded as the fundamental set for the scalar variables. The boundary conditions for these quantities will be considered next.

Let the dimensionless coordinate in the direction of U be denoted as x, and let $\vec{k}$ be the unit vector in the x direction; i.e. $\vec{k} \cdot \vec{r} = x$. The cylindrical radial coordinate is denoted by z; i.e. $z = |\vec{r} - \vec{k}x|$. Then, $r = \sqrt{x^2 + z^2}$, where the spherical polar coordinate r is defined by $r = |\vec{r}|$ (see Figure 1). The helium is released uniformly over the sphere surface with a total mass flux denoted by $\dot{m}$. Then, the boundary conditions on ξ and Φ take the following form: At the surface $r = 1$;

$$\frac{\partial\Phi}{\partial r} = \frac{M}{Re} \tag{10}$$

$$\frac{\partial\xi}{\partial r} = -M(1 - \xi)Sc \tag{11}$$

Here, the blowing parameter M is given by $M = \dot{m}\alpha/4\pi\mu a$, where μ is the viscosity of the gas. Far from the sphere, as $r \to \infty$:

$$\nabla\Phi = \vec{k} \tag{12}$$

$$\xi = 0 \tag{13}$$

Equations (10) thru (13) uniquely determine Φ and ξ.

The solution for Φ can be readily obtained without approximation as the sum of a source of strength M/Re and the potential flow past a sphere.

$$\Phi = \frac{-M}{Re}\frac{1}{r} + \vec{k} \cdot \vec{r}(1 + \frac{1}{2r^3}) \tag{14}$$

This solution ensures that the normal velocity component vanishes at the sphere surface. Since the component of Φ parallel to the sphere surface is of order 1 in Re, $\vec{v}$ can be expanded in ascending powers of Re at distances of order unity from the sphere as follows:

$$\vec{v} = \vec{v_0}(\vec{r}) + O(Re) \tag{15}$$

Similarly, the scalar ξ is of order unity in this region, and can be expanded in a corresponding manner:

$$\xi = \xi_0(\vec{r}) + O(Re) \tag{16}$$

When these expansions are inserted into equation (9), $\vec{v}$ makes no contribution to the convective terms at lowest order, and the solution for ξ_0 can be readily found to be:

$$\xi_0(\vec{r}) = 1 - exp(\frac{-ScM}{r}) \tag{17}$$

Expansions of the form defined in equations (15) and (16) break down when r is sufficiently large for $\vec{y}$ defined by $\vec{y} = Re\vec{r}$ to be of order one in Re. Then the velocity associated with the potential Φ takes the form:

$$\nabla\Phi(\vec{y}) = \vec{k} + O(Re) \tag{18}$$

Similarly, rewriting the inner solution $\xi_0(\vec{r})$ in terms of $\vec{y}$ shows that the correct form for the outer expansions for ξ and $\vec{v}$ are given by:

$$\xi(\vec{y}) = Re\xi^0(\vec{y}) + O(Re)^2 \tag{19}$$

$$\vec{v}(\vec{y}) = Re\vec{v^0}(\vec{y}) + O(Re)^2 \tag{20}$$

Substitution of the expansions defined by equations (19), (20), and (18) into equation (9) yields the following equation for $\xi^0(\vec{y})$:

$$\vec{k} \cdot \tilde{\nabla}\xi^0 = \frac{1}{Sc}\tilde{\nabla}^2\xi^0 \tag{21}$$

The tilde appearing in equation(21) denotes vector operations with $\vec{y}$ as the independent variable.

The required solution to equation (21) vanishes at infinity and matches the inner solution ξ_0 when y, defined as $y = |\vec{y}|$, is $O(Re)$. It is readily found by writing ξ^0 in the form:

$$\xi^0(\vec{y}) = exp[\frac{Sc}{2}(\vec{k} \cdot \vec{y})]\lambda(y) \tag{22}$$

The quantity $\lambda(y)$ is a solution to the equation

$$\frac{1}{y^2}\frac{\partial}{\partial y}(y^2\frac{\partial\lambda}{\partial y}) = \frac{(Sc)^2}{4}\lambda \tag{23}$$

The solution to equation (23) satisfying the conditions stated above is then given by:

$$\lambda = \frac{ScM}{y} exp(-\frac{Sc}{2}y) \tag{24}$$

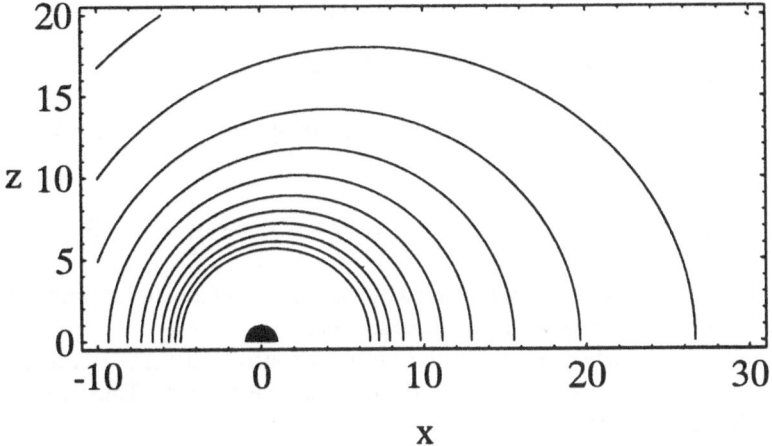

Fig. 2: Helium mass fraction contours for $Re = 0.05, Sc = 1, M = 2$. The porous sphere (solid disk) is shown to scale.

Finally, a composite solution for ξ valid everywhere that is correct to first order in Re can be written by replacing the pre-exponential term in equation (24) by the inner solution for ξ_0 given by equation (17). Then:

$$\xi(\vec{r}, Re) = [1 - exp(\frac{-ScM}{r})] \, exp(\frac{ScRe}{2}[\vec{k} \cdot \vec{r} - r]) \qquad (25)$$

Equation (25) (and the corresponding solution to the combustion problem given by equation (50) below) is the most important result of the paper. Note that no information about the nature of the solenoidal component of the velocity was used to derive this result. Only the fact that the potential flow was the dominant mode of convection was needed. Given the solution for ξ, the density ρ and mass fraction Y can be obtained using equations (3) and (8). These results, together with the solution for Φ given by equation (14) determine all the scalar quantities. Note that ϕ is found through the use of equation (6). Contours of constant mass fraction are displayed in Figure 2 for a flow corresponding to a value of $Re = 0.05$ with $M = 2$, $Sc = 1$, and the molecular weight ratio $\alpha = 7$. Note the transition from the effectively spherical contours near the surface, where the blowing velocity is dominant, to the the far field. Here the balance between convection by the uniform flow and the transverse diffusion produces the classical downstream Gaussian wake profile.

3   The Solenoidal Velocity

Now consider the solenoidal component of the velocity. The goal is not to calculate the complete solution, but to determine enough of the structure of $\vec{v}$ to demonstrate that the formal ordering of the terms used in the analysis up to this point also leads to a consistent solution for the vorticity induced flow. For this purpose, it is convenient to work with the the dimensionless vorticity $\vec{\omega}$ defined as:

$$\vec{\omega}(\vec{r}) = \nabla \times \vec{v}(\vec{r}) \tag{26}$$

The vorticity is determined from the curl of the momentum equation, which takes the form:

$$\nabla\rho \times [\frac{1}{2}\nabla(u^2) - \vec{u} \times \vec{\omega}] - \rho\nabla \times (\vec{u} \times \vec{\omega}) = -\frac{1}{Re}\nabla \times (\nabla \times \vec{\omega}) \tag{27}$$

Equations (26) and (27) are supplemented by the solenoidal condition, equation (5), and the solution for ρ obtained in the previous section.

The inner region, where r is of order unity, is considered first. Recall that in this region, ρ is a function of r only, as is the dominant term in ϕ. Thus, all coefficients of $\vec{\omega}$ in the equation for the lowest order terms in the inner expansion of equation (27) are spherically symmetric. The symmetry is broken only by the inhomogeneous baroclinic vorticity generation terms which are linear in $\vec{k}$ and the need to satisfy the no slip condition on the surface. This requires that $\vec{v}$ cancel the velocity generated by terms proportional to $\vec{k} \cdot \vec{r}$ in equation (14), which are also linear in $\vec{k}$. Under these circumstances, the arguments given by Landau and Lifshitz [11] about the form of $\vec{v}$ can be carried over directly. The velocity is required to be divergence free, linear in $\vec{k}$, and be a polar vector made up only of $\vec{k}$ and $\vec{r}$. Hence, the solenoidal velocity $\vec{v}_0$ must be of the form:

$$\vec{v}_0(\vec{r}) = \nabla \times [\nabla f(r) \times \vec{k}] \tag{28}$$

Expanding the terms in equation (28) then yields the following recipes for $\vec{v}_0$ and $\vec{\omega}$ in terms of $f(r)$:

$$\vec{v}_0 = -\vec{k}(2g + r\frac{dg}{dr}) + \vec{r}(\vec{k} \cdot \vec{r})\frac{1}{r}\frac{dg}{dr} \tag{29}$$

$$\vec{\omega} = \vec{k} \times \vec{r}(\frac{4}{r}\frac{dg}{dr} + \frac{d^2g}{dr^2}) \tag{30}$$

Here, the quantity $g(r)$ is given in terms of f through the relation:

$$g(r) = \frac{1}{r}\frac{df}{dr} \tag{31}$$

Rather than derive the full equation governing $g(r)$, it suffices to note that for large r, equation (27) assumes the constant density form. This leads to the asymptotic form for g as $g(r) = g_\infty/r$ and thus to the following asymptotic expressions for $\vec{v}_0$ and $\vec{\omega}$:

$$\vec{v}_0 = -g_\infty(\frac{\vec{k}}{r} + \frac{(\vec{r} \cdot \vec{k})\vec{r}}{r^3}) \tag{32}$$

$$\vec{\omega} = -\frac{2g_\infty}{r^3}(\vec{k} \times \vec{r}) \tag{33}$$

The constant g_∞ can be determined only by solving the full equation for $g(r)$, a task which will not be pursued here. Instead, note that when equations (32) and (33) are written in terms of $\vec{y}$, the form of the expansion for $\vec{v}$ given by equation (20) is confirmed to leading order. Moreover, for internal consistency, $\vec{\omega}$ in the outer region takes the form:

$$\vec{\omega} = (Re)^2 \vec{k} \times \tilde{\nabla}\Omega(\vec{y}) \tag{34}$$

When the outer expansions for all quantities are substituted into equation (27) it is easy to show that $\Omega(\vec{y})$ satisfies the same equation as $\xi^0(\vec{y})$ with $Sc = 1$. The solution which matches equation (33) is readily found to be:

$$\Omega(\vec{y}) = \frac{2g_\infty}{y} \, exp(\frac{1}{2}[\vec{k} \cdot \vec{y} - y]) \tag{35}$$

The similarity between the solutions for ξ and Ω is not accidental. In an axially symmetric flow without swirl, there is no vortex stretching. Moreover, in the outer region there is no vorticity generation. Hence, the vorticity and the scalar quantities are controlled by the same balance between convection and diffusion. This qualitative similarity can be extended to the uniformly valid composite solution for Ω.

$$\vec{\omega} = (\vec{k} \times \vec{r}) \, exp(\frac{Re}{2}[\vec{k} \cdot \vec{r} - r])(\frac{4}{r}\frac{dg}{dr} + \frac{d^2g}{dr^2}) \tag{36}$$

The physical differences between the vorticity and scalar quantities are buried in the last factor in equation (36) and the prefactor in equation (25).

Finally, the solenoidal velocity in the outer region can be found using equation (34) to note that:

$$\tilde{\nabla} \times (\vec{v^0} + \vec{k}\Omega) = 0 \tag{37}$$

Thus, $\vec{v^0}$ must have the form:

$$\vec{v^0} = -\vec{k}\Omega + \tilde{\nabla}\Lambda \tag{38}$$

Applying the solenoidal condition to equation (38) and using the equation governing $\Omega(\vec{y})$ yields the following equation for $\Lambda(\vec{y})$:

$$\tilde{\nabla}^2(\Lambda - \Omega) = 0 \tag{39}$$

The solution to equation (39) matching equation (32) as y vanishes then yields the final result for $\vec{v^0}$:

$$\vec{v^0}(y) = -\vec{k}\Omega + \tilde{\nabla}(\Omega - \frac{2g_\infty}{y}) \tag{40}$$

4 Combustion of Gases from a Porous Sphere

The physical configuration is the same as that studied above, with a flow of oxidizer in the x direction and radial blowing of fuel from the sphere. The combustion is assumed to be diffusion controlled . In the absence of flow, this problem has a considerable literature, which is summarized in Williams [12]. Here, the focus is on the extent to which the velocity field must be considered when the Reynolds number of the approach flow is small.

To this end, the combustion problem is posed in terms of a mixture fraction $Z(\vec{r})$, defined as the fraction of the gas at any point which at one time was fuel. Williams

[12] presents an extended discussion of this concept, as well as detailed relationships between Z and all other scalar variables. For the present purpose, all that is needed is that $Z = 1$ corresponds to pure fuel, while $Z = 0$ is pure oxidizer. Then, with the density and velocity non-dimensionalized as before, the mass and mixture fraction equations can be written in the form:

$$\nabla \cdot (\rho \vec{u}) = 0 \tag{41}$$

$$\nabla \cdot (\rho \vec{u} Z - \frac{1}{ReSc} \rho D \nabla Z) = 0 \tag{42}$$

These equations are supplemented by the "state relationships" $\rho(Z)$ and $D(Z)$. Note that the dimensionless diffusivity D is defined so that $D = 1$ in the oxidizer stream. The Reynolds and Schmidt numbers are based on properties defined in the oxidizer stream. Equations (41) and (42) are solved subject to the boundary conditions $Z = 1$ on the surface of the sphere and $Z = 0$ in the oxidizer stream far from the sphere. The velocity is again decomposed as in equation (4), and the boundary conditions on $\phi(\vec{r})$ are given by equation (12) as $r \to \infty$. At the surface of the sphere:

$$\frac{\partial \phi}{\partial r} = \frac{M_c}{Re} \tag{43}$$

Here, the combustion blowing parameter M_c is given by $M_c = \dot{m} \rho_\infty / 4 \pi \mu_\infty \rho_s a$, where ρ_s is the density at the surface of the sphere.

Following the procedure used in the isothermal problem, the inner region is considered first. The velocity components are expanded as in equations (14) and (15). In particular, $\phi(\vec{r})$ is written in the form:

$$\phi(\vec{r}) = \frac{\Phi_0(r)}{Re} + \vec{k} \cdot \vec{r}(1 + \frac{1}{2r^3}) \tag{44}$$

The density and mixture fraction are also expanded so that the leading terms are spherically symmetric.

$$\rho(\vec{r}) = \rho_0(r) + O(Re) \tag{45}$$

$$Z(\vec{r}) = Z_0(r) + O(Re) \tag{46}$$

Substitution of these expansions into equations (41) and (42) leads to the immediate integrals:

$$\rho_0 \frac{d\Phi_0}{dr} r^2 = M_c \tag{47}$$

$$\int_0^Z \frac{\rho(t) D(t) dt}{Z_\infty - t} = \frac{Sc M_c}{r} \tag{48}$$

The constant Z_∞ is determined from the condition that $Z = 1$ when $r = 1$. The solution can be evaluated explicitly only when a choice of state relationships specifying $\rho(Z)$ and $D(Z)$ has been made. However, for any physically plausible choice, the requirement that Z vanish as $r \to \infty$ leads to the asymptotic formula for the mixture fraction:

$$Z_0 = Z_\infty \frac{Sc M_c}{r} \tag{49}$$

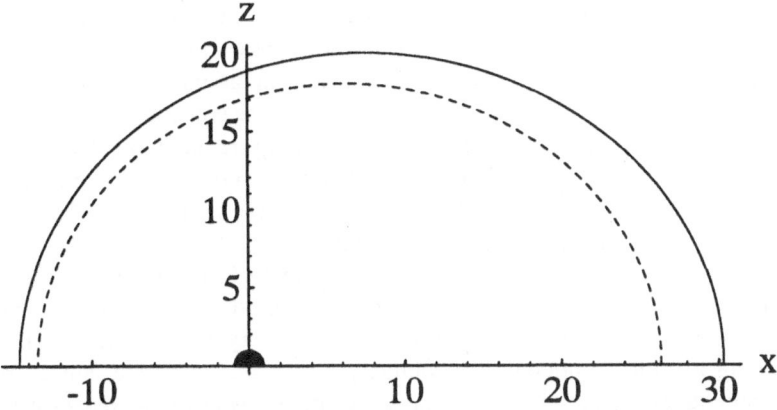

Fig. 3: Flame sheet location for Acetylene with 30 percent radiation loss (dashed line), and 60 percent radiation loss (solid line). The surface to ambient temperature ratio is 2, $Re = 0.05$, $Sc = 1$, and $M_c = 2$.

Finally, the solution in the outer region for $Z(\vec{y})$ is obtained by the same procedure as that for ξ^0 and Ω with the same result. Hence, a composite expansion for $Z(\vec{r})$ valid everywhere can be obtained by replacing the point source solution on the right hand side of equation (48) appropriate to the Laplace operator by that corresponding to the Oseen equation. The result is:

$$\int_0^Z \frac{\rho(t)D(t)dt}{Z_\infty - t} = \frac{ScM_c}{r} \, exp(\frac{ReSc}{2}[\vec{k} \cdot \vec{r} - r])) \tag{50}$$

Thus, the combustion problem has been solved to leading order in Reynolds number everywhere with only the potential flow used to obtain the desired result.

A set of state relations is chosen to complete the analysis. It is based on the empirical observation that for many fuels the specific volume is approximately a piecewise linear function of the mixture fraction [13]. The diffusivity is specified by assuming that $\rho^2 D = 1$. This is sufficient information to evaluate the integral on the left hand side of equation (50) explicitly . The flame sheet locations are plotted in Figure 3 for two cases corresponding to a sphere to ambient density ratio $\rho_s/\rho_\infty = 0.5$ and ambient to flame density ratios of 4.1 and 6. The flame values are chosen to be in agreement with values displayed in [13] corresponding to Acetylene with 60 percent and 30 percent respectively of the chemical energy release radiated away from the flame. All other parameters are identical with those used in Figure 2.

5   Generalizations and Conclusions

While the analyses presented above are restricted to a simple geometry in a steady state, the velocity decomposition and the simplification obtained using the potential flow approximation are quite general. Indeed, using the low Mach number form of the energy equation and an isobaric equation of state , it is easy to show that the velocity potential still satisfies a Poisson equation, with the source terms proportional to the net chemical heat release minus the radiative losses [14]. This is true whether or not the phenomena are in steady state, and makes no assumption about the relative importance of the vorticity induced flow. Of course, if the solenoidal component of the flow is important, the convection of species and energy requires the determination of this velocity. However, the calculation of the irrotational field induced by the thermal processes still depends only on the chemical and radiative sources.

There is a further advantage to the potential flow model when the low Reynolds number microgravity problem must be solved numerically in an unbounded domain. The determination of computational far field boundary conditions for velocity components is often a non-trivial task. Simply requiring that a velocity component or its derivative vanish along a finite computational boundary is not very accurate, since velocities and/or gradients decay slowly with a small negative power of the distance from the combustion zone. Thus, unless the computational boundaries are set far from the region of interest, significant errors can occur when boundary conditions of this type are employed. However, the Poisson equation can be solved formally in terms of suitably chosen Greens Functions for unbounded domains, even if source and surface terms are only known numerically. These formulae can be readily employed to generate accurate numerical open boundary conditions, with relatively little loss in computational efficiency ([6], [7]), [8]). Finally, the potential flow is efficient in a conceptual sense. It addresses the microgravity low Reynolds number thermal science problem of interest with the minimum technical apparatus required to obtain the desired results.

References

[1] Fendell, F.E., Sprankle, M.L., and Dodson, D.S., "Thin-Flame Theory for a Fuel Droplet in Slow Viscous Flow", *J. Fluid Mechanics*, vol. 26, pp. 267-280, (1968).

[2] Gogos, G., Sadhal, S.S., Ayyaswamy, P.S., and Sundararajan, T., "Thin-flame theory for the combustion of a moving liquid drop: effects due to variable density", *J. Fluid Mechanics*, vol. 171, pp. 121-144, (1986).

[3] Fendell, F.E., Coats, D.E., and Smith, E.B., "Compressible Slow Viscous Flow Past a Vaporizing Droplet", *AIAA Journal*, vol. 6, pp.1953-1060, (1968).

[4] Kaplun, S., and Lagerstrom, P.A., "Asymptotic Expansions of Navier-Stokes Solutions for Small Reynolds Number", *J. Mathematics and Mechanics*, vol. 8, pp. 585-593, (1957).

[5] Proudman, I., and Pearson, J.R., "Expansions at Small Reynolds Numbers for the Flow Past a Sphere and a Circular Cylinder", *J. Fluid Mechanics*, vol. 2, pp. 237-262, (1957).

[6] Kushida, G., Baum, H.R., Kashiwagi, T., and di Blast, C., "Heat and Mass Transport From Thermally Degrading Thin Cellulosic Materials in a Microgravity Environment", *ASME Journal of Heat Transfer*, vol. 114, pp. 494-502, (1992).

[7] Yamashita, H., Baum, H.R., Kushida, G., Nakabe, K., and Kashiwagi, T., "Heat Transfer From Radiatively Heated Material in a Low Reynolds Number Microgravity Environment", *ASME Journal of Heat Transfer* , vol. 115, pp. 418-425, (1993).

[8] Nakabe, K., McGrattan, K.B., Kashiwagi, T., Baum, H.R., Yamashita, H., and Kushida, G., "Ignition and Transition to Flame Spread over a Thermally Thin Cellulosic Sheet in a Microgravity Environment", *Combustion and Flame*, vol. 98, pp. 361-374, (1994) .

[9] de Ris, J.N., "Spread of a Laminar Diffusion Flame", *Twelfth Symposium (International) on Combustion*, The Combustion Institute, Pittsburgh, pp. 241-252, (1969).

[10] Wichman, I.S., Williams, F.A., and Glassman, I., "Theoretical Aspects of Flame Spread in an Opposed Flow over Flat Surfaces of Solid Fuels", *Nineteenth Symposium (International) on Combustion*, The Combustion Institute, Pittsburgh, pp. 835-845, (1983).

[11] Landau, L.D., and Lifshitz, E.M., *Fluid Mechanics*, Addison-Wesley, Reading, MA, pp. 64-67, (1959).

[12] Williams, F.A., *Combustion Theory*, Second Edition, Benjamin Cummings, Menlo Park, CA, pp. 52-76, (1985).

[13] Baum, H.R., Rehm, R.G., and Gore, J.V., "Transient Combustion in a Turbulent Eddy", *Twenty-Third Symposium (International) on Combustion*, The Combustion Institute, Pittsburgh, pp. 715-722, (1990).

[14] Baum, H.R., Ezekoye, O.A., McGrattan, K.B., and Rehm, R.G., "Mathematical Modeling and Computer Simulation of Fire Phenomena", *Theoret. Comput. Fluid Dynamics*, vol. 6, pp. 125-139, (1994).

Numerical Modeling of Graphite Combustion using Elementary, Reduced and Semi-global Heterogeneous Reaction Mechanisms

H.K. Chelliah, Department of Mechanical, Aerospace and Nuclear Engineering
University of Virginia, Charlottesville, VA 22903, USA

1. Introduction

The combustion of coal is known to occur in two stages - rapid pyrolysis producing volatiles, tar and char, followed by (or sometimes simultaneously with) slow heterogeneous char oxidation. Because of the slow oxidation of char, which can be an order of magnitude slower than burning the same mass in gas phase, efficient coal combustion heavily depends on the char burnout time [1]. As char mainly consists of carbon and since carbon can be graphitized at high temperatures [2-4], numerous experimental, analytical and computational studies have been performed to study the *graphite* burnout times in different flow configurations, over a wide range of oxidizing environments. Although there is a good qualitative understanding of the overall oxidation process, truly intrinsic heterogeneous surface reaction mechanisms are yet to be well developed.

One major difficulty in validating the estimated elementary surface reaction rates is the inability to measure the concentration of adsorbed surface intermediates at the high temperatures and pressures of interest. The explicit dependence of the concentration of surface intermediates on the chemical rate expressions, however, can be eliminated by introducing the steady-state approximation for these species, as originally shown by Langmuir [5,6]. The resulting reduced semi-global rate expressions still require information on the specific reaction rate constants of the elementary surface mechanism. Thus, the traditional approach in analyzing surface reaction rates has been to model the semi-global rate constants based on experimental data, for a limited range of conditions. In contrast, the elementary rate data of gas-phase homogeneous kinetics, are relatively well established. For example, an elementary homogeneous mechanism of hydrocarbon-air mixture can typically include about 30 species in over 100 reactions. However, in engineering applications, not all these elementary homogeneous reactions are rate controlling. Methods of reducing the complexity of such homogeneous elementary mechanisms have received significant attention during the last decade. Systematic application of steady-state approximation to gaseous reaction intermediates has been a popular approach [7-9], although the similarity with the Langmuir's approach in heterogeneous kinetics has not been recognized in the literature. In homogeneous combustion problems, the accuracy of the steady-state approximation introduced directly depends on the characteristic transport and chemical times involved. In the case of heterogeneous kinetics, the steady-state approximation is in fact exact because the convection and diffusion of surface intermediates are nonexistent (mobile adsorbed species can jump from one site to another site, but in a global sense, these concentrations are assumed to be uniform on the surface). Thus, production and consumption of the surface intermediates must balance each other exactly. In the present chapter, starting with an elementary surface reaction mechanism proposed recently by Mitchell et al. [10] for graphite oxidation, steady-state approximations have been applied systematically to obtain a reduced surface reaction mechanism. By performing numerical integration of the governing equations, a comparison of the mass burning rate predictions with the reduced surface mechanism developed here and the existing semi-global surface reaction mechanisms is presented to indicate the uncertainties of the elementary rate constants that appear in the reduced surface mechanism.

Recent theoretical investigations aimed at better understanding the heterogeneous kinetics of carbon combustion under flame environment have considered two reacting flow configurations [10-19]. They are (a) combustion of graphite particles in entrained flow reactors or thermogravimetric analyzers and (b) combustion of heated graphite surfaces in oxidizer streams. Numerical calculations have been performed recently in both these configurations to demonstrate the effect of surface reaction models employed [18,19]. In the studies on particulate burning in flame environments, the particle oxidation process has been approximated by that of a single particle burning in a quiescent oxidizing atmosphere. Assuming quasi-steady and spherically symmetric burning conditions, conservation equations reduce to a set of ordinary differential equations and the solution procedure becomes considerably simplified. In the second flow configuration, combustion of a heated graphite rod of the order of 1cm placed in a uniform flow field, has been investigated. In this configuration, for radius of curvature of the graphite rod much larger than the thickness of the boundary layer, the species and temperature profiles across the mixing layer have been assumed to be planar, while boundary layer equations for mass continuity and momentum have been used. In this chapter, numerical calculations of graphite for the two flow configurations discussed above are presented. For both cases, an elementary reaction mechanism for the homogeneous kinetics and two semi-global mechanisms for the heterogeneous kinetics are employed. The differences in the two semi-global surface mechanisms employed are related to the porosity of the graphite used in the experiments to determine the rate constants.

2. Governing Equations

The external flow field near a burning cylindrical graphite rod and a spherically symmetric graphite particle, with a detached gas-phase CO-flame are illustrated in Figs. 1 and 2, respectively. By decreasing the local diffusion time, ie. by increasing the strain rate of flow near the graphite surface or by decreasing the graphite particle size, the CO-flame can be extinguished. Once the flame extinction occurs, the CO produced at the surface by reaction with O_2 is simply convected and diffused to the ambience.

2.1 Graphite Rod

The flow over the heated graphite rod is assumed to be steady and laminar. If x and y are the coordinates tangential and normal to the graphite surface, respectively, and u and v are the corresponding velocity components, then the outer, inviscid, oxidizer flow can be described by $u_\infty = ax$ and $v_\infty = -ay$, where the subscript ∞ identifies the conditions in the outer flow and a is the strain rate, defined as the velocity gradient dv/dy in the oxidizer stream, just upstream of the mixing layer. The strain rate a can be related to a local diffusion time [20].

Introducing the notation $f' = u/u_\infty$, the governing boundary layer equations for mass, momentum, species and energy in the inner viscous region can be transformed into a system of ordinary differential equations along the stagnation-point stream line (x=0), and can be written as [21]

$$\frac{d}{dy}(\rho v) + a\rho f' = 0, \tag{1}$$

$$\frac{d}{dy}(\mu \frac{df'}{dy}) - (\rho v)\frac{df'}{dy} + (a/2)(\rho_\infty - \rho(f')^2) = 0, \tag{2}$$

$$(\rho v)\frac{dY_i}{dy} + \frac{d}{dy}(\rho Y_i V_i) = W_i \dot{\omega}_i, \quad i = 1, 2, ..., N, \tag{3}$$

$$c_p(\rho v)\frac{dT}{dy} + \sum_{i=1}^{N} \rho Y_i V_i c_{pi}\frac{dT}{dy} + \frac{d}{dy}(\lambda\frac{dT}{dy}) = -\sum_{i=1}^{N} W_i \dot{\omega}_i h_i. \tag{4}$$

Here ρ is the density in the gas phase, T the temperature, Y_i the mass fraction of the ith species, h_i the specific enthalpy of the ith species, V_i the diffusion velocity of the ith species in y direction, c_{pi} the heat capacity of the ith species, $\dot{\omega}_i$ the molar rate of production of ith species by homogeneous reactions, W_i the molecular weight of the ith species, c_p the heat capacity of the mixture, μ the viscosity of the mixture, and λ the thermal conductivity of the mixture.

The boundary conditions at the surface ($y = y_s$) are

$$f' = 0; \quad (\rho v)_s = \sum_{i=1}^{N} \dot{s}_i; \quad [\rho Y_i(v + V_i)]_s = \dot{s}_i, \quad i = 1, ..., N; \quad T = T_s, \tag{5}$$

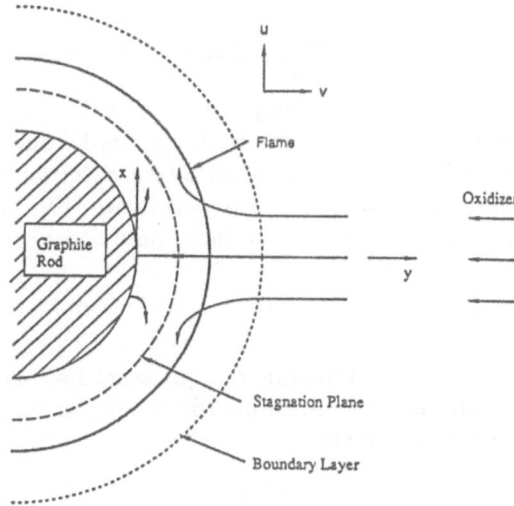

Figure 1: Illustration of the stagnation-point flow field over a reacting cylindrical graphite rod.

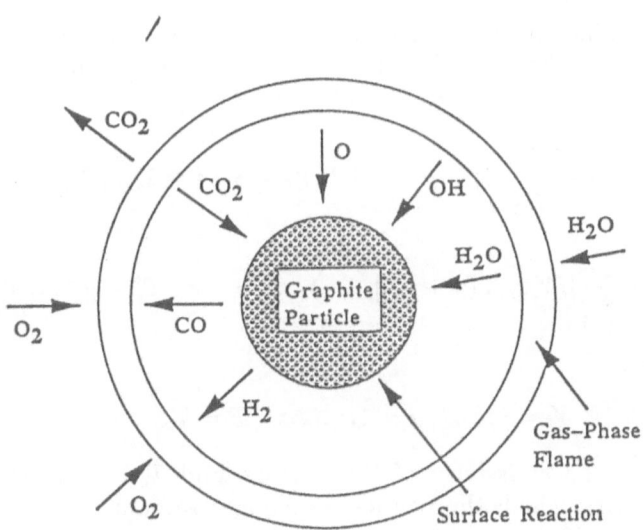

Figure 2: Illustration of the flow around a reacting graphite particle.

and, at $y = y_\infty$, are

$$f' = 1; \quad Y_i = Y_{i,\infty}, \quad i = 1, ..., N; \quad T = T_\infty. \tag{6}$$

In this formulation the mass burning rate of graphite is equivalent to $(\rho v)_s$. Here $\dot{s}_i$ is the mass production rate of ith species by surface reactions. Since the surface temperature T_s is controlled by an external electric current and held fixed for a particular run in the experiments [17], heat lost by thermal radiation does not appear in the boundary condition expression for T.

2.2 Graphite Particle

For a single graphite particle burning in a quiescent hot oxidizing environment under spherically symmetric conditions, the quasi-steady governing equations for mass, species and energy are

$$\frac{\partial(r^2 \rho v)}{\partial r} = 0, \tag{7}$$

$$\rho v \frac{\partial Y_i}{\partial r} + \frac{1}{r^2} \frac{\partial}{\partial r}(r^2 \rho Y_i V_i) = W_i \dot{\omega}_i, \quad i = 1, ..., N, \tag{8}$$

$$\rho v c_p \frac{\partial T}{\partial r} + \sum_{i=1}^{N} \rho Y_i V_i c_{pi} \frac{\partial T}{\partial r} - \frac{1}{r^2} \frac{\partial}{\partial r} \left(r^2 \lambda \frac{\partial T}{\partial r} \right) = -\sum_{i=1}^{N} W_i \dot{\omega}_i h_i. \tag{9}$$

Here, r is the radial distance from the center of the particle and v is the velocity in the radial direction. The boundary conditions at the surface of the particle, $r = r_s$, are

$$(\rho v)_s = \sum_{i=1}^{N} \dot{s}_i; \quad [\rho Y_i(v + V_i)]_s = \dot{s}_i, \quad i = 1, ..., N;$$

$$\lambda \left(\frac{\partial T}{\partial r} \right)_s - \sum_{i=1}^{N} [\rho Y_i c_{pi} T(v + V_i)]_s - \sum_{i=1}^{N} \dot{s}_i h_i - \sigma \epsilon (T_s^4 - T_e^4) = 0, \tag{10}$$

and at $r = \infty$ are

$$Y_i = (Y_i)_\infty, \quad i = 1, ..., N; \quad T = T_\infty. \tag{11}$$

In Eq. (10), ϵ is the emissivity of the particle and T_e is the temperature of the medium with which the particle exchanges radiant energy. Unlike in the graphite rod oxidation, there is no strain rate involved in this problem to indicate the variation of the local diffusion time. However, appropriate nondimensionalization of the governing equations shows that the characteristic diffusion time is proportional to the square of the particle size. In numerical calculations, the inverse of the scalar dissipation rate defined as τ [20],

$$\tau = [2\lambda/(\rho c_p)|dZ/dr|^2]^{-1}, \tag{12}$$

provides a more convenient and rigorous method of obtaining a characteristic diffusion time of a burning graphite particle. For the overall reaction $CO + (1/2)O_2 \rightarrow CO_2$, the mixture fraction, Z, in Eq. (12) is defined as [19]

$$Z = \frac{Z_{C,F}/W_C + Z_{O,F}/W_O - 2Z_{O,OX}/W_O + (2Z_{O,OX}/W_O)_2}{(Z_{C,F}/W_C + Z_{O,F}/W_O)_1 + (2Z_{O,OX}/W_O)_2}, \tag{13}$$

where $Z_{j,F}$ the mass fraction of element j in the mixture originating from the fuel stream, and $Z_{j,OX}$ the mass fraction of element j in the mixture originating from the oxidizer stream. Subscripts 1 and 2 identify the conditions in the fuel and oxidizer stream, respectively [22].

2.3 Porosity Effects

The overall carbon reactivity can be estimated by $R = (\psi A_e + \eta A_g) \sum_{i=1}^{N} \dot{s}_i$, where A_e is the external surface area, A_g the internal surface area, ψ the surface roughness factor, and η an effectiveness factor which is a measure of the depth of penetration of gases into the carbon [2,3]. The semi-global rate constants $\dot{s}_i$ reported in the literature for porous graphite, absorb the surface area terms into the frequency factors in general, making them porosity dependent. Thus, the semi-global rate constants of porous graphite are not truly intrinsic. The flux rate at the surface of the porous graphite specimen depends on the reaction rates, both heterogeneous and homogeneous, taking place within the bulk carbon. In the present one-dimensional formulation, short of solving for the coupled conservation equations for the gas-phase within the porous graphite, there is only one simple method to incorporate the porosity effects, ie. by changing the boundary conditions of Eqs. (5) and (10) as $(\rho v)_s = \phi \times \sum_{i=1}^{N} \dot{s}_i$ to take into account the increased surface area accessible for combustion. Here, ϕ is some effective surface area factor which depends on the porosity of graphite employed.

3. Reaction Mechanisms

3.1 Heterogeneous Elementary Mechanisms

The overall surface reaction $2C + O_2 \rightarrow 2CO_2$, for example, is postulated to occur through the following sequence of elementary surface reactions, $2C_f + O_2 \rightarrow 2C'(O)$ and $C'(O) + C_b \rightarrow CO + C_f$, where subscripts f and b identify the free-surface and bulk carbon sites, C(O) identifies adsorbed O-atom, and superscript ′ identifies a mobile adsorbed site. At high temperature, the additional desorption path $C'(O) \rightarrow C(O)$ and $C(O) + C_b \rightarrow CO + C_f$ can contribute to the overall desorption rate. Xxidation reaction pathways of carbon have been extensively reviewed in Refs. [2-3]. Based on the published rate data, Mitchell at al. [10] have recently compiled an elementary adsorption-desorption surface reaction mechanism involving 16 species in 18 elementary reactions. This mechanism is listed in Table 1 and is used here as the starting mechanism for the systematic development of the reduced surface reaction mechanism. Rate constants of reactions 1, 5-7 in Table 1 have been modified in Ref. [10] to get agreement with their experimental data, while estimated values were used for the unknown rate constants. Conservation of surface sites require an additional input value for surface site density (S) which was assumed to be 10^{18} sites/cm^2 [10].

Table 1: Elementary surface reaction mechanism [10], where specific rate constants in the form of $k_j = A_j T^{\alpha_j} \exp(-E_j/RT)$

j	Reaction	A_j	α_j	E_j
1	$2C_f + O_2 \rightarrow 2C'(O)$	2.3×10^{16}	0	15000
2	$C_f + CO_2 \rightarrow C'(O) + CO$	4.8×10^9	0	30000
3	$C_f + H_2O \rightarrow C'(O) + H_2$	6.2×10^8	0	25000
4	$C'(O) \rightarrow C(O)$	5.0×10^{12}	0	70000
5	$C(O) + C_b \rightarrow CO + C_f$	2.0×10^{11}	0	60000
6	$C'(O) + C_b \rightarrow CO + C_f$	2.0×10^7	0	40000
7	$2C'(O) + C_b \rightarrow CO_2 + 2C_f$	4.0×10^{13}	0	35000
8	$C_f + O \rightarrow C'(O)$	2.0×10^9	0	1000
9	$C_f + H \rightarrow C'(H)$	2.0×10^8	0	1000
10	$C_f + OH \rightarrow C'(OH)$	2.0×10^8	0	1000
11	$C_f + HO_2 \rightarrow C'(HO_2)$	2.0×10^8	0	1000
12	$C'(H) + H \rightarrow H_2 + C_f$	2.0×10^9	0	1000
13	$C'(H) + OH \rightarrow H_2O + C_f$	2.0×10^9	0	1000
14	$C'(OH) + H \rightarrow H_2O + C_f$	2.0×10^9	0	1000
15	$C'(OH) + OH \rightarrow H_2 + O_2 + C_f$	2.0×10^9	0	1000
16	$C'(HO_2) + H \rightarrow H_2 + O_2 + C_f$	2.0×10^9	0	1000
17	$C'(HO_2) + OH \rightarrow H_2O + O_2 + C_f$	2.0×10^9	0	1000
18	$C'(OH) + C_f \rightarrow C'(O) + C'(H)$	1.0×10^{14}	0	20000

Units: mole, cubic centimeters, seconds, Kelvin and calories per mole

3.2 Reduced Heterogeneous Mechanisms

As discussed in the Introduction, the validity of the steady-state approximation introduced to homogeneous systems depends on the flow field involved. However, in heterogeneous kinetics, the surface intermediates are neither convected or diffused, therefore under steady burning conditions their production and consumption should balance each other exactly. Thus, the reduced surface reaction mechanisms obtained with the application of steady-state approximation for surface intermediates is an exact representation of the the elementary surface mechanism, and provides significant reduction in computational effort.

For the heterogeneous elementary mechanism listed in Table 1, by introducing the steady-state approximation for surface species $C'(O)$, $C(O)$, $C'(H)$, $C'(OH)$ and $C'(HO_2)$, a eight-step reduced surface reaction mechanism can be obtained, which is represented here by

$$2C_b + O_2 \rightarrow 2CO, \qquad (I)$$

$$C_b + CO_2 \rightarrow 2CO, \qquad (II)$$

$$C_b + H_2O \rightarrow CO + H_2, \qquad (III)$$

$$C_b + O \rightarrow CO, \qquad (IV)$$

$$HO_2 + OH \rightarrow O_2 + H_2O, \qquad (V)$$

$$OH + OH \rightarrow O_2 + H_2, \qquad (VI)$$

$$OH + H \rightarrow H_2O, \qquad (VII)$$

$$H + H \rightarrow H_2, \qquad (VIII)$$

where the molar rates of the semi-global reactions I-VII are related to the elementary molar reaction rates ω_j as

$$\omega_I = \omega_1,$$

$$\omega_{II} = \omega_2 - \omega_7,$$

$$\omega_{III} = \omega_3 + \omega_{18},$$

$$\omega_{IV} = \omega_8,$$

$$\omega_V = \omega_{11},$$

$$\omega_{VI} = \omega_{15},$$

$$\omega_{VII} = \omega_9 + \omega_{10} - \omega_{15} - \omega_{12} - \omega_{16} + \omega_{18},$$

$$\omega_{VIII} = \omega_{12} + \omega_{16} - \omega_{18}.$$

Steady-state relationships provide algebraic expressions for the concentration of surface intermediate species. Coupling between these expressions generally requires introduction of truncation approximations [9] in order to obtain explicit expression in terms of the known gas-phase species concentrations. Introducing such truncations, the steady-state surface concentrations (moles/cm^2) of $C'(O)$, $C(O)$, $C'(H)$, $C'(OH)$ and $C'(HO_2)$, can be written as

$$c_{C'(O)} = \frac{k_2 c_{CO_2} + k_3 c_{H_2O} + k_8 c_O}{k_4 + k_6 c_{C_b}} c_{C_f} + \frac{2k_1 c_{O_2} + k_{18} B}{k_4 + k_6 c_{C_b}} c_{C_f}^2, \qquad (14)$$

$$c_{C(O)} = \frac{k_4}{k_5 c_{C_b}} c_{C'(O)}, \qquad (15)$$

$$c_{C'(H)} = \frac{k_9 c_H}{k_{12} c_H + k_{13} c_{OH}} c_{C_f} + \frac{k_{18} B}{k_{12} c_H + k_{13} c_{OH}} c_{C_f}^2, \qquad (16)$$

$$c_{C'(OH)} = \frac{k_{10} c_{OH}}{k_{14} c_H + k_{15} c_{OH} c_{C_f}} = B c_{C_f}, \qquad (17)$$

$$c_{C'(HO_2)} = \frac{k_{11} c_{HO_2}}{k_{16} c_H + k_{17} c_{OH} c_{C_f}}, \qquad (18)$$

where B is given by

$$B = \frac{k_{10} c_{OH}}{k_{14} c_H + k_{15} c_{OH}}. \qquad (19)$$

If the total number of active sites is conserved, then the following constraint must be satisfied

$$\sum_{i=1}^{N_s} Z_i = 1, \tag{19}$$

where Z_i is the fraction of sites occupied by species i. Since the surface concentration of ith species is given by $c_i = (S/N)Z_i$ where S is the surface density (sites/cm^2) and N the Avagadro number (molecules/mole), the concentration of free-sites at the surface is given by

$$c_{C_f} = \frac{S}{N} - \sum_{\substack{i=1 \\ i \neq C_f}}^{N_s} c_i. \tag{20}$$

Substitution of c_i's from Eqs. (14)-(18) to Eq. (20) yields a quadratic equation for c_{C_f}, which can be then solved to obtain explicit expressions for the concentration of surface intermediate species.

In the reduced surface mechanism, reactions (I)-(IV) correspond to consumption of carbon and production of CO, while reactions (V)-(VIII) correspond to recombination of gaseous reaction intermediates through adsorption-desorption reactions with carbon. In the elementary mechanism, the production of CO_2 by surface reactions is included in reaction 7, however, in the reduced surface mechanism the reaction 7 has been absorbed into ω_{II}. Hence, no explicit reaction for production of CO_2 is represented. Another important feature is that, for the starting elementary heterogeneous mechanism implemented, there is no reduced reaction of carbon with OH that leads to production of CO. Reaction 18 however does provide a reaction pathway to produce CO and, in fact, appears in the rate expression for ω_{III}. In contrast, ad-hoc semi-global surface mechanisms listed below have assumed explicit reaction of carbon with OH producing CO.

3.3 Semi-global Heterogeneous Mechanisms

Because of the uncertainties of the elementary heterogeneous rate constants, previous studies have used experimental data to obtain fits for the semi-global rates of reactions (I)-(IV). In addition, a rate constant for $C_b + OH \rightarrow CO + H$ has also been used. As discussed in section 2.3, when experimentally measured mass burning rate data are employed, the semi-global rate constants obtained will depend on the porosity of the graphite used in the experiments. Two sets of semi-global rate data reported in the literature for nonporous graphite [11] and porous graphite [17] are used here and are listed in Tables 2 and 3, respectively.

Table 2: Surface-reaction rate constants for *nonporous* graphite from Ref. [11]

$\dot{s}_i$	Reaction	k_j	B_j	n_j	E_j
1	$C_s + OH \rightarrow CO + H$	1	6.65×10^2	-0.5	0.0
2	$C_s + O \rightarrow CO$	2	3.61×10^2	-0.5	0.0
3	$C_s + H_2O \rightarrow CO + H_2$	3	9.0×10^3	0.0	68100
4	$C_s + CO_2 \rightarrow 2CO$	4	4.8×10^5	0.0	68800
5	$C_s + (1/2)O_2 \rightarrow CO$	5	2.4×10^3	0.0	30000
		6	2.13×10^1	0.0	-4100
		7	5.35×10^{-1}	0.0	15200
		8	1.81×10^7	0.0	97000

In Table 2, the surface flux rates $\dot{s}_i$ (kg/m^2/s) are given by $\dot{s}_1 = k_1 P_{OH}$, $\dot{s}_2 = k_2 P_O$, $\dot{s}_3 = k_3 P_{H_2O}^{0.5}$, $\dot{s}_4 = k_4 P_{CO_2}^{0.5}$, $\dot{s}_5 = [k_5 P_{O_2} Y/(1 + k_6 P_{O_2}) + k_7 P_{O_2}(1-Y)]$, where $Y = [1 + k_8/(k_7 P_{O_2})]^{-1}$, $k_j = B_j T^{n_j} \exp(-E_j/RT)$ are the specific reaction rate constants and P_i is the partial pressure of species i (atm). Units of E_j and T are in cal/mole and Kelvin, respectively.

Table 3: Surface-reaction rate constants for *porous* graphite from Ref. [17]

$\dot{s}_i$	Reaction	B_i	n_i	E_i
1	$C_s + OH \rightarrow CO + H$	1.65	0.5	0
2	$C_s + O \rightarrow CO$	3.41	0.5	0
3	$C_s + H_2O \rightarrow CO + H_2$	2.00×10^8	0.0	64750
4	$C_s + CO_2 \rightarrow 2CO$	6.00×10^7	0.0	64300
5	$C_s + (1/2)O_2 \rightarrow CO$	4.40×10^6	0.0	43000

In Table 3, the surface flux rates $\dot{s}_i$ (kg/m^2/s) are expressed in the form $\dot{s}_i = \nu_i W_i (\rho Y_i/W_i) B_i T^{n_i} \exp(-E_i/RT)$. Reaction rate constants of steps 1 and 2 in Table 3 are the same as those in Table 2, but the rate constants appear different because of the different form used.

3.4 Homogeneous Mechanisms

The gas-phase reaction mechanism of CO oxidation in the presence of H_2O is well established. A mechanism proposed by Yetter et al. [23], consisting of 12 species in 28 elementary reactions, is used in the present calculations.

4. Numerical Procedure

The governing equations of steady diffusion flames have been integrated previously by employing a global finite difference scheme [24]. A similar approach is employed here by modifying Smooke's diffusion flame code described in Ref. [21] to solve for the graphite rod oxidation and spherically symmetric graphite particles with heterogeneous kinetics. The numerical method uses a damped modified Newton iteration scheme, with an adaptive grid algorithm to refine the grid where the gradients of the dependent variables are large [25]. The thermodynamic and transport properties of the gas-phase species have been evaluated using the CHEMKIN thermodynamic data base [26] and the molecular parameters [27], respectively.

The solution of a graphite rod oxidation at extremely low strain rate was used as the starting estimate for numerical integration of the spherically symmetric governing equations of a graphite particle. The modified spherically symmetric governing equations were found to converge to the steady solution readily. The spatial domain of burning particles was increased to more than 50 particle diameters to assure that the gradients at $r = \infty$ are small.

5. Results and Discussions

5.1 Graphite Rod Oxidation

For the uniform air stream conditions of $Y_{O_2,\infty} = 0.233$, $Y_{N_2,\infty} = 0.762$, $Y_{H_2O,\infty} = 0.005$, $T_\infty = 300$ K and $a=200$ s^{-1}, Fig. 3 shows a comparison of the predicted mass burning rate of a porous graphite rod as a function of the surface temperature with the experimental results of Makino [17]. The density of the graphite rod used in the experiments was estimated to be 1.82×10^3 gm/cm^3, corresponding to a porosity of about 19%. The symbols are from experiments of Makino et al. [17], while the dashed and solid lines are from present numerical calculations using the semi-global rate constants listed in Tables 2 and 3, respectively. The rate constants of reactions 3-5 in Table 3 are, in fact, based on experimental data of Makino et al. [17] for porous graphite rods, hence the close agreement between the predictions (solid line) and experiments. A comparison of the predictions with the nonporous (dashed line) and porous (solid) rate constants clearly indicates the influence of the surface area effects.

Because carbon reaction with O-atoms has the highest reaction probability [28], gas-phase flame structure results have indicated that most of the radicals are consumed right at the surface for the temperature range 1200-2000 K [18]. Therefore, for surface temperatures below 2000 K (ie. below diffusion control limit) with water present in trace amounts in the air stream, only the semi-global reactions $C + (1/2)O_2 \rightarrow CO$ and $C + CO_2 \rightarrow 2CO$ are rate controlling. Under these conditions, to get agreement between the two semi-global mechanisms, reactions 4 and 5 of Table 2 had to be multiplied by factors 100 and 550, respectively. Based on the porosity of the graphite used, two orders in magnitude difference correspond to a reactant penetration depth of about 10 μm [18].

For the same oxidizer and graphite conditions used in Fig. 3, plots of the calculated surface flux rates $\dot{s}_i$ for the nonporous and porous semi-global surface mechanisms are shown in Figs. 4 and 5, respectively. For the nonporous

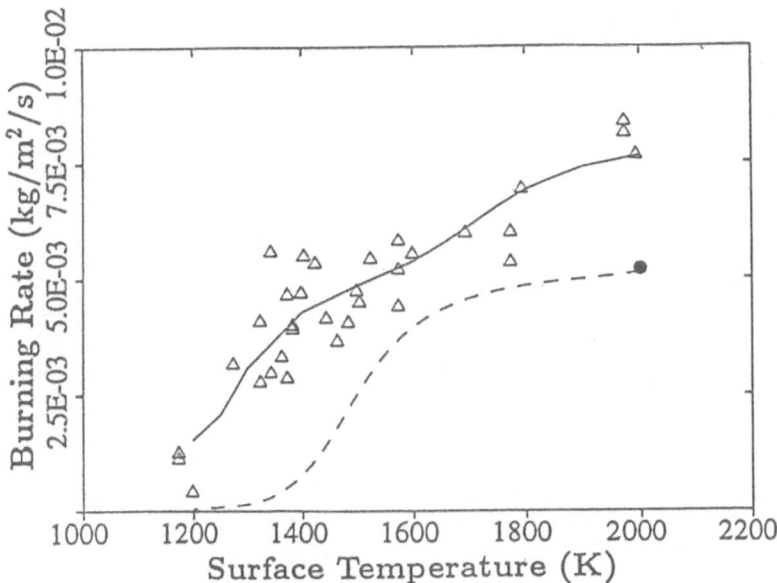

Figure 3: The mass burning rate of a graphite rod as a function of the surface temperature in air ($Y_{O_2,\infty} = 0.233$, $Y_{N_2,\infty} = 0.762$, $Y_{H_2O,\infty} = 0.005$, p=1 atm, $T_\infty = 300$ K, a=200 s^{-1}); $\triangle$ = experiments [17]; $- - -$ = nonporous rate data; $\underline{}$ = porous rate data; and $\bullet$ = reduced surface mechanism.

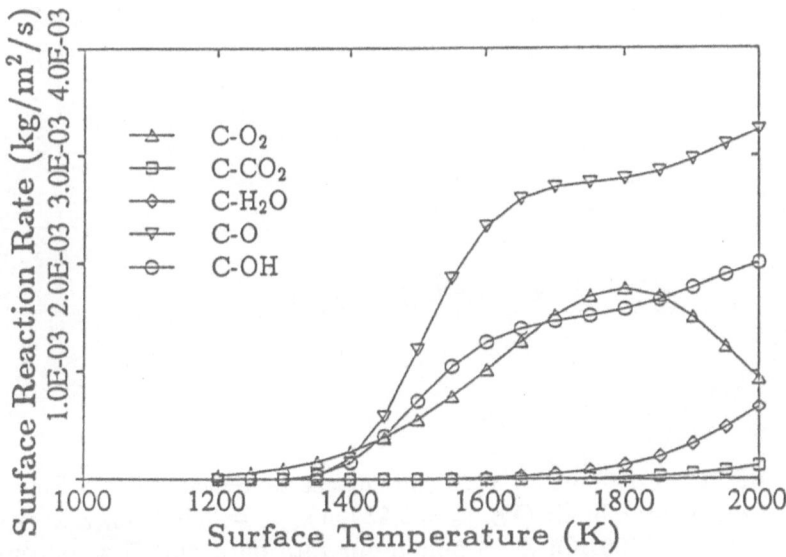

Figure 4: Surface reaction rates as a function of the surface temperature for nonporous rate constants.

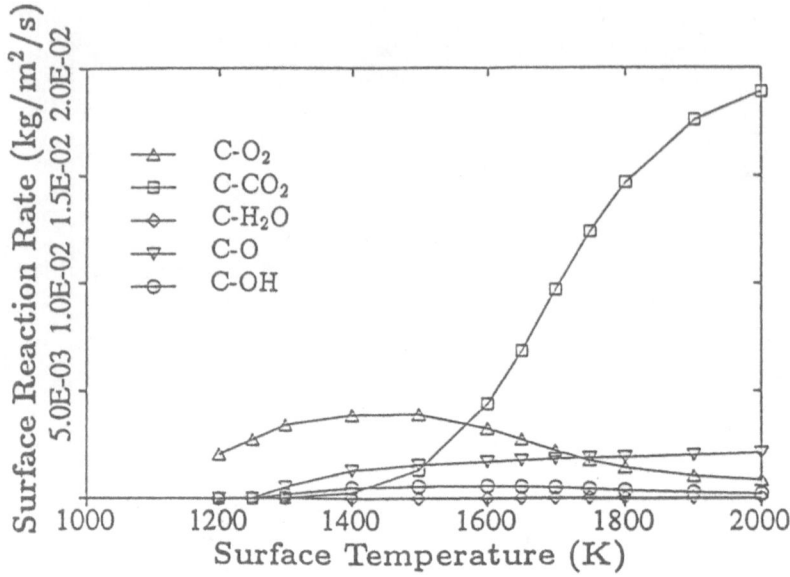

Figure 5: Surface reaction rates as a function of the surface temperature for porous rate constants.

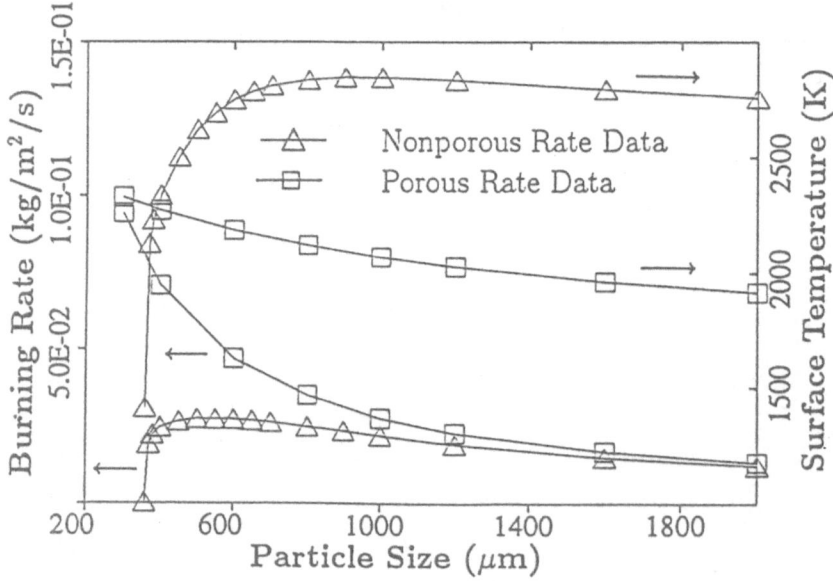

Figure 6: The mass burning rate and surface temperature as a function of particle size, burning in air ($Y_{O_2,\infty} = 0.232$, $Y_{N_2,\infty} = 0.762$, $Y_{H_2O,\infty} = 0.006$, p=1 atm, $T_\infty = 1400$ K); $\triangle$ = nonporous rate data and $\square$ = porous rate data.

rate constants, the dominant reaction is carbon with O-atoms and for porous rate constants, carbon reaction with O_2 dominates for temperatures below 1600 K and carbon reaction with CO_2 dominates for temperature above 1600 K.

When the 8-step reduced reaction mechanism developed in section 3.2 with the elementary rate constants listed in Table 1 and total surface concentration (S/N) of 1.0×10^{-6} moles/cm^2 is employed, the predicted mass burning rate of graphite is even lower than that predicted with the nonporous mechanism listed in Table 2. Because of the uncertainty of the surface site density (S), it is meaningless to fine tune the rate constants of the surface elementary mechanism. For example, for $S/N=1.0 \times 10^{-6}$, it was found that the reduced surface reaction rates $\omega_I - \omega_{IV}$ must be multiplied by 2.6, 35.0, 0.8 and 200.0, respectively, to get agreement with the species concentration obtained with the nonporous semi-global mechanism listed in Table 2. For a surface temperature of 2000 K, the mass burning rate predicted with these modified rates of the reduced reaction mechanism is shown in Fig. 3 by a filled circle. Because of the current arbitrariness in specifying the surface concentration, it is premature to make any attempts to validate the elementary surface reaction rate constants. However, a systematic method has been developed that could be used in the future to validate the elementary rate constants rather than the semi-global rate data, once the surface sites that take part in combustion are known.

5.2 Graphite Particle Oxidation

As shown in the previous section, graphite rods can be useful in validating surface reaction models below the diffusion control limit, provided that the porosity of graphite can be accounted for accurately. In this section the two sets of semi-global rate constants listed in Tables 2 and 3 are applied to model the mass burning rate of spherically symmetric graphite particles. Figure 6 shows the mass burning rate as a function of the particle size, based on the numerical integration of the quasi-steady conservation equations described in section 2.2. The ambient conditions were assumed to be $Y_{O_2,\infty} = 0.232$, $Y_{N_2,\infty} = 0.762$, $Y_{H_2O,\infty} = 0.006$, p=1 atm, and $T_\infty = 1400$ K. Unlike in graphite rods, the particle temperature cannot be controlled externally and hence will depend on heat transfer by radiation. Assuming that the temperature within the particle is uniform and thermal radiation is negligible, the predicted particle temperature as a function of particle size is also shown in Fig. 6. The most interesting feature of these predictions is that for the nonporous rate constants listed in Table 2, the mass burning rate predictions drops dramatically for particle sizes below 360 μm, indicating an extinction of the gas-phase flame and subsequent weak combustion of carbon with ambient O_2 at low ambient temperature. The porous surface rate constants, however, show much stronger combustion phenomena down to particle size of about 10 μm before the burning rate starts to drop dramatically, as observed for nonporous rate constants. When thermal radiation is included, the predictions with nonporous rate constants change significantly [19].

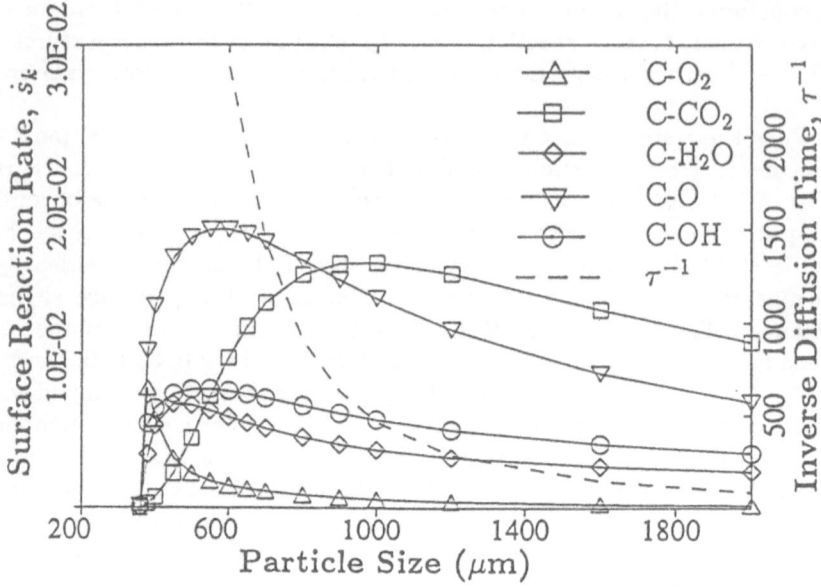

Figure 7: Surface reaction rates and inverse diffusion time as a function of particle size for nonporous rate constants.

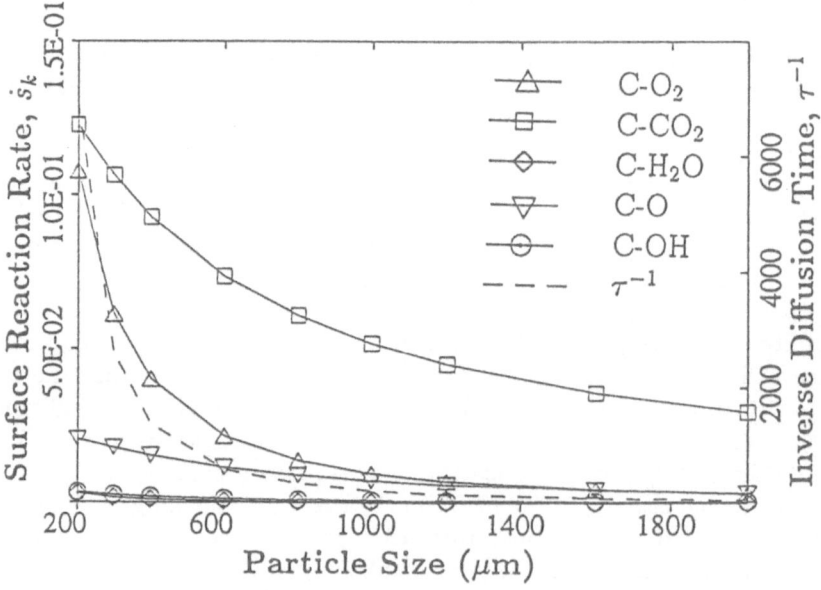

Figure 8: Surface reaction rates and inverse diffusion time as a function of particle size for porous rate constants.

Once again the variation of the surface reaction rates as a function of the particle size can be plotted to obtain information on the dominant surface reactions controlling the combustion of graphite particles. Figures 7 and 8 show such plots when nonporous and porous rate constants are employed, respectively. The ordering of the reaction flux rates is similar to those shown in Figs. 4 and 5. Also shown is the diffusion time evaluated using Eq. (12) at the stoichiometric mixture fraction of $Z_{st}=0.54$. As the particle size is reduced, the diffusion time decreases, while the chemical time associated with the porous surface reactions follows a similar trend, at least for the particle size range of 200 to 2000 μm. However, the chemical time associated with the nonporous surface reactions shows an increase as the particle size is decreased, leading to a further decrease in the Damköhler associated with the surface reactions and subsequent extinction. Further detailed analysis is needed to illustrate the uncertainties and effects of heterogeneous chemistry and the interaction with the homogeneous chemistry on graphite particle combustion.

6. Conclusions

Numerical calculations have been performed for two graphite combustion configurations. It has been shown that under a chemically controlled regime, the graphite rod experiments can, in fact, be used to validate surface reaction mechanisms. Calculations have clearly indicated that the internal surface area of the graphite has a major influence on the rate constants so determined. The graphite particle calculations have shown that the vigorous particle combustion regime heavily depends on the porosity of graphite used. This can have practical applications in terms of determining the optimal combustion conditions for a given class of graphite or any other porous particles.

A method of implementing a systematically reduced surface reaction mechanism, without additional computational effort, was described. This approach may be used in the future to factor out the surface area effects and validate truly intrinsic elementary surface reaction rate constants.

References

[1] Howarth, J.B., *Twenty-Third Symposium (International) on Combustion*, The Combustion Institute, p. 1107, 1990.

[2] Laurendeau, N.M., *Prog. Energy Comb. Sci.* 4, p. 221, 1978.

[3] Essenhigh, R.H., in *Chemistry of Coal Utilization*, Second Supplementary Volume, M.A. Elliot (Ed.), Wiley, New York, p. 1153, 1981.

[4] Davis, K., Hurt, R.H., Yang, N.Y.C., and Headley, T.H., "Evolution of Char Chemistry, Crystallinity, and Ultrafine Structure during Pulverized-Coal Combustion," to appear in *Twenty-Fifth Symposium (International) on Combustion*, The Combustion Institute, Pittsburgh, 1994.

[5] Langmuir, I, *J. Am. Chem. Soc.* Vol. 37, No. 5, p.1139, 1915.

[6] Hayward, D.O. and Trapnell, M.W., *Chemisorption*, Butterworths, Washington, 1964.

[7] Peters, N., *Numerical Simulation of Combustion Phenomena*, Lecture Notes in Physics **241**, p. 90, 1985.

[8] Smooke, M.D, *Reduced Kinetic Mechanisms and Asymptotic Approximations for Methane-Air Flames*, Lecture Notes in Physics, Springer-Verlag, Vol. 384, 1991.

[9] Peters, N. and Rogg, B., *Reduced Kinetic Mechanisms for Applications in Combustion Systems*, Lecture Notes in Physics, Springer-Verlag, Vol. M15, 1993.

[10] Mitchell, R.E., Kee, R.J., Glarborg, P., and Coltrin, M.E., *Twenty-third Symposium (International) on Combustion*, The Combustion Institute, p.1169, 1991.

[11] Bradley, D., Dixon-Lewis, G., Habik, S.E., and Mushi, E.M.J., *Twentieth Symposium (International) on Combustion*, The Combustion Institute, p. 931, 1984.

[12] Henriksen, K., *Twenty-second Symposium (International) on Combustion*, The Combustion Institute, p. 47, 1988.

[13] Matsui, K., and Tsuji, H., *Comb. Flame* **70**, p. 79, 1987.

[14] Makino, A. and Law, C.K., *Comb. Sci. and Tech.* **73**, p. 589, 1990.

[15] Cho, S.Y., Yetter, R. and Dryer, F.L., *J. Comp. Phys.* **102**, p. 160, 1992.

[16] Lee, C.Y., Yetter, R.A., Cho, S.Y. and Dryer, F.L., *Central and Eastern States Section of the Combustion Institute* New Orleans, March 1993.

[17] Makino, A., Araki, N. and Mihara, Y., *Comb. and Flame* **96**, p. 261, 1994.

[18] Chelliah, H.K., Makino, A., Araki, N. and Mihara, Y., and Law, C.K., "Effects of Porosity and Carbon-Radical Reactions on Graphite Oxidation," submitted to *Comb. and Flame*.

[19] Chelliah, H.K., "The influence of Heterogeneous Kinetics and Thermal Radiation on the Oxidation of Graphite Particles," submitted to *Comb. and Flame*.

[20] Chelliah, H.K., and Williams, F.A., *Combustion and Flame* **80**, p. 17-48, 1990.

[21] Smooke, M.D., Puri, I.K., and Seshadri, K., *Twenty first Symposium (International) on Combustion*, The Combustion Institute, p. 1783-1792, 1988.

[22] Peters, N., *Comb. Sci. and Tech.* **30**, p. 1, 1983.

[23] Yetter, R.A., Dryer, F.L., and Rabitz, H., *Comb. Sci. and Tech.*, **79**, p. 129, 1991.

[24] Smooke, M.D., *J. Comp. Phy.*, **48**, No. 1, p.72, 1982.

[25] Keyes, D.E. and Smooke, M.D., *J. Comp. Phys.* **73**, p. 267, 1987.

[26] Kee, R.J., Miller, J.A., and Jefferson, T.H., "Chemkin: A general purpose, problem-independent, transportable, fortran chemical kinetics code package," Sandia Report, SAND 80-8003.

[27] Kee, R.J., Warnatz, J., and Miller, J.A., "A fortran computer code package for the evaluation of gas-phase viscosities, conductivities, and diffusion coefficients," Sandia Report, SAND 83-8209.

[28] Rosner, D.E. and Allendorf, H.D., *AIAA J.* **6**, p. 650, 1968.

Atomistic Approach on Diamond Growth from Gaseous Phases

Seishiro Fukutani, Nílson Kunioshi
Okayama Prefectural University

and

Takayuki Iwasaki
Hitachi, Ltd.

Introduction

Diamond film deposition is performed with various methods such as the CVD method, thermal filament method [1] and arc–plasma method [2]. All these methods are based on the dissociation of hydrocarbons in gas phases, and their subsequent deposition on diamond crystal surfaces. The conditions imposed on the gas phase and diamond substrate, therefore, influence in a great extent the production of diamond films. The best conditions are difficult to be predicted experimentally. Then, the individual chemical and physical events involved in the diamond growth process have not been clearly analyzed yet, though the phenomenological data have been already accumulated. For example, Spytin *et al* [3] found that hydrogen atoms play an important role on the growth process of diamond, Matsumoto *et al.* [4].

The purpose of this work is to elucidate the relationship between the properties of the reactive gases, such as composition and temperature, and those of the growing diamond crystal, as for example the direction of growth.

Two atomistic approaches, molecular orbital method and molecular dynamics, were used. The molecular orbital method was applied for the static phenomena, and the molecular dynamics for the dynamical phenomena.

Diamond Crystal

Perfect crystals can be described as the repetition of their unit cells. The unit cell of diamond crystal is shown in Fig. 1. There are eight carbon atoms at the corners, each of which having 1/8 of an atom inside the cell, six on the surfaces, each one having half an atom inside the cell, and four completely inside the cell; then, the unit cell contains a total of eight carbon atoms.

In a unit cell, we can imagine several hypothetical planes, consisting of constituent particles located on them. The planes in a unit cell are identified by

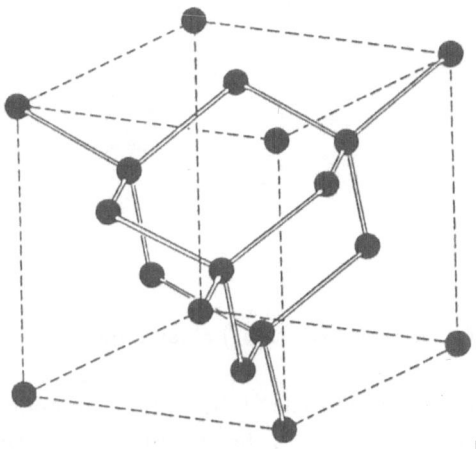

Fig. 1. Unit cell of diamond crystal

using the Miller indices. In this work, the growth of (100) and the (111) planes were studied.

Molecular Orbital Approach

Chemical reactions are processes where chemical bonds rearrange into new forms. Therefore, they can be studied by analyzing the states of the electrons involved in those bonds.

The states of electrons in atoms or molecules are described by solving the Schrödinger equation. But, unfortunately, this equation can be analitically and strictly solved only for atoms containing one electron, the so-called hydrogen-like atoms. For the case of common atoms and molecules, where many electrons are involved, the application of some approximate methods is necessary.

The molecular orbital method is used for the description of states of electrons which contribute to build up chemical bonds in a molecule. In this method, the molecular orbital ψ is constructed with linear combinations of the so-called basis functions, which are based on the wave functions for one-electron atoms. Several methods, differing according to the aproximations introduced, are available to the estimation of the molecular orbitals. Here, the $X\alpha$ method[5,6] is used.

The $X\alpha$ method is one of the approximate methods. It simplifies the integrals of the wave equation by approximating the Hamiltonian, and has a reliability between that of the *ab initio* method and of the semi-empirical method.

Using the $X\alpha$ method, the energy of a system for given configurations of atoms contained in the system can be estimated.

First, the growing process of a (100) plane of diamond crystal is discussed. The molecular orbital methods cannot treat molecules consisting of many

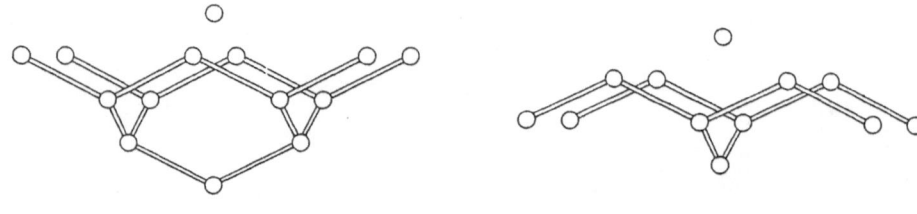

Fig. 2. Bridge site (left) and hollow site (right) cluster models.

atoms, so that cluster models for the growth of the (100) plane of diamond are assumed. When a species in a gas phase approaches the (100) plane, two typical depositing positions are possible; a bridge site and a hollow site. Figure 2 shows the two cluster models correspondent to each possibility. If a carbon-containing molecule falls toward the bridge site, a new diamond layer can be formed, but if it falls into the hollow site and deposits there, it originates a structural deficiency there.

The $X\alpha$ calculation was performed to estimate potential energies of the system under various conditions.

The dashed line in Fig. 3 shows the variation of the energy of the cluster when a methylene radical approaches a bridge site of the (100) plane. The two neighboring surface carbon atoms are separated by 2.51 Å, which is the distance in a standard configuration of diamond crystal. The abscissa indicates the distance between the falling methylene radical and the surface carbon atoms.

Similarly, the solid line in Fig. 3 also shows the variation of the energy of the system when a methylene radical approaches a bridge site of the (100) plane, but now the surface, except the point at which the methylene radical will deposit, is assumed to be terminated by hydrogen atoms.

In the case of hydrogen-free surface, that is, the case of carbon atoms at the surface having dangling bonds, the methylene radical penetrates down to 0.05 Åbelow the surface and only a graphite-like structure is given if the falling species cannot be removed afterwards. When the carbon atoms are linked to hydrogen atoms, however, the methylene radical is stabilized at 0.12 Å above the surface. This distance is still too small comparing with that in a normal diamond crystal, but the obtained result suggests that hydrogen atoms linking to surface carbon have important roles in diamond growth processes. This is supported by experiments, which showed that too high temperatures in the gas phase result in graphite-like crystals because hydrogen atoms are eliminated from the surface by increasing active species produced in the gas phase [7].

The energy gradient in the region correspondent to methylene penetration

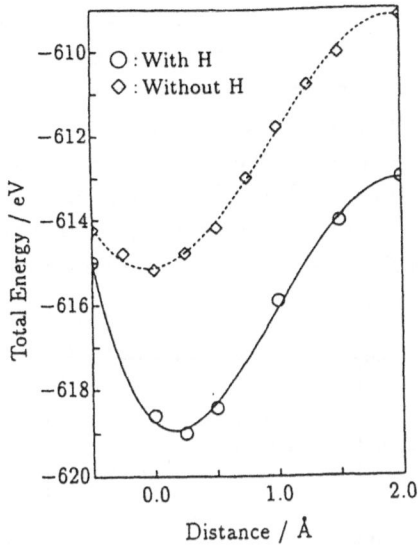

Fig. 3. Variation of the cluster energy when a methylene radical approaches a bridge site.

into the crystal (negative distance) is much steeper for the case with hydrogen atoms linking to the surface carbons, so that the falling methylene radical hardly penetrates into the crystal when the crystal surface is covered by hydrogen atoms.

The too short equilibrium distance between the falling methylene radical and the surface carbon atoms was improved by the reduction of the surface carbon distances. The two curves in Fig. 4 indicate the total energy of the clusters where the distances between neighboring surface carbon atoms are reduced to 1.51 Å. The falling methylene is stabilized at 0.59 Å above the surface.

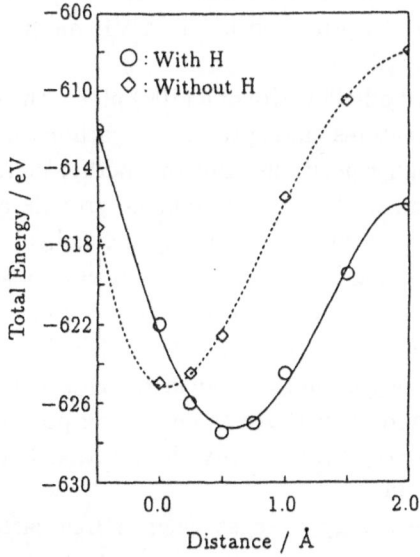

Fig. 4. Variation of the cluster energy when a methylene radical approaches a bridge site. The distance between two neighboring carbon atoms on the surface is reduced to 1.51 Å.

The reduction of the distances between two neighboring surface carbon atoms can be justified as follows: the amplitude of vibrations of surface carbon atoms depends on the temperature of the crystal, and the shorter distances result from excited vibrations due to high temperatures. This temperature effect is more remarkable in the (100) planes than in the (111) planes because the (100) planes have a more sparse structure and then the opening space admits falling methylene to penetrate inside the crystal. This interpretation also agrees with experimental observations, which show that the (100) plane growth becomes predominant with increasing temperature.

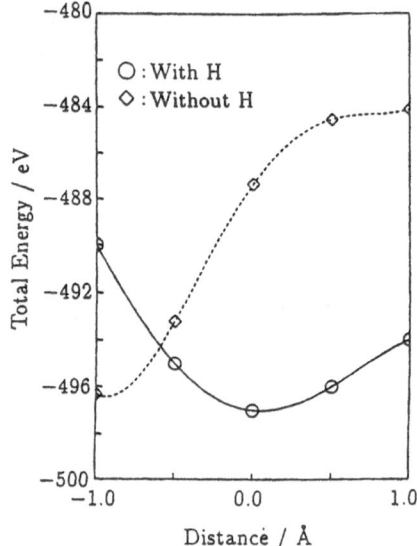

Fig. 5. Total energy of the cluster when a methylene radical approaches a hollow site.

In addition to the temperature effect, it has also been ascertained by experiments that the size of the diamond unit cell produced by the CVD method is smaller than that of natural diamond crystal.

When a methylene radical falls toward a hollow site of a (100) plane, the energy changes as shown in Fig. 5. Hydrogen atoms linking to surface carbon atoms can again hold the falling methylene at higher positions. But the methylene radical depositing at a hollow site cannot lead to the normal diamond growth, and then it should be removed anyhow. Elimination of the methylene radical depositing on a hollow site, when the surface is not covered by hydrogen atoms, is difficult because of its deep penetration. Figure 6 illustrates the top view of a (111) diamond plane. Carbon atoms in the third layer from the surface lay just below those of the second layer, so that they do not appear in the figure. Four representative sites, designated A, B, C and D, were set as depositing positions of species from the gas phase; among these positions, only the position D can lead to the normal growth of the (111) plane.

The four curves in Fig. 7 show the energy changes when a free carbon

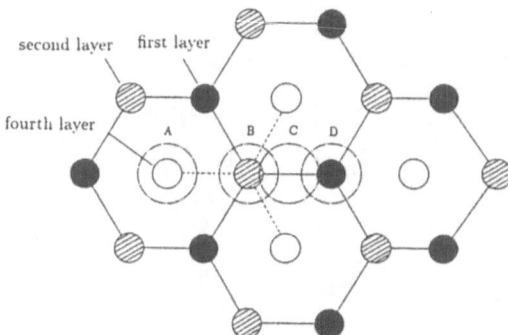

Fig. 6. Top view of the diamond lattice

atom falls down from the gas phase toward each of the four sites. As descibed previously, the site D is the position leading to normal diamond growth, but the free carbon atom approaching the site A gives the lowest energy. This result comes from the fact that the free carbon atom can make interactions with three surface carbon atoms which surround equivalently the site A; the formation of three bonds linking the falling carbon atom to the surface is the most stable possibility for equilibrating the atom into the surface.

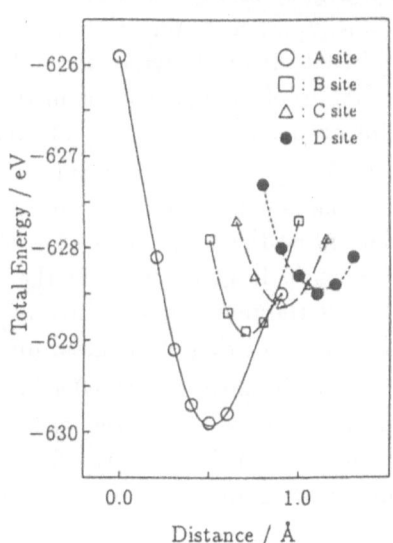

Fig. 7. Energy changes as a carbon atom approaches the sites A, B, C and D of Fig. 6.

Free carbon atoms produced in the gas phase easily deposit on the crystal surfaces due to their high reactivity, but unfortunately they have only low

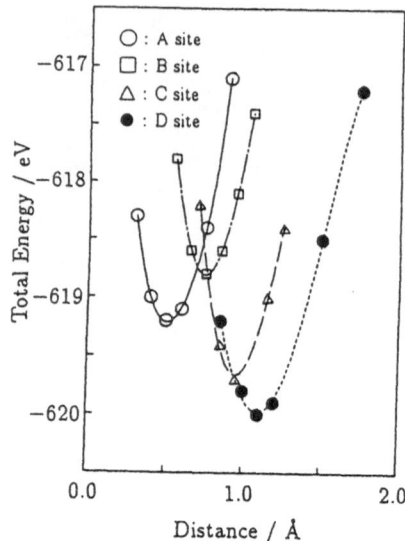

Fig. 8. Energy changes as a methyl radical approaches the sites A, B, C and D of Fig. 6.

position-selectivity, and then can hardly lead to the normal diamond growth.

On the other hand, when a methyl radical falls toward the (111) plane, the D sites gives the lowest energy, as shown in Fig. 8. Methyl radicals, consisting of three hydrogen atoms linked to a carbon atom, have direction-dependent properties, and therefore their position selectivity is much higher than that of a simple free carbon atom. Then, the production of methyl radicals in the gas phase is expected to be very important for the normal growth of the (111) planes.

However, the kind of the falling species cannot be specified only by the melecular orbital calculations. It depends on the amounts of species present in the gas phase. Chemical equilibrium calculations on the gas mixture of methane with air show that methyl radicals are present in large amounts in the temperature range from 2,000K to 2,500K, which is the temperature range usually employed as the reaction temperature in actual production of diamond films.

Hydrogen atoms linking to carbon atoms on the surface work effectively in favor of the diamond growth. But they must be removed from the surface for allowing gaseous species to deposit on the diamond surface. In the following, the efficiency of hydrogen atoms and methyl radicals for the elimination of the hydrogen atoms from surfaces is discussed.

Figure 9 shows the change in the length of the bond between a hydrogen atom and the carbon atom to which it is linked, as a function of the distance to the surface of an approaching hydrogen atom. The bond length increases from 1.1 Å to 1.5 Å when the distance between free hydrogen atom and the hydrogen atom on the surface comes down to about 0.7 Å. The steep increase in the C–H bond length suggests that the hydrogen atom attached to the surface is removed by the approaching hydrogen atom, with the formation of a hydrogen molecule.

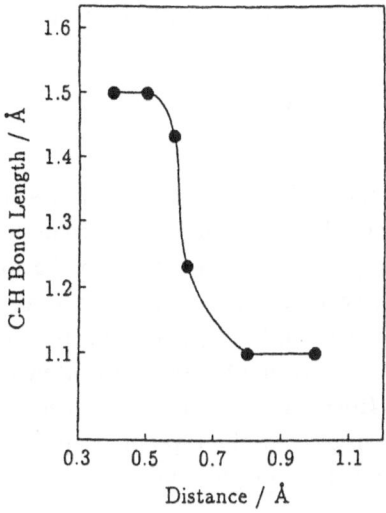

Fig. 9. C–H bond length at the surface as a function of the distance H–H.

When a methyl radical falls down, the distance between the attached hydrogen atom and the surface carbon does increase, but not so remarkably as in the case of an approaching hydrogen atom. The conclusion that the efficiency of methyl radicals as a remover of hydrogen atoms from the crystal surface is lower than that of hydrogen atoms immediately comes out. The lower efficiency of methyl radicals is ascribed to the fact that the unpaired electron in these species is not localized in a determined orbital, but is stabilized through a resonance state.

Free hydrogen atoms in the gas phase work efficiently as a remover of hydrogen atoms linking to the carbon atoms on the surface, so that they are important for the formation of dangling bonds at the crystal surface. On the other hand, if the concentration of free hydrogen atoms exceeds an optimum amount, the elimination of hydrogen atoms from the surface can become too frequent and too many dangling bonds can be formed. This can result in a lack of hydrogen atoms on the surface, which, as described previously, prevents the normal growth of the diamond crystal. Controll of the amount of free hydrogen atoms in the gase phases is then of great importance for the normal diamond growth.

Molecular Dynamics Approach

The time-dependent behavior of the carbon atoms on the diamond lattice can also influence the formation of new layers. The motions of particles can be perfectly described when the initial positions and velocities of the particles, and the potential among them are known. The motions of atomistic particles such as atoms and ions can, therefore, be solved according to the classical mechanics by using molecular dynamics.

The potential Φ of a system containing N interacting particles can be formulated as

$$\Phi(r_1, ..., r_N) = \sum_i \phi_1(r_i) + \sum_{i,j} \phi_2(r_i, r_j) + \sum_{i,j,k} \phi_3(r_i, r_j, r_k) + \cdots,$$

where ϕ_1, ϕ_2 and ϕ_3 express, respectively, one-body, two-body and three-body potentials. The one-body potential terms equal zero when any external forces do not work on the system. The two-body potential corresponds to the force which is based on the interactions between two particles, and the summation must be taken for all the possible pairs of particles in the system. This potential can be successfully applied to simulate the behavior of rare gases or metals, which have the closest packing structures. However, it cannot reproduce more sparse structures such as the diamond crystal, so that the three-body potential terms given as functions of the angles among three neighboring particles must be added.

The systems for the molecular dynamics simulation are composed of a (100) plane of diamond crystal having initially the standard configuration and a carbon atom approaching the surface of the crystal. All the carbon atoms in the system have initial velocities given by the Maxwellian distributions corresponding to various temperatures.

The $Z-$axis was taken to be perpendicular to the (100) plane, and 2–D periodic boudary conditions were applied for the $X-$ and $Y-$ directions, assuming that the crystal was infinitely wide in both directions.

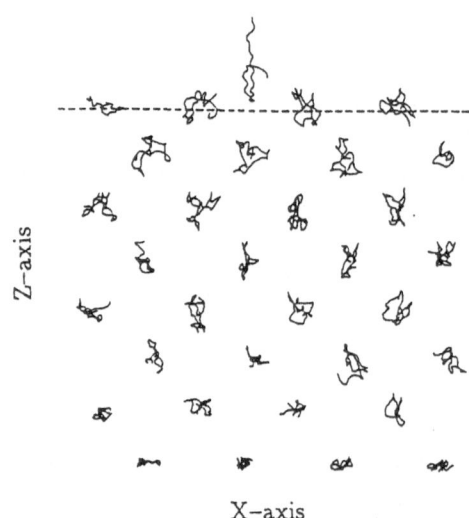

Z–axis

X–axis

Fig. 10. Projection of the particle trajectories in the $X - Z$ plane.

Figure 10 illustrates the motions of the carbon atoms as their trajectories are projected on the $X - Z$ plane. When a carbon atom with the kinetic energy correspondent to a temperature of 2200 K falls into a diamond surface kept at 1800 K, the falling atom first tend to penetrate into the crystal structure, but after strong interactions with the surface carbon atoms, it forms new bonds with two surface carbon atoms and is stabilized at a position favoring the normal growth of the crystal. The amplitudes of thermal vibrations are larger at the surface carbons than at the inner carbons, because of difference in the freedom of motion.

In the case of a carbon atom falling with the same kinetic energy as that of the case illustrated in Fig. 10 but starting from a deviated position, the trajectory was found to change gradually toward the same equilibrium position shown in Fig. 10.

When the temperature of the falling carbon atom is reduced to 1800 K, its kinetic energy was found to be too small and the carbon atom is repeled before coming close to the crystal surface.

On the other hand, when the temperature of the crystal is set as low as 1500 K, the falling carbon atom was found to penetrate easily into the crystal, threading through the slow motions of the carbon atoms in the crystal.

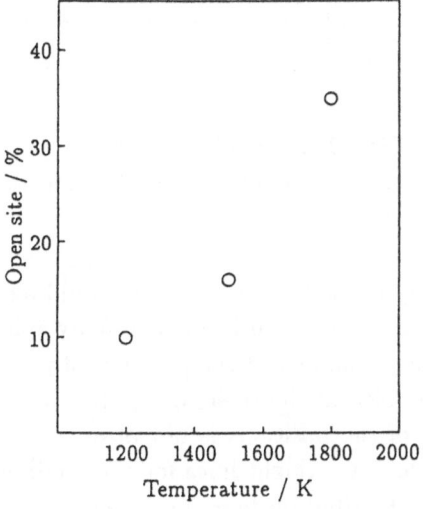

Fig. 11. Percentage of open sites on the surface varying with crystal temperature.

There are three possible bonding states on the surface carbon: the open site, or the dangling bond, the linkage to hydrogen atoms, and the dimer site. The first two types have already been discussed. The third one is a chemical bond produced between two adjacent surface carbon atoms, which prevents the formation of new bonds with the falling species. Then, dimers must be broken before the deposition of the gaseous species. Dimer formation occurs more often on the (100) planes than on the (111) planes due to their structures.

The percentage of open sites as a function of the temperature of the diamond crystal is shown in Fig. 11. The number of open sites increase with increasing temperature, dimer bonds being broken at high temperatures. This can also be pointed out as a reason for explaining why the growth of the (100) planes are accelerated as the crystal temperature is raised.

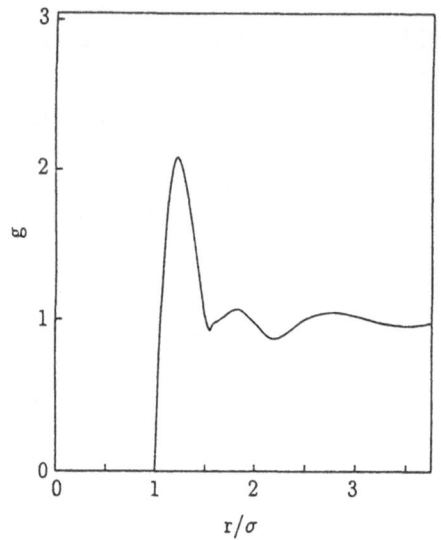 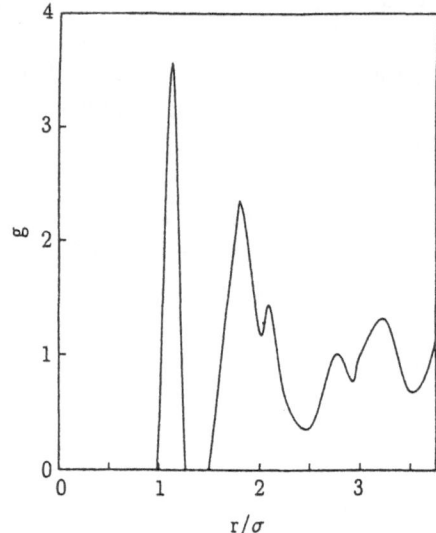

Fig. 12. Two-body particle distribution function for crystal temperature of 1500 K.

Fig. 13. Two-body particle distribution function for crystal temperature of 1900 K.

The crystallographic state of a newly-formed diamond layer should also be discussed. The crystal state is often expressed by using a two-body particle distribution function, which decribes the probability of the presence of particles at certain positions when one particle is fixed as the reference. When a crystal has a perfect structure, the fluctuation of the positions is not observed at all, and then the distribution function is given by straight lines for each individual position. Figure 12 shows the two-body distribution function when the crystal temperature is 1500 K; it does show a peak around the radius 1 σ, but, the peaks in Fig. 13, which shows the distribution function for crystal temperature of 1900 K, are much more pronounced, confirming the result that imposition of high temperatures on crystals gives highly-ordered structures.

Conclusions

According to the results obtained with this investigation on the diamond growth processes through the molecular orbital method and molecular dynamics, the conditions necessary for the normal growth of diamond crystal are:

1. Surface carbon atoms be covered with hydrogen atoms in order to prevent penetration of the approaching species inside the crystal.
2. Temperature of the crystal be high enough (1) to hold the approaching species to the normal distance from the crystal surface, (2) to produce dangling bonds by dissociation of dimers, (3) to prevent the penetration of an approaching species into the crystal by increasing the amplitude of the thermal vibrations among carbon atoms on the surface.
3. Temperature of the gas phase not exceed a critical value because (1) too many hydrogen atoms in the gas phase would remove the hydrogen atoms from the diamond surface, and (2) increase of the number of carbon atoms, which have low position-selectivity and penetrate into the crystal structure, would introduce structure defects.

References

1. Matsumoto, S., Sato, Y., Kamo, M. and Setaka, N., *Jap. J. Apll. Phys.* 21, 183(1982)
2. Suzuki, K., Yasuda, H., Sawabe, A. and Inuzuka, T., *Appl. Phys. Lett.* 50, 782(1981)
3. Spytin, D. V., Bouilov, L. L. and Derjaguin, B. V., *J. Crys. Growth* 52, 219(1981)
4. Matsumoto, S., Sato, Y., Tsutsumi, M. and Setaka, N., *J. Mate. Sci.* 17, 3106(1982)
5. Adachi, Y., *J. Phys. Soc. Jap.* 45, 875(1978)
6. Adachi, Y., *J. Phys. Soc. Jap.* 45, 1333(1978)
7. Kamo, M., Sato, Y., Matsumoto, S. and Setaka, N., *J. Crys. Growth* 62, 642(1983)

4. Flame - Pressure Interactions

Flame Propagation in Closed Vessels

Moshe Matalon *

McCormick School of Engineering and Applied Science
Northwestern University
Evanston, Illinois 60208-3125.

1 Introduction

When a combustible gas mixture contained in an open tube is ignited at one end of the tube, a nearly planar flame propagates throughout the gas at approximately a constant speed. The burned gas expands freely during the combustion process; that is the process takes place at constant pressure. If the chemical reaction goes to completion under adiabatic conditions, the temperature of the burned gas reaches the *adiabatic flame temperature* T_a. Let T_0 be the initial temperature of the fresh mixture, the change in enthalpy associated with the process yields

$$T_a = T_0 + QY_0/c_p \tag{1}$$

where Y_0 is the mass fraction of the deficient reactant in the mixture, c_p is the specific heat at constant pressure, assumed constant, and Q is the heat of combustion per unit mass of reactant. In a closed tube the burning process is accompanied by a rise in pressure. The confinement prevents a free expansion of the burned gas so that the *total* heat released by the chemical reaction goes into raising the gas temperature (none being used for the work of expansion). The temperature of the burned gas is therefore significantly higher than the adiabatic flame temperature for the same mixture. If the reaction proceeds uniformly throughout the entire volume the final temperature, evaluated from the change in internal energy, is

$$T_b = T_0 + QY_0/c_v \tag{2}$$

where c_v is the specific heat at constant volume, assumed constant. In practice there are temperature differences within the vessel, because adiabatic compression may convey energy from the gas elements which burn first to those which burn later. Thus, the flame temperature does not necessarily reach T_b. Nevertheless, because of the higher temperatures encountered, the burning velocity increases during the burning and the flame no longer travels at a constant speed

* This work has been partially supported by NASA's microgravity Combustion Science program and by the National Science Foundation.

as it does in an open vessel. Fluid dynamical effects appear to play an important role especially when the flame deforms near the walls or in the presence of hydrodynamic instabilities. All these effects complicate the burning process and necessitate a detailed analysis.

Studies of premixed flames in closed vessels have significant technological importance. They are directly applicable to internal combustion engine processes and are relevant to the initial development of detonation waves. Closed spherical bombs are used often for flame speed measurements.

In this paper a mathematical model that describes the propagation of a premixed flame in a closed vessel will be presented. Results concerning the burning velocity, pressure and temperature rise, flame front position and total burnout time will be reviewed. Recent results pertaining flame front instability and the development of a *tulip flame*, a flame shape often observed during flame propagation in closed tubes, will also be discussed.

2 Mathematical Model

A premixed combustible mixture of density ρ_0, temperature T_0 and pressure P_0 occupies the volume of a closed vessel. The mixture consists of a deficient reactant with an initial mass fraction Y_0 and a molecular weight W. When ignited, a flame propagates throughout the vessel. The present discussion is concerned with the post ignition events. It is therefore assumed that at time $t = 0$ a flame is already established near the ignition source. The model is based on the following main assumptions (cf. [14],[20]):-

1. The propagation velocity (typically $\sim 50 - 100$cm/s) is much smaller then the speed of sound. This suggests that acoustic disturbances propagate relatively fast so that the pressure is almost instantaneously equalized throughout the vessel. The pressure can therefore be expressed as $P(t) + p(\mathbf{x}, t) + \cdots$ where $P(t)$ represents the mean pressure level and p accounts for the *small* spatial variations.

2. The combustible mixture is treated as an inviscid ideal gas with constant specific heats c_p and c_v (with $\gamma = c_p/c_v$) and an average molecular weight $\bar{W}$. The thermal conductivity of the mixture λ and the mass diffusivity of the deficient reactant $\mathcal{D}$ are, in general, temperature dependent.

3. The chemical activity is modeled by a one step irreversible overall reaction. The reaction rate is assumed to be of Arrhenius type with an activation energy E a pre-exponential factor $\mathcal{B}$.

4. The combustion process occurs under adiabatic conditions.

The governing equations are

$$\frac{D\rho}{Dt} + \rho \nabla \cdot \mathbf{v} = 0 \qquad (3)$$

$$\rho \frac{D\mathbf{v}}{Dt} = -\nabla p \qquad (4)$$

$$\rho c_p \frac{DT}{Dt} = \frac{dP}{dt} + \nabla \cdot \lambda \nabla T + QB \left(\frac{\rho Y}{W}\right)^n \exp(-E/R^oT) \tag{5}$$

$$\rho \frac{DY}{Dt} = \nabla \cdot \rho \mathcal{D} \nabla Y - B \left(\frac{\rho Y}{W}\right)^n \exp(-E/R^oT) \tag{6}$$

$$\rho R^o T / \bar{W} = P \tag{7}$$

where $\mathbf{v}(\mathbf{x}, t), \rho(\mathbf{x}, t), Y(\mathbf{x}, t)$ and $T(\mathbf{x}, t)$ are the velocity, density, mass fraction and temperature fields, R^o is the universal gas constant and n the reaction order. The boundary conditions along the walls of the vessel are

$$\hat{\mathbf{n}} \cdot \mathbf{v} = \hat{\mathbf{n}} \cdot \nabla T = \hat{\mathbf{n}} \cdot \nabla Y = 0 \tag{8}$$

where $\hat{\mathbf{n}}$ is a unit vector normal to the walls.

The diffusion length $L_D = \lambda^o/\rho_0 c_p S_L \sim 10^{-2}$cm, is typically much smaller than the characteristic dimension of the vessel so that the flame may be treated as a surface of discontinuity (λ^o is a reference value). Across the flame front, described by $F(\mathbf{x}, t) = 0$, the Rankine-Hugoniot jump relations

$$[\rho(\mathbf{v} \cdot \mathbf{n} - V_f)] = 0 \quad [\mathbf{v} \times \mathbf{n}] = 0$$
$$\tag{9}$$
$$[p + \rho(\mathbf{v} \cdot \mathbf{n} - V_f)(\mathbf{v} \cdot \mathbf{n})] = 0 \quad [T] = QY_0/c_p$$

must be satisfied. Here $\mathbf{n} = -\nabla F/|\nabla F|$ is a unit vector normal to the flame front (pointing toward the burned gas) and $V_f = |\nabla F|^{-1}(\partial F/\partial t)$ is the normal velocity of the front. The jump in a quantity is defined as its value in the burned side minus that in the unburned side.

Diffusion and chemical reaction are negligible everywhere but in the flame zone. Thus, on either side of the flame the energy equation then simplifies to

$$\frac{DS}{Dt} = 0, \quad S = c_p \ln \left\{ \left(\frac{T}{T_0}\right)\left(\frac{P}{P_0}\right)^{(1-\gamma)/\gamma} \right\} \tag{10}$$

implying that the entropy S of the mixture is conserved along particle paths. Since the state of the fresh mixture is initially uniform, the entropy of the unburned gas remains constant. Consequently the temperature in the unburned gas remains uniform and rises in time according to the law of adiabatic compression; i.e.

$$T = T_0 \left(\frac{P}{P_0}\right)^{(\gamma-1)/\gamma} \quad \text{for } F > 0. \tag{11}$$

The flame temperature can now be determined by applying the appropriate jump from (9). One finds

$$T_f = T_a + T_0 \left\{ \left(\frac{P}{P_0}\right)^{(\gamma-1)/\gamma} - 1 \right\}. \tag{12}$$

Hence the entropy of a gas element, which is constant ahead of the flame, rises in the flame zone to the value S^*, given by $S(T = T_f)$, and remains constant thereafter. It is apparent from (10) and (12) that S^* decreases continuously in time, so that the gas element which burns first has the largest entropy. This implies that the highest temperature in the vessel is not reached at the flame front but rather at the point where ignition occurred; the maximum temperature is $T_{\max} = T_a(P/P_0)^{(\gamma-1)/\gamma}$. A temperature gradient is thus formed in the burned gas region causing re-illumination of the gas near the point of ignition. This is often seen in photographic records of flames in closed vessels [9], but not in open vessels. The temperature distribution in the burned gas can be expressed in the form

$$T = T_0 \left(\frac{P}{P_0}\right)^{(\gamma-1)/\gamma} \psi(\mathbf{x}, t) \qquad \text{for } F > 0 \tag{13}$$

where the function ψ is found by solving $D\psi/Dt = 0$ subject to $\psi = \exp(S^*/c_p)$ at $F = 0$. Note that the adiabatic condition (8) at the walls of the vessel is automatically satisfied at the cold boundary. Near the walls which are in contact with the hot burned gas, a thermal boundary layer must exist to ensure that the adiabatic condition is satisfied.

The flow field in the burned gas region is also markedly different than that in the unburned gas region. In the absence of vorticity in the fresh mixture, the unburned gas remains irrotational. The velocity field can therefore be deduced from a potential $\mathbf{v} = \nabla \Phi$ which must satisfy

$$\nabla^2 \Phi = -\frac{1}{\gamma P} \frac{dP}{dt}.$$

In the burned gas the flow is, in general, rotational. There are two possible sources of vorticity: (i) a curved flame in a nonuniform flow generates vorticity because, as a result of thermal expansion, the flow is refracted on passage through the curved front; and (ii) spatial nonuniformity in the density and pressure fields produce a nonzero baroclinic vector $\nabla\rho \times \nabla p$, which appears as a source term on the right hand side of the vorticity equation

$$\frac{D}{Dt}\left(\frac{\Omega}{\rho}\right) - \left(\frac{\Omega}{\rho} \cdot \nabla\right)\mathbf{v} = \frac{\nabla\rho \times \nabla p}{\rho^3}.$$

Clearly both are absent for one-dimensional configurations because the flow field remains uniform, albeit unsteady and the baroclinic vector vanishes identically.

To complete the system one needs equations for $P(t)$ and $F(\mathbf{x}, t)$; the latter can be also expressed in terms of the mass burning rate $M \equiv \rho(\mathbf{v} \cdot \mathbf{n} - V_f)$. Both depend on the details of the flame structure and are discussed next.

3 Burning Rate and Pressure Buildup

A standard analysis of the flame structure for a large activation energy yields an expression for the mass burning rate of the form

$$M = \sqrt{\frac{2\lambda^o \mathcal{B} Y_0^{n-1} \Gamma(n+1) P_0^n}{c_p Le^{-n} R^{o n} \beta^{n+1}}} \left(\frac{P}{P_0}\right)^a \left(\frac{T_f}{T_0}\right)^b \exp(-E/2R^o T_f). \tag{14}$$

Here $Le = \lambda/\rho c_p \mathcal{D}$ is the Lewis number and $\beta = E(T_a - T_0)/R^o T_a^2$ is the dimensionless activation energy parameter. The exponents a and b depend on the overall reaction order n and on the assumptions that one adopts regarding the thermal conductivity λ and the mass diffusivity $\mathcal{D}$. For example, when λ and $\rho\mathcal{D}$ are both assumed constants one finds that $a = n/2$ and $b = (n+2)/2$; but if one assumes [13] that $\lambda \sim T$ and $\rho^2\mathcal{D} = const$ then $a = n/2$ and $b = (n+3)/2$. If instead $\lambda\rho$ and $\rho^2\mathcal{D}$ are both taken constants, then the exponents are $a = (n-1)/2$ and $b = (n+3)/2$.

For isobaric conditions, $T_f = T_a$, and (14) reduces to the corresponding relation for a freely propagating flame [21]; i.e. $M = \rho_0 S_L$. Hence, in a closed vessel $M \sim \exp(E/2R^o T_a - E/R^o T_f)$, which implies that an $O(1)$ change in the flame temperature from the adiabatic value T_a causes an exponentially large change in the mass burning rate (except during the initial phase when $P \sim P_0$). The asymptotic analysis for large β is therefore not uniformly valid in time; the limitation arises of course because of the strong temperature dependence of the reaction rate. An $O(1)$ change in M can be obtained by further assuming that the heat release is large [13] [17]. Alternatively, one can adopt (14) as being suggested by the asymptotics and treats the activation energy as a finite parameter, albeit large. Formally, the latter can be *derived* by modeling the reaction rate in the governing equations by a delta function with a strength $\sim \exp(-E/2R^o T_f)$. Finally, it should be noted, that expressions for the burning rate that include a power law dependence on the pressure and temperature have been previously written (see the review in [1]); the exponents a and b have been determined empirically.

An equation for the mean pressure P can be obtained by combining (3) and (5) and integrating the resulting equation throughout the volume of the vessel. One finds

$$\frac{1}{P_0}\frac{dP}{dt} = \left(\frac{QY_0}{c_v T_0}\right)\frac{\mathcal{A}_f}{V}\frac{M}{\rho_0} \tag{15}$$

where V is the total volume and $\mathcal{A}_f$ the flame surface area. If the reactant is totally consumed in the vessel, the ending pressure P_e is found to be independent of the dynamics and can be determined a-priori as follows. By appropriately adding (5) and (6) an enthalpy equation, free of the reaction term, is derived. Integrating this equation throughout the volume while using the no flux boundary conditions at the walls yields

$$\frac{d}{dt}\int_V \rho(c_p T + QY)\,dV = \frac{dP}{dt}V.$$

When use is made of the initial and end conditions, this relation implies

$$P_e = P_0\left\{1 + QY_0/c_v T_0\right\}. \tag{16}$$

Note that P_e is characteristic of a given mixture. For typical values of the parameters the ending pressure is found to be nearly ten times its initial value, in accord with experimental results (cf. [19]). The pressure level in the vessel is determined by solving (15) subject to $P = P_0$ at $t = 0$. The condition (16) then determines the ending time, t_e, when the whole mixture has burned out.

4 Planar Flames

For definiteness, we assume that the combustible gas is ignited at the left end of a tube of length L. As the flame propagates throughout the gas the fresh mixture is compressed and pushed to the right while the burned gas moves away toward the ignition end. The axial velocity is given by

$$u = \begin{cases} -\dfrac{1}{\gamma P}\dfrac{dP}{dt}x & 0 \le x < x_f \\[3mm] \dfrac{1}{\gamma P}\dfrac{dP}{dt}(L - x) & x_f < x \le L \end{cases}$$

with the flame front located at

$$x_f(t) = L\left\{1 - \frac{P_e - P}{P_e - P_0}\left(\frac{P}{P_0}\right)^{-1/\gamma}\right\}.$$

The pressure level is obtained by integrating (15) which reduces to

$$\frac{1}{P_0}\frac{dP}{dt} = \frac{\gamma q}{\rho_0 L}M$$

where $q \equiv QY_0/c_p T_0$ is the dimensionless heat release parameter. Figure 1 shows the variations in the pressure level for various values of q and representative values of the remaining parameters: $E/R^o T_0 = 20$, $\gamma = 1.4$, $a = 1/2$ and $b = 3/2$.

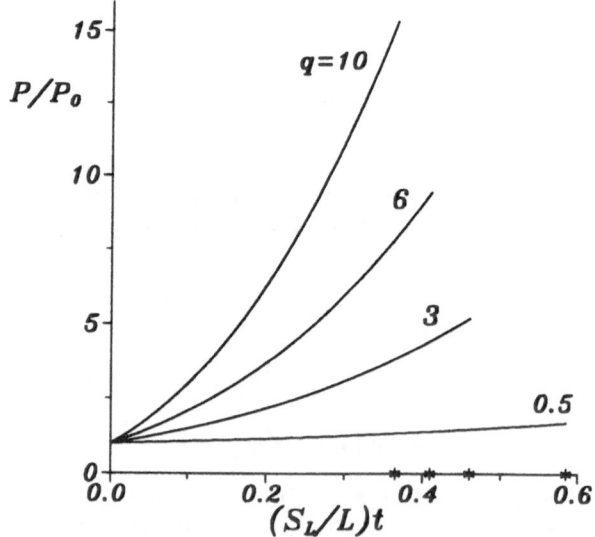

Fig. 1. The variation in the pressure level as a function of the scaled time $\tau = (S_L/L)t$

Note that the total burnout time, which is identified on the abscissa of the graph, is significantly shorter than the time it would take for the flame to travel steadily at the laminar flame speed S_L. The burning rate increases monotonically during the whole process, according to (14).

Temperature profiles at selected times are shown in Fig. 2 for the same choice of parameters and for $q = 6$. The temperature in the burned gas is expressed in the form (13) with $\psi = \psi(xP^{1/\gamma})$. The functional form of ψ is determined by the requirement that $S = S^*$ at $x = x_f(t)$. As pointed out earlier the finite gradient in T at $x = 0$ is smoothed out in a thin thermal boundary layer in which heat conduction dominates.

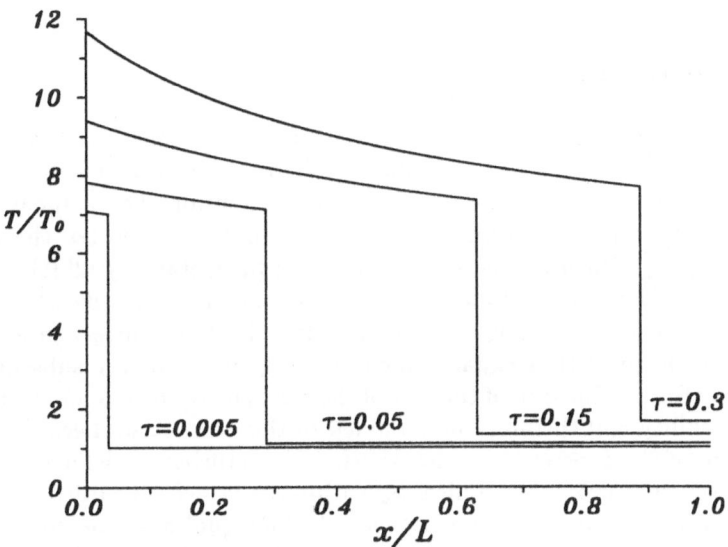

Fig. 2. Temperature profiles at selected times.

It is particularly instructive to examine these results in the limits of smallheat release and large activation energy, because explicit expressions can then be written for the physical quantities of interest. One finds

$$P \sim P_0 \left[1 + \gamma q(x_f/L)\right]$$
$$M \sim \rho_0 S_L \exp(\alpha x_f/L)$$
$$x_f \sim -\frac{1}{\alpha} \ln\left[1 - \alpha(S_L/L)t\right] L$$

where $\alpha = \frac{1}{2}(\gamma - 1)\beta q \sim O(1)$. Thus, although the pressure variations are small, they cause an $O(1)$ increase in the burning rate. The flame front propagates through the mixture at a continuously increasing speed and reaches the end of the tube at time $t_e \sim \frac{1}{\alpha}[1 - \exp(-\alpha)](L/S_L)$.

5 Lewis Number Effects

As pointed out earlier, a consistent asymptotic analysis for large β can be carried out if one assumes that the heat release is also large; but this is of no great advantage unless the Lewis number is sufficiently distinct from one. The analysis in this case [13] yields an equation for M in lieu of (14). It is found that qualitatively distinct behaviors in flame propagation may result, depending on whether the Lewis number is greater/less than one. When $Le < 1$ the burning rate increases monotonically in time as indicated earlier. However, for mixtures with $Le > 1$ the evolution is very sensitive to the initial conditions. Although the burning rate increases in time during the initial phase, a point is reached beyond which there can be either a bulk explosion of the remaining unburned mixture or a slow down in the burning process leading to extinction.

6 Flame Instability

Two well known phenomena that are responsible for intrinsic instabilities of premixed flames are hydrodynamic effects and diffusive-thermal effects. The hydrodynamic instability is a result of thermal expansion. For a freely propagating flame, thermal expansion causes a drop in density across the flame, $\rho_{burned} < \rho_{unburned}$, which is a necessary condition for instability [2],[8]. Disturbances of *all* wavelength grow and, furthermore, the short wavelength disturbances grow more rapidly than the long wavelength disturbances. In a closed vessel, the gas elements that expand upon crossing the flame are subsequently compressed by the combustion of the rest of the gas. The condition for the hydrodynamic instability will therefore be modified in this case. Also, because of the finite duration of the process, the long wavelength disturbances which grow relatively slower cannot develop significantly within the vessel. The other source of instability of premixed flames arises as a result of unequal diffusion coefficients. In particular, when the effective mass diffusivity (associated with the deficient reactant) is significantly larger than the thermal diffusivity of the mixture or, when the effective Lewis number $Le < 1$, a diffusive-thermal instability arises. Otherwise [2], i.e. when $Le \geq 1$, the flame is stable and diffusive-thermal effects have stabilizing influences. For mixtures with $Le > 1$ diffusive-thermal effects may often exert stabilizing influences that can offset the hydrodynamic instability. These stabilizing influences are associated with effects that occur within the flame zone and are therefore expected to play a similar role for combustion in closed vessels.

The stability of a nominally flat flame propagating in a finite tube of length L to non planar disturbances has been discussed in [11], [14] based on a similar model as the one described above. In order to account for hydrodynamic as well as diffusive-thermal effects, the expression (14) has been modified by adopting a Markstein-like model for the dependence of the burning rate on the local flame

---

[2] Indeed, for Le sufficiently bigger than one there is also a traveling waves or pulsating instability, which we do not discuss here.

front curvature. Specifically, the expression for the burning rate has been taken to be

$$M = M_0(1 + \mu\nabla \cdot \mathbf{n})$$

where M_0 stands for the right hand side of (14). The parameter μ, referred to as the Markstein length, has a magnitude comparable to the diffusion length L_D, so that $\mu/L \sim 10^{-3}$. An expression for the Markstein length has been derived for the case of a freely propagating flame [10],[16] showing an explicit dependence of μ on the effective Lewis number of the mixture. Generally speaking μ was found to be positive/negative depending on whether Le is greater/less than one, respectively. A similar dependence of μ on the mixture composition is also anticipated here.

The results of the stability analysis show that the long wavelength disturbances, which normally grow because of the hydrodynamic instability, are now damped out. The shorter wavelength disturbances which grow more rapidly can only be stabilized by diffusive-thermal effects . Thus, for mixtures with $\mu > \mu^*$ the planar flame is stable; otherwise an instability develops. Neutral stability curves are shown in Fig. 3 for various values of μ/L. For a given μ in the range $\mu_* < \mu < \mu^*$, there exists a band of wavenumber of disturbances that become unstable after x_f reaches the position x_f^* (the value of x_f at the turning point of that curve). The disturbance with wavenumber $k = k_c$, grows first and is therefore the most dangerous mode which is likely to be manifested first as an instability in an experiment. The dependence of k_c on x_f^* is identified by the dashed curve in the figure. Also from the figure, $\mu_* \simeq 0.001$ and $\mu^* \simeq 0.006$. Finally, for $\mu < \mu_*$ the planar flame is unstable to all disturbances.

7 Tulip Flames

Observations of combustion in closed tubes reveal that the flame often acquires a peculiar shape in which its center part forms a cusp pointing toward the burned gas. This flame shape is often referred to as a *tulip flame*. Experiments [5][7][19][4] reveal that the tulip flame forms only in tubes of sufficiently large aspect ratio (length/diameter); typically > 2. It has been also noted [7] that in extremely long tubes, of aspect ratio ~ 20, the inversion of the flame front can reverse itself. After a short transient, the center part of the inverted flame which points toward the burned gas starts accelerating and overtakes the outer edges of the flame thus forming a shape which is convex toward the unburned gas. This process repeats itself up to the end of the propagation. The experiments also show that the formation of tulip flames depend on the mixture composition and on the initial pressure in the tube.

There have been also numerous numerical investigations [18][15][6] aimed at reproducing the tulip flame. Based on these studies and on the above mentioned experimental observations various possible explanations have been suggested for the formation of the tulip flame. The actual cause for its formation, however, has not been conclusively determined.

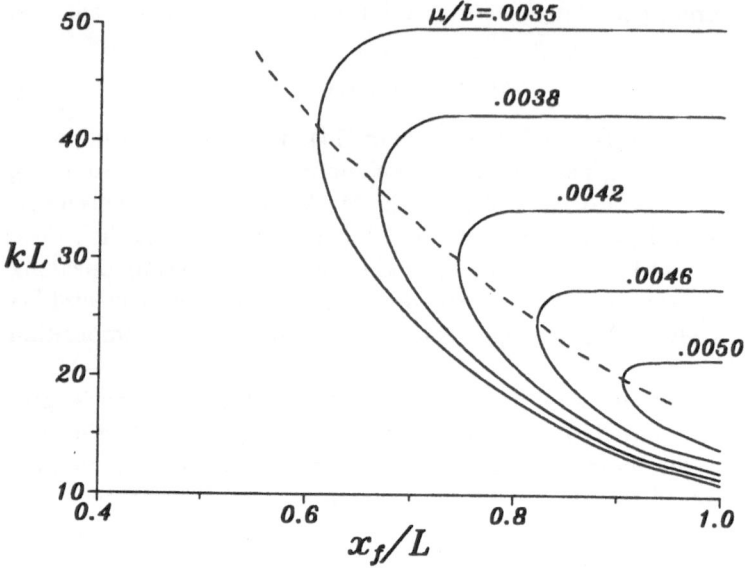

Fig. 3. Stability curves in terms of the disturbances wavenumber kL and the mean flame location x_f/L

It is possible to identify the physical mechanisms responsible for the initial development of the tulip flame from the stability results of the previous section. Consider a mixture with $\mu > \mu_*$ in a tube of aspect ratio $\mathcal{R} = kL/2\pi$. The flow field between two streamlines that are $2\pi/k$ apart correspond to an idealization of that within the tube. Thus, when $\mathcal{R} < k_c L/2\pi$, no instability will be observed and the flame front remains planar until it reaches the end of the tube. In longer tubes, however, the initially planar flame evolves into a corrugated front identified by a wavenumber $k_* \geq k_c$, where $k_* = 2m\pi L^{-1}\mathcal{R}$, with m an integer. A one cell structure indented toward the burned gas at the center of the tube that resembles the initial form of the tulip flame may result when $m = 1$. At the onset of instability, the perturbed flow field in the unburned gas region (Fig. 4) consists of two vortices, one on each side of the mid plane. The sense of the circulation in the upper plane is clockwise in agreement with experimental measurements [19]. Thus, as disturbances grow, the perturbed flow at the center of the tube opposes the gas motion and advects the flame into a shape whereby its center part becomes increasingly more convex with respect to the burned gas. Consistent with that one finds that the flame temperature, and hence the burning rate, decreases at the convex portion of the flame front.

The linear stability results discussed above are limited to the determination of the conditions for the development of an instability. In order to describe the new flame structure that evolves beyond the onset of instability, and in

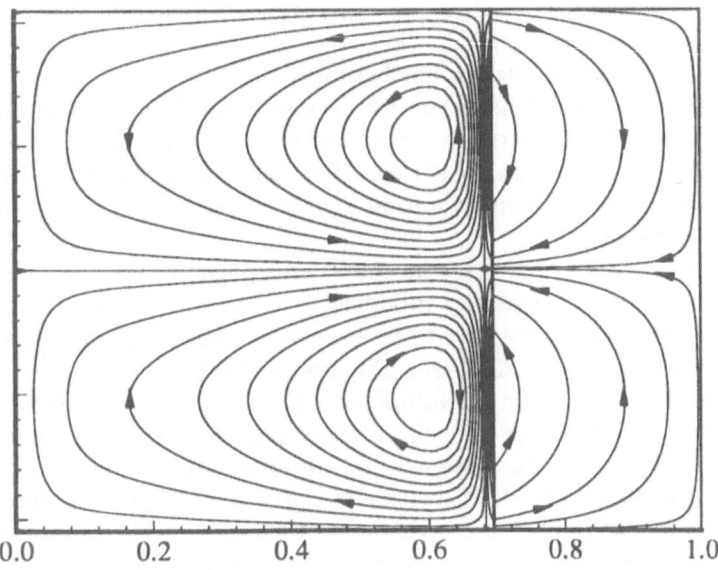

0.0   0.2   0.4   0.6   0.8   1.0

Fig. 4. Streamlines of the perturbed velocity field in the tube at the onset of instability.

particular the formation of the tulip flame and the oscillatory phenomenon of flame inversion, a nonlinear theory has been developed. Assuming that the heat release parameter is small, the problem reduces to solving a single evolution equation for the flame front perturbations. If $\varphi(y,t)$ denotes the deviation from a planar front, expressed in terms of the appropriately scaled transverse coordinate y and time variable t, the evolution equation is

$$\frac{\partial \varphi}{\partial t} = q \left\{ \hat{\mu} \frac{\partial^2 \varphi}{\partial y^2} + \frac{1}{2} \left(\frac{\partial \varphi}{\partial y} \right)^2 - \varphi + U(y,t) \right\} \tag{17}$$

where

$$U = \frac{1}{2\pi} \int\limits_{-\infty}^{\infty} \int\limits_{-\infty}^{\infty} \int\limits_{0}^{\tau} \frac{\kappa^2 \sinh[\kappa(1-t)] \cosh \kappa \tau}{\sinh \kappa} \varphi(\eta, \tau) \, e^{i\kappa(y-\eta)} \, d\tau \, d\kappa \, d\eta$$

and $\hat{\mu}$ is the scaled Markstein parameter. The derivation of this equation, its numerical simulation and the relation of the results to tulip flames are discussed in [12]. It should be pointed out that an important element of this equation is the memory effect appearing in the integral on the right hand side of the equation. This term has its origin in the vorticity production at the flame front which is *essential* for the description of the inversion phenomenon of tulip flames. In the absence of vorticity it can be readily verified that if at a given instant, $t = t_0$ say, the flame becomes flat, i.e. $\varphi(y, t_0) = 0$, it must remain flat thereafter since the equation in this case implies that $\varphi_{yt}(y, t_0) = 0$. Equation (17) on the other hand shows that $\varphi_{yt}(y, t_0) \neq 0$.

The inversion of tulip flames has been also discussed in [3] by considering an evolution equation that contains no memory term but has a second time derivative φ_{tt} instead. Mathematically, this term has a similar effect to the memory term in (17). However, unlike (17) which has been systematically *derived* from the conservation laws, their evolution equation has been written down using plausible arguments.

In summary, it appears that the tulip flame phenomenon is a manifestation of the hydrodynamic instability modified so as to account for the effects due to confinement, nonlinearities and flame front curvature. Ourpredictions suggest that the formation of the tulip flame depends on the mixture composition via the Lewis number (or μ) and on the initial pressure in the tube P_0 (which affects the burning rate as well as the diffusion coefficient), as experiments indicate. Also in agreement with the observations, the tulip flame forms only after the mean position of the flame exceeds x_f^*, with x_f^*/L greater than 1/2 and does not form in short tubes. The predicted critical wavenumbers (Fig. 4) correspond to $\mathcal{R}$ in the range $3 - 8$ which are typical in the reported experiments.

References

1. Bradley, D. and Mitcheson, A. Combustion and Flame, **26** (1976) 201.
2. Darrieus, G. unpublished manuscript (1938); also presented at the Sixth International Congress of Applied Mechanics.
3. Dold, J.W. and Joulin, G. this proceedings.
4. Dunn-Rankin, D., Barr, P.K. and Sawyer, R.F. Twenty-First Symposium (International) on Combustion, The combustion Institute, Pittsburgh (1986) 1291.
5. Ellis, O.C. DE C. , Fuel in Science and Practice, **7** (1928) 502.
6. Gonzalez, M., Borghi, B. and Saouab A., Combustion and Flame **88** (1982) 210.
7. Guénoche, H. in *Nonsteady Flame Propagation*, Markstein, G.H. editor, The Macmillan Company, New York (1964).
8. Landau, L.D. Acta Physicochimica URSS **19** (1944) 77.
9. Lewis, B. and von Elbe, G. *Combustion, Flames and Explosion of Gases*, third edition, Academic Press, Orlando (1987).
10. Matalon, M. and Matkowsky, B.J. Journal of Fluid Mechanics **124** (1982) 239.
11. Matalon, M. and McGreevy J.L. Twenty-Fifth Symposium (International) on Combustion, The combustion Institute, Pittsburgh (1994).
12. Matalon, M. and Metzener, P. submitted for publication (1995).
13. McGreevy, J.L. and Matalon, M. Combustion and Flame **91** (1992) 213.
14. McGreevy, J.L. and Matalon, M. Combustion Science and Technology, to appear (1994).
15. N'konga, B. Fernandez, G. Guillard, H. and Larrouturou, B. Combustion Science and Technology **87** (1992) 69.
16. Pelce, P. and Clavin, P. Journal of Fluid Mechanics **124** (1982) 219.
17. Peters, N. and Ludford, G.S.S. Combustion Science and Technology **34** (1983) 331.

18. Rotman, D.A. and Oppenheim A.K. Twenty-First Symposium (International) on Combustion, The combustion Institute, Pittsburgh (1986) 1303.
19. Starke, R. and Roth, P. Combustion and Flame **66** (1986) 249.
20. Sivashinsky, G.I. Acta Astronautica **6** (1979) 631.
21. Williams, F.A. *Combustion Theory*, second edition,. The Benjamin/Cummings Publishing Company (1985).

Pressure-driven disturbances in fluid dynamic interactions with flames

A.C. McIntosh

Department of Fuel and Energy, University of Leeds, LEEDS LS2 9JT, U.K.

Abstract. This paper discusses the progress of premixed flames after undergoing strong interactions with pressure driven disturbances. Even in one dimensional models of these interactions, if a pressure drop is fast enough and sufficiently great, extinction of the flame can result.

For two-dimensional interactions, a further very important mechanism is the baroclinic effect where the force due to the imposed pressure gradient acts on a region of fluid with variable density (i.e. the flame) which is at an angle to it, thus producing a vorticity field which can break up the flame. This was clearly demonstrated by the classic experimental work of Markstein. An overview is presented of recent numerical work that shows the dependence of the vorticity generated in the fluid on the viscosity and the strength of the reaction.

Keywords. Pressure interactions, acoustics, flames, burning velocity, Rayleigh-Taylor instability, shock-waves, baroclinic effect.

1. The role of pressure changes in premixed flames

The importance of pressure interactions with combustion processes cannot be underestimated. It is well known that small acoustic disturbances can affect the characteristics of the flame. The so called "singing flames" studied by Toepler [1] and Rayleigh [2] showed that oscillations and highly non-linear alterations in flame shape could occur for both diffusion and premixed combustion (see the book by Gaydon and Wolfhard [3], in particular chapter 7).

Since premixed flames are far more "active" and hazardous due to the preferred direction for burning, it is for this type of combustion that pressure effects become most relevant. For a given mixture strength there are two possible burning velocities - one small and of the order of centimetres sec^{-1} and the other large and of the order of kilometres sec^{-1}. For steady propagation, the former involves only a slight change in pressure across the combustion zone, and is termed a deflagration, whereas the latter case involves a large jump in pressure and is termed a detonation. For further details of the structure of these combustion waves the reader is referred to an introductory text on this subject (e.g. Strehlow [4], in particular chapters 5, 8 and 9). Our concern here is to demonstrate that there are a number of mechanisms which can cause a steadily propagating flame to accelerate. It is well known that turbulence increases the flame area so that the overall burning is increased. However, what is not generally appreciated is that pressure oscillations of a fraction of a bar are all that is needed to severely distort the flame and cause a turbulence flame to emerge. Thus although undoubtedly a flame encountering turbulent eddies will become stretched and under certain circumstances increase its overall burning rate, this may not be the primary cause of flame break-up. It can in fact be the pressure field itself (which may be created by obstacles) which causes the flame to break up into a turbulent flame brush. Fig. 1. illustrates schematically this type of effect.

The interaction of a plane pressure wave with a slightly rippled flame leads to the onset of the Rayleigh-Taylor instability (where regions of different density in the combustion region are accelerated by different amounts due to the same fluid force. This instablility is normally associated with acceleration due to gravity acting on an interface between a heavier fluid sitting above a lighter one [5]. For the purposes of the study of similar behaviour in flame propagation the gravity force is replaced by an imposed pressure gradient. When the gradient of pressure is more acute and the incident pressure wave is on a much shorter length scale, then the vorticity generation is more significant and the same effect (but on a larger scale) is known as the Richtmeyer-Meshkov instability [6].

For pressure disturbances of a very short length scale (on the scale of a typical shock thickness) the interaction can be very severe, as demonstrated by the experiments of Markstein [7]. Though completed 3 decades ago, these experiments have only recently been simulated to any accuracy, first by Picone, Oran, Boris and Young [8] and more recently by Batley et al [9,10]. It is shown in these recent simulations that the break up of the flame into a turbulent flame brush can take place if the relative combustion rate is slow.

Even for one-dimensional interactions there can be considerable alteration of the local mass flux due to a sharp pressure change. In section 4 we consider the possibility of flame extinction due to steep pressure drops and secondly a mechanism for flame acceleration by passing a shock wave through a flame from the hot side. Under certain conditions this can invoke the Clarke equation which describes the progress of acoustic waves in an explosive mixture [11], which is then coupled with a "fast flame" which emanates from the original accelerated flame.

2. Increase in flame surface area due to Rayleigh-Taylor instability

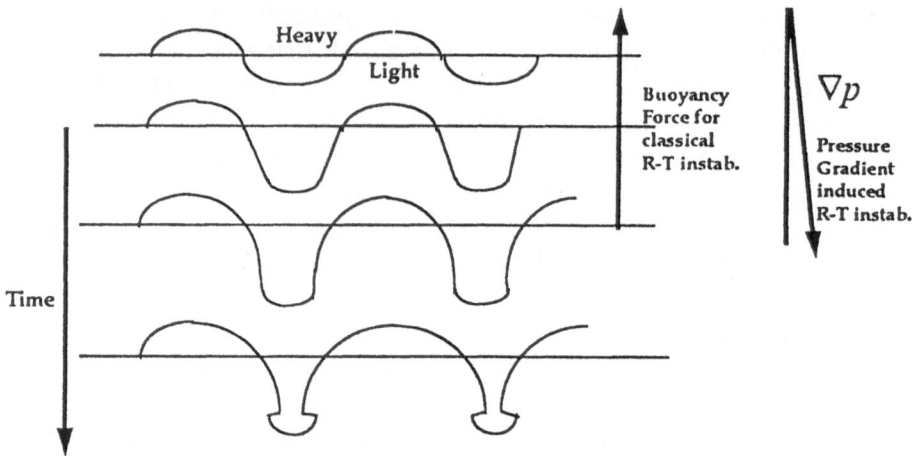

Fig. 1 Schematic of the Rayleigh-Taylor Instability applied to premixed flames.

Figure 1 is a schematic of the Rayleigh-Taylor instability. In our case the acceleration due to gravity is replaced by the pressure gradient. From the momentum equation $-\nabla p / \rho$ will have an equivalent effect in accelerating the fluid by different amounts because of the disparity in density between the hot and cold gases. Such an effect has been observed by Tsuruda and Hirano

[12]. This reference includes photographs, the essence of which have been reproduced numerically through the recent work of Liu et al [13] modelling the flame as a discontinuity, and Edwards et al [14] modelling the flame with a distributed reaction zone.

If a linear gradient in pressure is introduced from the burnt (light density) side, then this induces (by differential acceleration) a larger velocity near the peak than the trough. This then causes a spike to develop such that a "tongue" of unburnt reactants is pushed into the hot (burnt products). One should recognise that the overall progress of the wrinled premixed flame in units of combustion time ($\sim 10^{-3}$ secs) is still in the direction of the heavy (unburnt) gases, but that during the passage time of the pressure wave ($\sim 10^{-8}$ secs) significant *relative* velocities are induced in the other direction (heavy to light) due to differences in the effect of the imposed pressure gradient. The differences in density at different lateral locations encountered by the planar pressure wave lead to differential induced velocities. If the pressure gradient is in the other direction (that is where the pressure gradient is introduced from the cold (heavy) side), then the instability is suppressed [13]. Initially the same spikes are found in numerical models which use a distributed reaction zone. However, it is only with the more realistic model of the reaction that one can properly simulate the spherical lobes of unburnt gas which form at the end of the spikes entering the lighter fluid. These investigations [14] show that after the initial stage, a Kelvin-Helmoltz type instability is observed due to the shear flow of the heavy spike of fluid against the light fluid surrounding it (see Fig. 2). This creates a secondary vortex system leading to the lobes of unburnt gas at the end of the initial unburnt protrusion into the hot products. The subsequent development of the flame will be much affected by the increased burning rate which is encouraged by the greater heat transfer from hot products to cold reactants. This instability will enhance the turbulent break-up of a laminar flame and so further increase the mass burning rate because of the increase in the flame area.

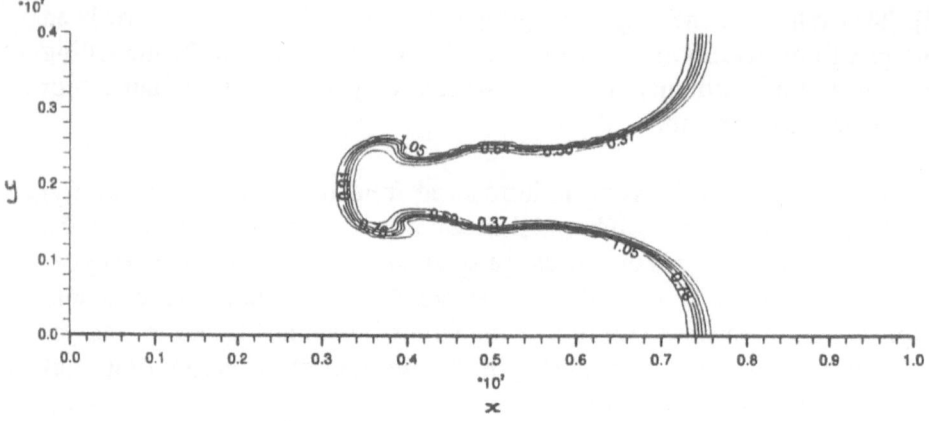

Fig. 2(a) Temperature contours of a rippled flame undergoing interaction with a 1 atm
m^{-1} pressure gradient, after .98 ms. Grid size : 21mm(x) × 17mm(y). (From ref. 14).

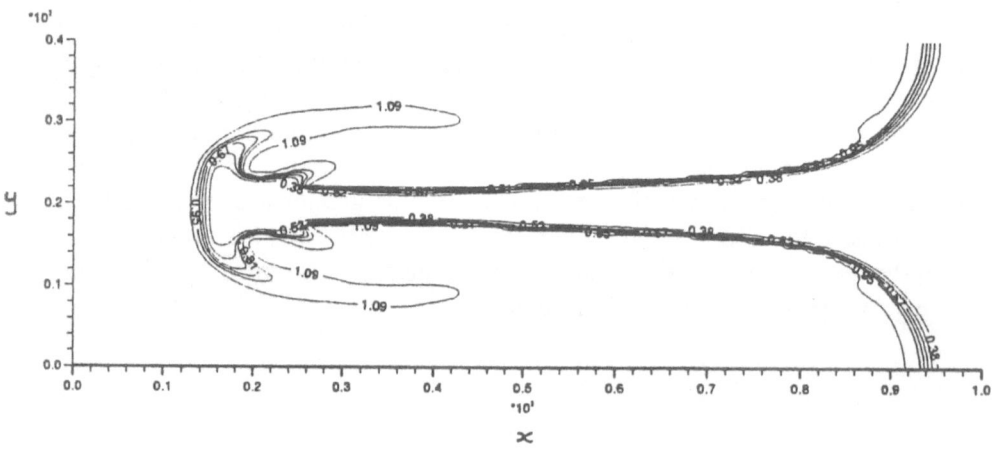

Fig.. 2(b) Temperature contours of a rippled flame undergoing interaction with a 1 atm
m^{-1} pressure gradient, after 1.2 ms. Grid size : 21mm(x) × 17mm(y). (From ref. 14).

3. Baroclinic distortion due to the interaction of a shock wave with a flame

If the premixed flame is strongly curved already and the pressure disturbance is sharp, then the $\nabla\rho \wedge \nabla p$ term in the vorticity generation equation,

$$\frac{D(\omega/\rho)}{Dt} = \frac{\nabla\rho \wedge \nabla p}{\rho^3} + \left(\frac{\omega}{\rho}.\nabla\right)\underline{u} + \nu\nabla^2\left(\frac{\omega}{\rho}\right)$$

induces strong roll-up of the flame surface. This effect is simply a more accentuated form of the Rayleigh-Taylor instability but with the generation of much more intense vorticity (see Fig.3).

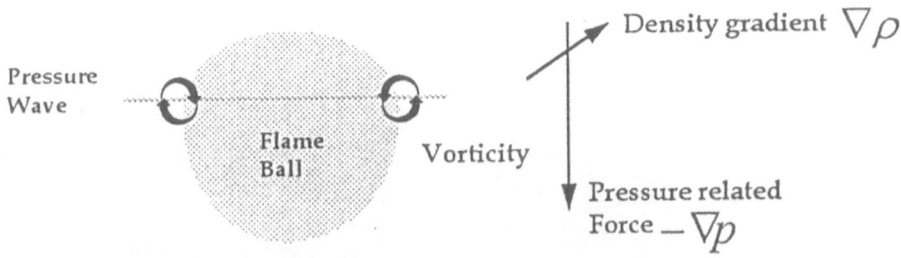

Fig. 3 Baroclinic generation of vorticity due to a shock wave passing through a curved flame

It is the baroclinic production of turbulence in premixed combustion which was first studied in detail by Markstein [7]. Since then there have been a number of attempts to simulate the vorticity production, treating the flame region as a prescribed density profile [8], and more recently including the reaction [9,10]. Fig. 4 shows the results of this work where a circular flame is simulated in a tube closed at one end. The shock wave passes through the flame once producing a strong distortion of the flame. The shock then reflects of the bottom end and passes through the flame a second time, further breaking up the flame. The amount of vorticity produced is found to depend heavily on the initial flame speed and reactivity. Faster, stronger flames (where the original reaction rates are high enough) have a greater tendency to recover from these interactions.

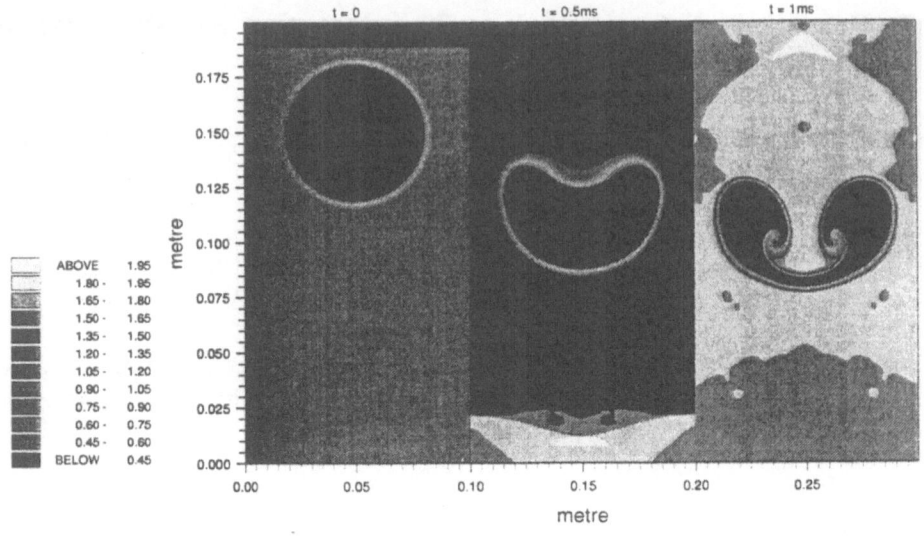

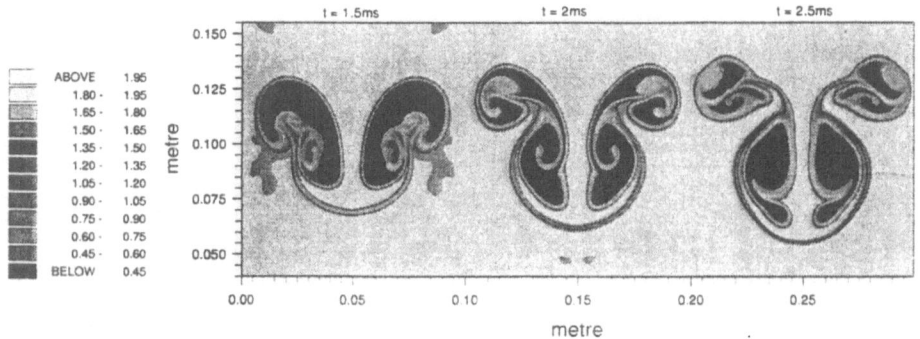

Fig. 4 The effect of a shock wave of strength 0.3 bar propagating across a flame ball with local laminar flame speed 1 ms⁻¹. The flame is in a tube closed at one end. The top centre and right-hand plots show the shock-wave reflecting off the closed end, simulating the experimental work of Markstein (1964). Upstream temperature 300K, downstream temperature 1500K. Reaction Rate constant $k_0 = 5 \times 10^{-7} s^{-1}$. Reaction Rate $= k_0 e^{-\theta/T}$, with $\theta = 10$. Thermal conductivity $\lambda \propto T$ with λ (300K) = 0.1Jm⁻¹s⁻¹K⁻¹. Upstream density ρ (300K) = 1.17kg m⁻³ and thermal diffusivity K (300K) = 9×10⁻⁵ m⁻¹s⁻¹. Lewis No. $Le = 1$ and Schmidt No. $Sc = 1$. The plots of density contours show the effect of baroclinicity in generating a strong vortex field as a result of the interaction (from refs. 9 and 10).

Fig. 4 clearly shows after the second shock passage has taken place, that the flame proceeds to break up into smaller and smaller segments. There are clearly four main "flamelets" visible by the time 2.5 ms have elapsed. The details of these effects are reported in the recent papers by Batley, McIntosh, Brindley and Falle [9] and Batley, McIntosh and Brindley [10]. Fig. 4 serves to illustrate the principle of these effects for relatively low reactivity, such rhat the pressure wave severely distorts the flame. There can be situations for fast, very reactive flames, where the flame "recovers" to some extent and the roll up of unburnt into burnt gases is not so pronounced.

Not only is this relevant to flame propagation in the presence of obstacles but also flame behaviour immediately after a vented explosion [15,16]. Immediately after a vented explosion there is a decompression wave which forces much of the hot gases out through the vent. Consequently the pressure gradient is operating in the unstable direction for the Rayleigh-Taylor instability and experiments confirm the rapid curling up of the deflagration, increase in flame area and consequent flame acceleration. Under cases of severe pressure buid up, the transient acceleration of the combustion wave can be very large (to values in excess of 100 ms^{-1}) and the expelled unburnt mixture undergoes a second ignition event.

4. One dimensional effects

The effects which are most visual and in some ways easier to be understood are those discussed in sections 2 and 3 concerning the rolling up of the flame surface area. These effects can be very pronounced and indeed on a macroscopic scale seem experimentally the most important phenomenon. However there are one dimensional effects which can be very significant when it comes to understanding (a) local flame extinction and (b) local acoustic coupling with precursor shock waves.

4.1 Flame extinction due to a pressure drop

Undoubtedly there will be flame extinction locally due to flame stretch (i.e. two-dimensional) effects. However extinction can take place even in one dimension. This can be done by applying a sufficiently large pressure drop across a flame so as to cause the flame combustion zone to dilate to such an extent that the overall mass burning rate drops and does not recover. This shown schematically in Fig. 5.

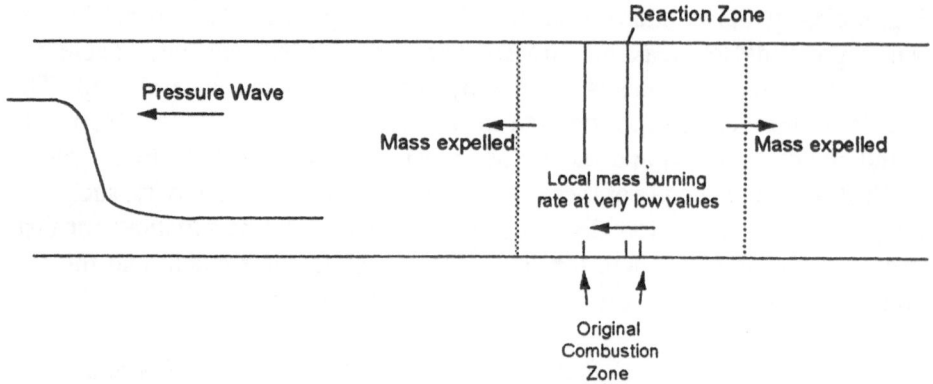

Fig. 5   Transient flame dilation after a sharp pressure drop.

For a single overall reaction rate model of the combustion, the unsteady flame evolution equations following the passage of a pressure wave [17,18,19,20,21,22] simplify, for unit Lewis number to :

$$\frac{\partial T}{\partial t} + m_0 \frac{\partial T}{\partial x} - p_0 \frac{\partial^2 T}{\partial x^2} = Q\Lambda C e^{\theta(1-1/T)} + (1-\gamma^{-1})\frac{T}{p_0}\frac{dp_0}{dt}$$

$$\frac{\partial C}{\partial t} + m_0 \frac{\partial C}{\partial x} - p_0 \frac{\partial^2 C}{\partial x^2} = -\Lambda C e^{\theta(1-1/T)}$$

where the thermal conductivity λ is assumed to be proportional to temperature T (non-dimensionalised with respect to the initial burnt temperature). The reaction rate is assumed to be of Arrhenius form with non-dimensional activation energy θ. The other variables are Q heat of reaction (non-dimensionalised with respect to $c'_p T'_b$, γ specific heats ratio, pressure p (only a function of time after the shock has gone through), t time, and x mass-weighted distance. C is the concentration of reactant. Noting that the steady-state eigen value Λ is given by

$$\Lambda \equiv \frac{A' D'_0 e^{-\theta}}{u'^2_0} \approx \frac{Q^2 \theta^2}{2}$$

(where D'_0 is the upstream mass diffusion coefficient, u'_0 is the initial flame burning velocity and the latter result is a deduction from classical asymptotic analysis), it can then be shown that the new steady mass burning rate m_{02} at the final pressure p_{min} is given by

$$m_{02} = p_{min}^{1/2} T_B^2 e^{\theta(1-T_B^{-1})/2}$$

where θ_1 is the activation energy non-dimensionalised with respect to the initial burnt gas temperature and T_B is the ratio of the final burnt gas temperature to the initial burnt gas temperature. Since there is a small but significant adiabatic heating term, the temperature ratio is given by

$$T_B = T_{01} p_{min}^{(1-1/\gamma)} + Q_1$$

where the notation Q_1 emphasises that the heat release is non dimensionalised with respect to the initial burnt temperature, and T_{01} is the ratio of initial unburnt to burnt temperature.

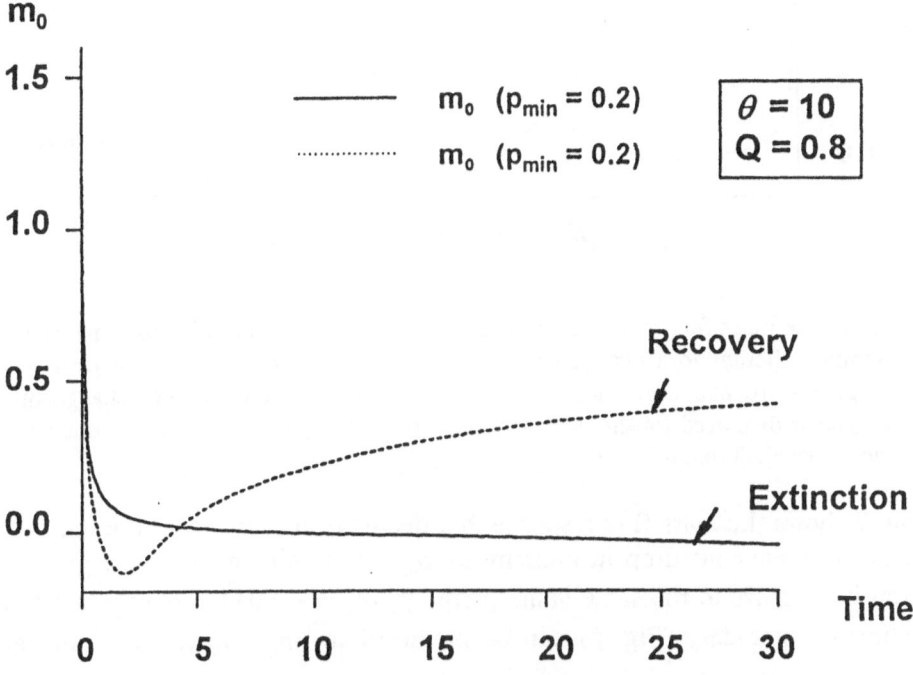

Fig. 6 The mass flux response of a premixed flame to a sudden decrease in pressure. Initial pressure $p = 1$, activation energy $\theta = 10$ and heat release $Q = 0.8$. The effect of pressure level is displayed for the two cases of recovery $(p_{min} = 0.4)$ and extinction $(p_{min} = 0.2)$. (From ref. 21).

Unlike the work of Ledder and Kapila [23], these studies consider pressure variations on time scales considerably faster than the typical diffusion time $(D_0'/u_0'^2)$. In Fig. 6 it can be seen that for the case of a sudden pressure drop to $p_{min} = 0.4$, the flame is able to recover its final steady m_{02} value just below 0.5. However when p_{min} is reduced to 0.2, the mass flux value m_{02} does not recover but continues to drop to extinction. These plots are for $\theta_1 = 10$ and $Q_1 = 0.8$.

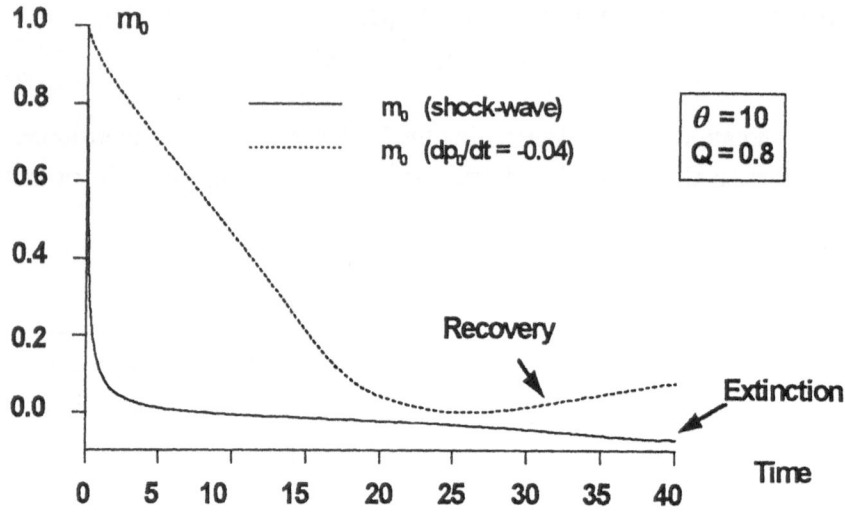

Fig. 7 The mass flux response (measured at the peak reaction rate position) of a premixed flame to a decrease in pressure. Initial pressure $p = 1$, activation energy $\theta = 10$ and heat release $Q = 0.8$. The effect of the rate of change of pressure is displayed for the two cases of recovery ($dp_0/dt = -0.04$) and extinction (shock-wave). (From ref. 21).

Fig. 7 shows the mass flux response to a decrease in pressure where the case of an instantaneous drop in pressure to $p_\infty = 0.2$ is compared to a gradual drop in pressure to the same value (with $dp_0/dt = -0.04$) Consequently an extinction boundary (Fig. 8) can be predicted in ($p_{min}, dp_0/dt$) parameter space.

By considering more than one reaction it can be verified that this is not an artefact of a simplified one-step scheme. Furthermore negative pressure pulses cause severe distortion of the reaction region such that locally the mass burning rate can become negative.

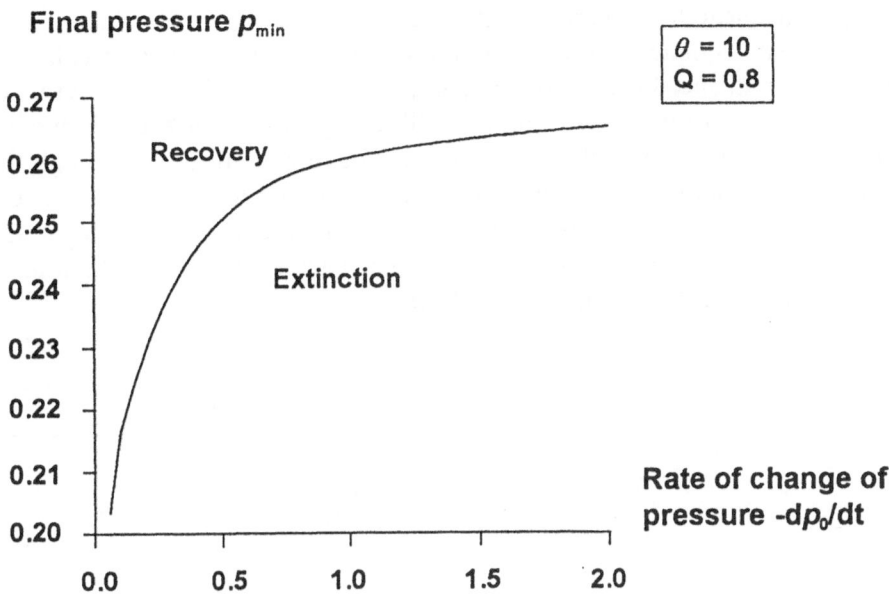

Fig. 8   The boundary between extinction and recovery in $(p_{min}, -dp_0/dt)$ parameter space for activation energy $\theta = 10$ and heat release $Q = 0.8$. (From ref. 21).

4.2 Acoustic coupling with precursor shock waves

A most interesting effect which has recently received attention by the author's research group at Leeds is the possibility of acoustic transmission after a strong interaction has taken place between a planar deflagration and a pressure wave. This can then lead to acceleration of deflagration even as a purely one-dimensional phenomenon.   Such studies follow on from the pioneering work of Clarke [11,24,25,26,27,28,29]. Those works have investigated the reactive-Euler equations and their solution prior to the formation of a detonation. In particular, for small perturbations they have explored the applicability of what is termed the Clarke equation which describes the acoustic/explosive events which can occur in the vicinity of a contact surface advancing through a combustible mixture (either a piston or shock wave). The work of Blythe and Crighton [30], Chue, Clarke and Lee [31], Dold, Kapila and Short [32] and Short and Dold [33] is also very relevant to these studies. There is a detailed discussion of the whole area in an excellent review article by Clarke [11]. For a briefer introduction to this subject, the reader is referred to the book by Buckmaster [34] and in particular Chapter 5 by Clarke : "Finite amplitude waves in combustible gases". The differential equation first derived by Clarke encapsulates the most important physics of acoustic transmission in the presence of an explosion. Such a non-linear acoustic equation can describe the thermodynamic behaviour behind a shock wave and ahead (by a large distance in terms of combustion lengths) of a traditional diffusion-driven "flame". This region is termed the induction zone. If the pressure disturbance is strong enough and advances through the flame from the hot side, the effect on the flame is first to severely distort the reaction zone, but also during the short but intense passage, to heat the preheat gases and send strong acoustic signals back to the flame which itself is readjusting after the passage of the shock.

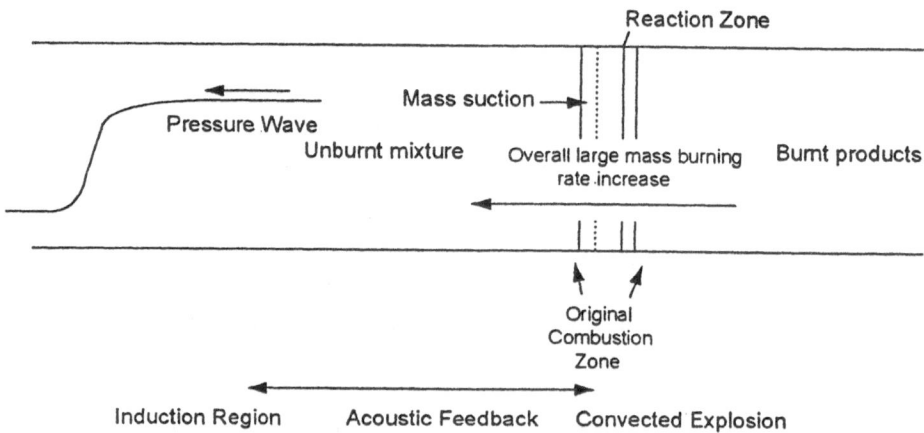

Fig. 9 Transient flame contraction and acceleration after a sharp
pressure increase.

After a shock wave has gone through a flame from the hot side into the unburnt mixture, if the amplitude of the shock is sufficiently strong, then at a comparatively large distance (in diffusion terms) from the flame, the Clarke equation is invoked (which describes acoustics in an explosive mixture) <u>at the same time</u> as the flame itself is still undergoing transient contractions and dilations as a result of the sharp (initial pressure change). For this case (Fig. 9) the flame experiences a sharp inrease in its burning rate m_0 so that under transient conditions one has a quasi-steady "fast flame" or "convected explosion" [24,25] obeying a set of equations of the form

$$m_0 \frac{\partial \hat{T}}{\partial \xi} = QR$$

$$R \sim AC \; \hat{\theta}^{\hat{s}} \exp(n \, \hat{\theta} - \hat{\theta}/\hat{T}) \quad 0 \leq n < T_{01}$$

$$m_0 \frac{\partial C}{\partial \xi} = -R$$

where $\hat{T}$ is temperature non-dimensionalised with respect to the initial upstream unburnt mixture temperature, and $\overset{\wedge}{\theta}$ likewise is non-dimensionalised using the same upstream temperature. T_{01} is the initial ratio of upstream to downstream temperature. The distance coordinate ξ is much smaller than the induction zone typical length $\hat{x}$ such that in this zone the pressure gradient is not significant, but ξ is appreciably larger than a typical diffusion lengthscale, since diffusion is relegated for these fast flames. In the induction zone far ahead of the fast flame the Clarke equation is obeyed

$$\left\{ \frac{\partial^2 (\phi_i)}{\partial \hat{t}^2} - \frac{\partial^2 (\phi_i)}{\partial \hat{x}^2} \right\} - \left\{ \gamma \frac{\partial^2 (e^\phi)}{\partial \hat{t}^2} - \frac{\partial^2 (e^\phi)}{\partial \hat{x}^2} \right\} = 0$$

where $\phi \equiv \overset{\wedge}{\theta} (\hat{T} - 1)$ is a reduced temperature.

The interaction between the induction zone and the fast flame provides an intriguing mechanism for acoustic transmission to the flame, such that a possible route for flame acceleration via flame/shock interactions exists which does not depend on two-dimensional effects.

References

1. Toepler, A., *Ann. Physics*. Leipzig, **128**, 126-.,1866.

2. Rayleigh, Lord, *Theory of Sound*, Macmillan, London, 1896.

3. Gaydon, A.G. and Wolfhard, H.G., *Flames, Their structure, radiation and temperature*, 3rd Ed., Chapnam and Hall, London., 1970.

4. Strehlow, R.A., *Combustion Fundamentals*, McGraw-Hill, New York, 1988.

5. Gardner, C.L., Glimm, J., Grove, J.W., McBryan, O., Zhang, Q., Menkoff, R. and Sharp, D.H., "A study of chaos and mixing in Rayleigh-Taylor and Richtmeyer-Meshkov unstable interfaces", Nuclear Physics B (Proc. Suppl.) **2**, 441-452., 1987

6. Cloutman, L.D. and Wehner, M.F., "Numerical simulation of Richtmeyer-Meshkov instabilities", *Phys. Fluids A* **4**(8), 1821-1830, 1992.

7. Markstein, G.H., *Non-steady flame propagation*, AGARDograph 75, Pergamon, Oxford, 1964.

8. Picone, J.M., Oran, E.S., Boris, J.P. and Young, T.R., "Theory of vorticity generation by shock wave and flame interactions", Prog. in Astronautics and Aeronautics : *Dynamics of Shock waves, explosions and detonations.* **94**, 429-448, 1984.

9. Batley G.A., McIntosh A.C., Brindley J. and Falle S.A.E.G., "A numerical study of the vorticity field generated by the baroclinic effect due to the propagation of a planar pressure wave through a cylindrical premixed laminar flame", Jnl. Fluid Mech. **279**, 217-237, Nov. 1994.

10. Batley, G.A., McIntosh A.C. and Brindley J., "Baroclinic distortion of laminar flames", (submitted to Proc. Roy. Soc. A July 1994).

11. Clarke, J.F., "Fast flames, waves and detonation", Prog. Energy Combust. Sci. **15**, 241-271 1989.

12. Tsuruda T. and Hirano T., "Local Flame Disturbance Development under Acceleration", *Comb. and Flame.* **84**, 66-72, 1991.

13. Liu, F., McIntosh. A.C. and Brindley, J., "A Numerical Investigation of Rayleigh-Taylor Effects in Pressure Wave-Premixed Flame Interactions", *Comb. Sci. and Tech.* **91**(4-6), 373-386, 1993.

14. Edwards, N.R., McIntosh A.C., and Brindley, J., "The development of pressure induced instabilities in premixed flames", *Comb. Sci. and Tech.*, **99**(1-3), 179-200, 1994.

15. Cooper, M.G., Fairweather, M. and Tite, J.P., "On the mechanisms of pressure generation in vented explosions", *Comb. and Flame* **65**, 1-14, 1986.

16. McIntosh, A.C., "The influence of pressure waves in the early stages of explosions", AIAA 94-0104, 32nd Aerospace Sciences Meeting & Exhibit Jan. 10-13 1994, Reno, NV, USA, 1994.

17. McIntosh, A.C., "Pressure Disturbances of Different Lengthscales Interacting with Conventional Flames", *Comb. Sci. Tech.* **75**, 287-309, 1991.

18. McIntosh, A.C., "The linearised response of the mass burning rate of a premixed flame to rapid pressure changes", *Comb. Sci. and Tech.* 91(4-6), 329-346, 1993.

19. Batley, G., McIntosh, A.C. and Brindley, J., "The time evolution of interactions between short lengthscale pressure disturbances and premixed flames", *Comb. Sci. and Tech.* 92, 367-388, 1993.

20. Johnson, R.G., McIntosh A.C., Batley G.A. and Brindley J., "Nonlinear oscillations of premixed flames caused by sharp pressure changes", *Comb. Sci. and Tech.*99(1-3), 201, 1994.

21. Johnson, R.G., McIntosh A.C. and Brindley J., "Extinction of premixed flames by pressure drops", *Comb. and Flame* (submitted March 1994).

22. Johnson, R.G.and McIntosh A.C., "On the phenomenon of local flow reversal in a premixed flame due to a large pressure pulse", *Comb. Sci. and Tech.*.(submitted May 1994).

23. Ledder, G. and Kapila, A.K., "The Response of Premixed Flames to Pressure Perturbations", *Comb. Sci and Tech.* 76, 21-44, 1991.

24. Clarke, J.F., "On changes in the structure of steady plane flames as their speed increases", *Comb. and Flame.* 50, 125-138, 1983.

25. Clarke, J.F., "Combustion in plane steady compressible flow : general considerations and gasdynamical adjustment regions", Jnl. Fluid Mech. 136, 139-166, 1983.

26. Clarke, J.F., "Chemical reactions in high-speed flows", Phil. Trans. R. Soc. Lond. A 335, 161-199, 1991.

27. Clarke, J.F. and Cant, R.S., "Non-steady gas dynamic effects in the induction domain behind a strong shock wave", Prog. Astronaut. Aeronaut. 95, 142-163, 1984.

28. Kassoy, D.R. and Clarke, J.F., "The structure of a steady high-speed deflagration with a finite origin", Jnl. Fluid Mech. 150, 253-280, 1985.

29. Singh, G. and Clarke, J.F., "Transient phenomena in the initiation of a mechnically driven plane detonation", Proc. Roy. Soc. A 438, 23-46, 1992.

30. Blythe, P. and Crighton, D.G., Proc. Roy. Soc. A **426**, 189-209, 1989.

31. Chue, R.S., Clarke; J.F. and Lee, J.H., "Chapman-Jouguet deflagrations", Proc. Roy. Soc. A **441**, 607-623, 1993.

32. Dold, J.W., Kapila, A.K. and Short M., *Dynamic Structure of Detonation in Gaseous and Dispersed Media*, Kluwer, 109-141, 1991.

33. Short, M. and Dold, J.W., Prog. in Astron. and Aeronaut. **154**, 59-74, 1993.

34. Buckmaster, J. D., "The mathematics of combustion", *Frontiers in Applied Mathematics* vol **3**, SIAM Monograph series in Applied Mathematics, 1985.

The Nonlinear Dynamics of Intrinsic Acoustic Oscillations in Combustion-Driven Systems

Stephen B. Margolis

Combustion Research Facility
Sandia National Laboratories
Livermore, California 94551-0969 USA

Abstract. The appearance of nonlinear acoustic oscillations in various types of unsteady combustion devices arises from combustion-driven instabilities that excite one or more classical acoustic modes of the system. For sufficiently strong acoustic driving relative to various damping processes, these disturbances grow to finite amplitudes and, owing to the nonlinear coupling between linearly unstable and stable modes, a stable limit-cycle oscillation is typically established. The nonlinear dynamics of these oscillations are formally governed by an infinitely-coupled system of evolution equations for the complex mode amplitudes. The linear terms determine the relative growth and decay rates of infinitesimal perturbations, while the nonlinear coupling terms determine the ultimate amplitude of each mode. The present work describes the typical nonlinear acoustic response of one such system by obtaining approximate analytical and numerical solutions of the dynamical system of amplitude equations. In particular, it is shown that, depending on the value of a reduced driving parameter, finite-mode approximations to the full infinitely-coupled system may be employed to describe the acoustic response of the model. The nature of acoustic mode interactions is also investigated, and it is formally demonstrated that a resonance-like coupling exists between any growing mode, whose frequency is ω_j, and its first resonant harmonic, which is the mode whose frequency is $3\omega_j$. Thus, the first acoustic bifurcation of the system occurs at a critical value of the driving parameter for which some mode achieves a positive linear growth rate, and is governed by a decoupled subsystem for the two modes corresponding to these frequencies.

1. Introduction

The principle underlying the operation of most unsteady combustion-driven devices is the persistence of naturally-occurring acoustic oscillations, resulting in an oscillating flow field that enhances efficiency in various types of heating, drying and propulsion applications [1]. These oscillations arise from intrinsic combustion-driven instabilities similar to those that occur in rocket motors [2], and may usually be represented as infinite summations over the classical acoustic eigenmodes of the system [2,3]. For sufficiently strong acoustic driving, these modes grow to finite amplitudes, whereupon the nonlinear coupling between growing (i.e., linearly unstable) and decaying (linearly stable) modes results in steady acoustic limit-cycle oscillations [4] that are commonly observed in experiments [5] and direct numerical calculations [6]. The present paper describes the nonlinear acoustic response of a model pulse combustion device as a function of certain driving parameters, providing additional insight into the nonlinear dynamics of acoustic mode interactions in such systems.

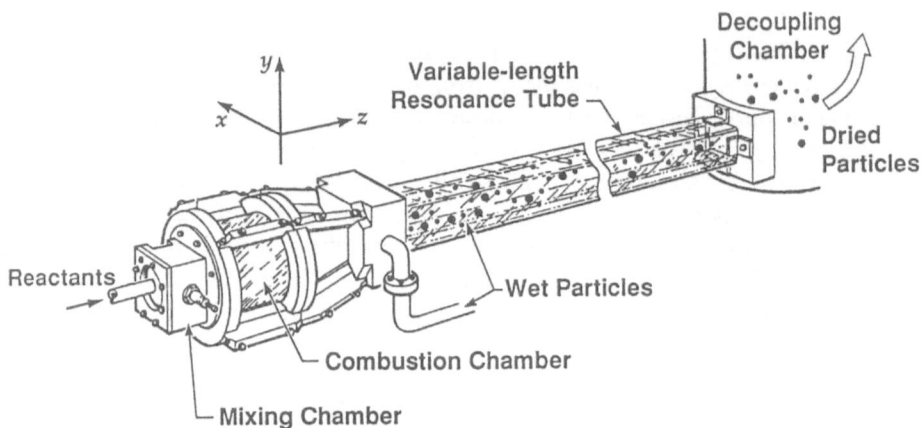

Fig. 1. *Geometry of the model pulse combustor considered here, illustrating a potential drying application. Wet particles are injected at the entrance to the resonance tube, which is fed by the flow from the combustion chamber. The drying process is enhanced by the presence of acoustic motions as the particles pass through the resonance tube, emerging in a dried or partially dried state and ready for further processing.*

The pulse combustor depicted in Fig. 1 is representative of a Helmholtz-type device in which the energy release in the combustion chamber drives the classical acoustic modes that are intrinsic to the geometry of the resonance tube. The injection of (wet) particles at or near the entrance to the resonance tube is incidental to the general discussion that follows, but serves to illustrate how such a device may be used in a typical drying application. A mathematical model of this device is given below, followed by a summary of the method by which the problem is reduced to an infinitely-coupled dynamical system of evolution equations for the complex mode amplitudes. Solutions to truncated versions of this system then reveal some features of the acoustic response, and also show how the number of modes required to give a correct approximation depends on parameters related to acoustic driving and damping. These results are supported by a formal analysis in the neighborhood of a simple acoustic bifurcation, where a two-mode subsystem for a single linearly unstable mode and its (stable) first resonant harmonic can be derived from the fully coupled system. This represents a key mechanism by which acoustic energy is transfered from a growing mode (one that is excited by the driving processes) to a decaying mode (one whose amplitude would otherwise decay to zero). Multiple growing modes, on the other hand, may be either strongly or weakly interacting, depending on whether or not one growing mode is also the first resonant harmonic of another. The first case corresponds to very strong modal resonance, while the second is more typical and leads to the prediction of secondary transition phenomena. In the present work, we focus on the general case of a simple acoustic bifurcation, corresponding to the passage through zero of the linear growth rate of a single acoustic mode. The latter cases are more complex and are treated elsewhere [7,8].

2. The Mathematical Model

The model employed here is similar to that introduced previously [3,4]. In particular, because combustion plays only a perturbative role in driving the classical acoustic oscillations dictated by the geometry of the device, it is sufficient for present purposes to confine the detailed analysis to the resonance tube by employing phenomenological admittance/impedance types of boundary conditions to model the acoustic driving and damping that occur at the ends and walls of the tube. Within the resonance tube, we have, for each phase, conservation of mass, momentum and energy, and an equation of state. In the small-particle limit considered here, these are given by

$$\frac{\partial \tilde{\rho}_\pi}{\partial \tilde{t}} + \tilde{\nabla} \cdot (\tilde{\rho}_\pi \tilde{\mathbf{u}}) = \tilde{r}(\tilde{p}, \tilde{T}) , \tag{1}$$

$$\frac{\partial \tilde{\rho}}{\partial \tilde{t}} + \tilde{\nabla} \cdot (\tilde{\rho} \tilde{\mathbf{u}}) = 0 , \tag{2}$$

$$\tilde{\rho}\frac{\partial \tilde{\mathbf{u}}}{\partial \tilde{t}} + \tilde{\rho}(\mathbf{u} \cdot \tilde{\nabla})\tilde{\mathbf{u}} = -\tilde{\nabla}\tilde{p} + \tilde{\mu}\tilde{\nabla}^2\tilde{\mathbf{u}} + \frac{1}{3}\tilde{\mu}\nabla(\nabla \cdot \tilde{\mathbf{u}}) , \tag{3}$$

$$\frac{\partial \tilde{p}}{\partial \tilde{t}} + \tilde{\mathbf{u}} \cdot \tilde{\nabla}\tilde{p} + \overline{\gamma}\tilde{p}\tilde{\nabla} \cdot \tilde{\mathbf{u}} = \tilde{p}\left(\frac{\partial \ln \overline{R}}{\partial \tilde{t}} + \tilde{\mathbf{u}} \cdot \tilde{\nabla} \ln \overline{R}\right) + (\overline{\gamma}-1)\left[\tilde{L}\tilde{r}(\tilde{p}, \tilde{T}) + \tilde{\mu}\tilde{\Phi}\right] , \tag{5}$$

$$\tilde{p} = \tilde{\rho}\overline{R}\tilde{T} , \quad \overline{R} \equiv (1 - \tilde{\rho}_\pi/\tilde{\rho})\tilde{R} . \tag{4}$$

Here, $\tilde{\rho}_\pi$ and $\tilde{\rho}$ are the particle and total densities, respectively, and $\tilde{\mathbf{u}}$, $\tilde{p}$ and $\tilde{T}$ are the velocity, pressure and temperature variables. The remaining quantities are defined in the Nomenclature.

To complete the specification of the problem, we prescribe admittance and impedance boundary conditions of the general form

$$\tilde{\mathbf{u}}|_{\tilde{z}=0} = (0, 0, \tilde{f}(\tilde{p}, \partial\tilde{p}/\partial\tilde{t}))|_{\tilde{t}-\tilde{t}_d} , \quad \tilde{p}|_{\tilde{z}=\tilde{H}} = \tilde{g}(\mathbf{n}_z \cdot \tilde{\mathbf{u}})|_{\tilde{t}-\tilde{t}_e} , \quad \mathbf{n} \cdot \tilde{\mathbf{u}}|_C = \tilde{h}(\tilde{p})|_{\tilde{t}-\tilde{t}_c} ,$$
$$(6a, b, c)$$

corresponding to imposed relationships, with time delays, between velocity and pressure at the ends and sides of the resonance tube. The first of these is a generalization of that used previously in that velocity is assumed to depend on the time rate of change of pressure as well as the pressure itself, similar to recent burning-rate models of combustion instabilities in rocket engines [9]. The parameters inherent in (6a) are phenomenologically related to the driving relationship between velocity and pressure due to combustion in the chamber upstream. For simplicity, we asssume a perfect decoupling chamber by setting $\tilde{h} = 0$, and set $\tilde{g} = 0$ as well. The overall effect of viscous wall damping is thus represented in a volumetric fashion by regarding the coefficient $\tilde{\mu}$ in (3) as a phenomenological damping parameter [4].

3. Reduction to an Infinite Dynamical System

When (1) - (6) are nondimensionalized using the speed of sound as a characteristic velocity, three small quantities emerge: a characteristic Mach number M of the unperturbed flow, a particle-loading parameter δ, and a characteristic acoustic amplitude ϵ associated with nonsteady perturbations. As a result [3], solutions for the nondimensional field variables may be sought as

$$
\begin{aligned}
p = \gamma^{-1} &+ M^2 p_s(z) + \wp(x,y,z,t) \\
&\sim \gamma^{-1} + M^2(\delta p_{s,1} + \cdots) + \epsilon(\wp_1 + \epsilon\wp_2 + \epsilon^2\wp_3 + \cdots),
\end{aligned} \tag{7a}
$$

$$
\rho = \rho_s(z) + \zeta(x,y,z,t) \sim 1 + \delta\rho_{s,1} + \cdots + \epsilon(\zeta_1 + \epsilon\zeta_2 + \epsilon^2\zeta_3 + \cdots), \tag{7b}
$$

$$
\rho_\pi = \delta\hat{\rho}_s(z) + \delta\hat{\zeta}(x,y,z,t) \sim \delta\hat{\rho}_{s,1} + \cdots + \delta\epsilon(\hat{\zeta}_1 + \epsilon\hat{\zeta}_2 + \epsilon^2\hat{\zeta}_3 + \cdots), \tag{7c}
$$

$$
T = T_s(z) + \theta(x,y,z,t) \sim 1 + \delta T_{s,1} + \cdots + \epsilon(\theta_1 + \epsilon\theta_2 + \epsilon^2\theta_3 + \cdots), \tag{7d}
$$

$$
\begin{aligned}
\mathbf{u} = M\mathbf{u}_s(z) &+ \mathbf{v}(x,y,z,t) \\
&\sim M\left[(0,0,w_0) + (0,0,\delta w_{s,1}) + \cdots\right] + \epsilon(\mathbf{v}_1 + \epsilon\mathbf{v}_2 + \epsilon^2\mathbf{v}_3 + \cdots), \tag{7e}
\end{aligned}
$$

where, in general, we may choose $w_0 = 1$ (which defines M) unless $\mathbf{u}_s \equiv 0$, in which case $w_0 = 0$ (and M is based on some appropriate hydrodynamic velocity scale). Here, we consider the regime $\delta \sim M$, which implies that particles and the basic flow affect the nonsteady acoustic waves at the same order in the analysis. This leads to the result that $\epsilon \sim M^{1/2}$ (i.e., acoustic perturbations are larger in magnitude than the Mach number of the unperturbed flow), whereupon a complete nonlinear perturbation analysis leads to a sequence of inhomogeneous problems for the coefficients in the nonsteady parts of the expansions. In particular, at $O(\epsilon^n)$, the nth order pressure and velocity fields are determined by

$$
\begin{aligned}
\frac{\partial \mathbf{v}_n}{\partial t} + \nabla\wp_n = \mathbf{r}_n, \qquad & \frac{\partial \wp_n}{\partial t} + \nabla \cdot \mathbf{v}_n = q_n, \\
\mathbf{v}_n|_{z=0} = (0,0,b_n), \qquad & \wp_n|_{z=1} = \mathbf{n} \cdot \mathbf{v}_n|_C = 0,
\end{aligned} \tag{8}
$$

where the inhomogeneous terms q_n, $\mathbf{r}_n$ and b_n are zero for $n = 1$ and, for $n > 1$, involve only lower-order coefficients. Thus, these problems may be solved in a sequential fashion.

The leading-order ($n = 1$) problem corresponds to classical linear acoustics in a tube that is closed at one end ($z = 0$) and open at the other ($z = 1$ in units of $\tilde{H}$). The solution for a rectangular resonance tube of transverse dimensions a and b ($0 \leq x \leq a$, $0 \leq y \leq b$) is given by

$$\wp_1 = \sum_{j=0}^{\infty} \sum_{k=0}^{\infty} \sum_{l=0}^{\infty} A_{j,k,l}(\tau)\, e^{i\omega_{j,k,l}t} \cos(j\pi x/a) \cos(k\pi y/b) \cos[(2l+1)\pi z/2] + \text{c.c.},$$

(9)

$$\mathbf{v}_1 = \sum_{j=0}^{\infty} \sum_{k=0}^{\infty} \sum_{l=0}^{\infty} \frac{A_{j,k,l}(\tau)}{i\omega_{j,k,l}} e^{i\omega_{j,k,l}t}$$

$$\times \left\{ \begin{array}{l} (j\pi/a)\sin(j\pi x/a)\cos(k\pi y/b)\cos[(2l+1)\pi z/2] \\ (k\pi/b)\cos(j\pi x/a)\sin(k\pi y/b)\cos[(2l+1)\pi z/2] \\ [(2l+1)\pi/2]\cos(j\pi x/a)\cos(k\pi y/b)\sin[(2l+1)\pi z/2] \end{array} \right\} + \text{c.c.},$$

(10)

where j, k and l are integers, $\omega_{j,k,l}^2 = \pi^2[(j/a)^2 + (k/b)^2 + (l+\frac{1}{2})^2]$, $\tau = \epsilon^2 t$ is a slow time, and "c.c." denotes the complex conjugate of the preceding term. The remaining first-order coefficients ζ_1, θ_1 and $\hat{\zeta}_1$ are then given in terms of $\wp_1$ and $\mathbf{v}_1$. Infinitely-coupled evolution equations for the complex mode amplitudes $A_{j,k,l}(\tau)$ are determined at a higher (third) order in the analysis by the application of solvability conditions on the inhomogeneous terms [3,4].

For resonance tubes used in pulse combustion applications, the transverse dimensions are typically small compared with the length of the tube ($a, b \ll 1$). Thus, the low-frequency modes are purely longitudinal ($j = k = 0$) and since damping increases with frequency [3], any finite-mode approximation to the problem will, for sufficiently long tubes, involve only the purely longitudinal complex mode amplitudes $A_{0,0,l} \equiv A_l$ corresponding to the lowest mode frequencies $\omega_{0,0,l} \equiv \omega_l = \pi(l+\frac{1}{2})$. Retaining N modes, the evolution equations for the A_l are given by

$$\frac{dA_l}{d\tau} = \left[(\alpha_l + \mu_0\beta_l)\hat{M} + \gamma_l\hat{\delta} \right] A_l$$

$$+ \sum_{l'=0}^{l-1\geq 0} \sum_{l''=0}^{l-1-l'} \sigma_0(\omega_l, \omega_{l'}, \omega_{l''}, \omega_{l'''}) A_{l'} A_{l''} A_{l'''} \Big|_{l'''=l-1-l'-l''}$$

$$+ \left(\sum_{l'=0}^{l} \sum_{l''=l-l'}^{N'} + \sum_{l'=l+1\leq N'}^{N'} \sum_{l''=0}^{N'+l-l'} \right) \sigma_1(\omega_l, \omega_{l'}, \omega_{l''}, \omega_{l'''}) A_{l'} A_{l''} A_{l'''}^* \Big|_{l'''=l'+l''-l}$$

$$+ pms + \sum_{l'=0}^{N'-l-1\geq 0} \sum_{l''=0}^{N'-l-1-l'} \sigma_2(\omega_l, \omega_{l'}, \omega_{l''}, \omega_{l'''}) A_{l'}^* A_{l''}^* A_{l'''} \Big|_{l'''=l+1+l'+l''}$$

$$+ pms$$

(11)

for $l = 0, 1, \ldots, N' = N - 1$. Here, "pms" denotes the permutations $l' \rightarrow l''$, $l'' \rightarrow l'''$, $l''' \rightarrow l'$ and $l' \rightarrow l'$, $l'' \rightarrow l'''$, $l''' \rightarrow l''$ in the preceding expressions, the scaled parameters $\hat{M}$ and $\hat{\delta}$ are defined by $M = \hat{M}\epsilon^2$, $\delta = \hat{\delta}\epsilon^2$, and μ_0 is the nondimensional viscous damping parameter (an effective inverse Reynolds number for the unsteady flow). The coefficients of the nonlinear terms, given explicitly in [4], are purely imaginary and depend only on the ratio of specific heats of the gas. The coefficients of the linear terms contain all remaining parametric dependencies of the model, and their real parts determine the linear growth or decay rate of each mode in the limit of infinitesimal perturbations. Defining $f_0' \equiv \partial f/\partial p|_{p_0}$ and $g_0' \equiv \partial f/\partial \dot{p}|_{p_0}$, where $\dot{p} \equiv \partial p/\partial t$ and $p_0 = \gamma^{-1}$ is the steady nondimensional reference pressure [cf. (7a)], we obtain $\alpha_l = (f_0' + i\omega_l g_0')\, e^{-i\omega_l t_d}$, $\beta_l = -2\omega_l^2/3$, and purely imaginary γ_l. Thus, the linear growth rates λ_l^r are given by

$$\lambda_l^r = f_0' \cos(\omega_l t_d) + g_0' \omega_l \sin(\omega_l t_d) - \frac{2}{3}\mu_0 \omega_l^2 , \qquad (12)$$

where the first two terms, which arise from the presumed pressure/velocity coupling due to combustion, determine the degree of acoustic driving. The last term is always negative and thus acts to damp all acoustic modes at a rate that increases with frequency, insuring that all but perhaps a few of the lower-frequency modes are damped [3]. In what follows, results will be obtained as a function of the empirically-determined parameters f_0', g_0' and t_d.

4. Limit-Cycle Solutions

Writing the N complex amplitudes in the polar form $A_l(\tau) = R_l(\tau)\, e^{i\phi_l(\tau)}$, and separating each amplitude equation into real and imaginary parts, yields a set of $2N$ equations for the real amplitudes R_l and the corresponding phases ϕ_l. From (11), these have the general form

$$\frac{dR_l}{d\tau} = \lambda_l^r R_l + F_{l,N}^{\sin}(R_0,\dots,R_{N-1},\phi_0,\dots,\phi_{N-1}) , \qquad (13)$$

$$R_l \frac{d\phi_l}{d\tau} = \lambda_l^i R_l + G_{l,N}(R_0,\dots,R_{N-1}) + F_{l,N}^{\cos}(R_0,\dots,R_{N-1},\phi_0,\dots,\phi_{N-1}) \quad (14)$$

for $l = 0,\dots,N-1$. Here, λ_l^r (λ_l^i) are the real (imaginary) parts of the complex linear growth rates λ_l, the $G_{l,N}$ are sums of triple products of the real amplitudes, and the $F_{l,N}^{\sin}$ $(F_{l,N}^{\cos})$ are sums of triple products of the real amplitudes multiplied by sines (cosines) of linear combinations of the phase variables. For example, the nonlinear terms for $N = 2$ are given by

$$F_{0,2}^{\sin} = \lambda_{\bar{0},\bar{0},1}^i R_0^2 R_1 \sin(3\phi_0 - \phi_1) , \quad F_{0,2}^{\cos} = \lambda_{\bar{0},\bar{0},1}^i R_0^2 R_1 \cos(3\phi_0 - \phi_1) , \qquad (15)$$

$$F_{1,2}^{\sin} = \lambda_{0,0,0}^i R_0^3 \sin(\phi_1 - 3\phi_0) , \quad F_{1,2}^{\cos} = \lambda_{0,0,0}^i R_0^3 \cos(\phi_1 - 3\phi_0) , \qquad (16)$$

$$G_{0,2} = \lambda_{0,0,\bar{0}}^i R_0^3 + \lambda_{0,1,\bar{1}}^i R_0 R_1^2 , \quad G_{1,2} = \lambda_{0,\bar{0},1}^i R_0^2 R_1 + \lambda_{1,1,\bar{1}}^i R_1^3 , \qquad (17)$$

where the triple subscripts on the coefficients refer to the particular complex amplitudes they multiply when like combinations of the A_l are grouped together in (11), with an overline denoting a complex conjugate factor A_l^*. Solutions for which the right hand sides of (13) vanish correspond to constant-amplitude limit-cycle oscillations that are sinusoidal in the slow time τ [since if $dR_l/d\tau = 0$, the right hand sides of (14) are constant, which implies that the $\phi_l(\tau) = \varphi_l \cdot \tau$]. In the context of the representations (9) - (10), such solutions correspond to a superposition of classical acoustic modes with time-independent amplitudes ϵR_l and perturbed frequencies $\omega_l + \epsilon^2 \varphi_l$, where, without loss of generality, we may set $\epsilon = M^{1/2}$ (and $\hat{M} = 1$).

A calculation of solutions to (13) - (14) requires a specification of the truncation order N, a proper choice for which depends on the λ_l^r. Physical considerations suggest that all growing modes and at least some decaying modes should be retained in order to provide the necessary nonlinear coupling and mechanism for energy transfer between growing and decaying modes. Thus, initial calculations with $N = 2$ and 3 were presented in [4] for a range of parameter values in which only a single linear growth rate (λ_0^r) achieved positive values. The results were similar to Fig. 2, which compares the analytically tractable limit-cycle solutions for $N = 2$ with those calculated numerically using $N = 4$ by computing the long-time trajectory of (13) - (14) from initial conditions. In this figure, which was obtained for $g_0' = 0$ and typical values for the particle-related parameters [4], the real amplitudes R_l have been plotted as a function of the reduced parameter $\mathcal{F} = f_0' \cos(\omega_0 t_d)/\mu_0$, which in this case represents the ratio of driving to damping for the zeroth mode. As $\mathcal{F}$ increases past the critical value $\mathcal{F}_0 = 2\omega_0^2/3 = \pi^2/6$ corresponding to the passage of λ_0^r through zero, a stable nontrivial solution corresponding to the steady limit-cycle oscillation described above bifurcates from the steady unperturbed state. It is clear from the figure that this bifurcation is accurately described by the two-mode approximation, a fact that can be deduced from the full infinitely-coupled (i.e., $N = \infty$) system, as discussed below. The two-mode approximation then predicts a turning point in the acoustic response, at which point stability of the limit-cycle is lost (only the lower branch is stable). For values of $\mathcal{F}$ beyond this point, solutions to the two-mode truncation grow unbounded in time, suggesting that it is possible to overdrive the system. The response obtained numerically for the four-mode approximation also gives a turning point, but at a larger value of $\mathcal{F}$, indicating that for stronger driving, additional decaying modes are crucial to the energy-transfer mechanism that limits the growth of the driven mode.

Although it has been conjectured [4], based on the relative magnitudes of the real ampltudes, that the turning point could be physical (i.e., the inclusion of additional modes would produce a convergence in the location of the turning point), further numerical calculations with larger numbers of modes indicate otherwise. Figure 3, employing 18 modes, shows that stronger driving can be accommodated (at least in the present model) through energy transfer to increasingly higher-frequency (and hence more strongly damped) modes. Indeed, Fig. 3 shows that stable limit-cycle oscillations can be obtained even when $\mathcal{F}$

surpasses the value $\mathcal{F}_1 = 3\pi^2/[2 - 8\sin^2(\pi t_d/2)]$ at which the second linear growth rate λ_1^r passes through zero. Calculations with 32 modes move the turning point well beyond the scale of Fig. 3. We thus conclude that the number of modes required to give an accurate prediction of the acoustic response of the system is dependent on parameters that determine the relative rates of driving and damping, although for given values of these parameters, a finite number of modes is sufficient. These conclusions are consistent with related studies of rocket-motor instabilities [10].

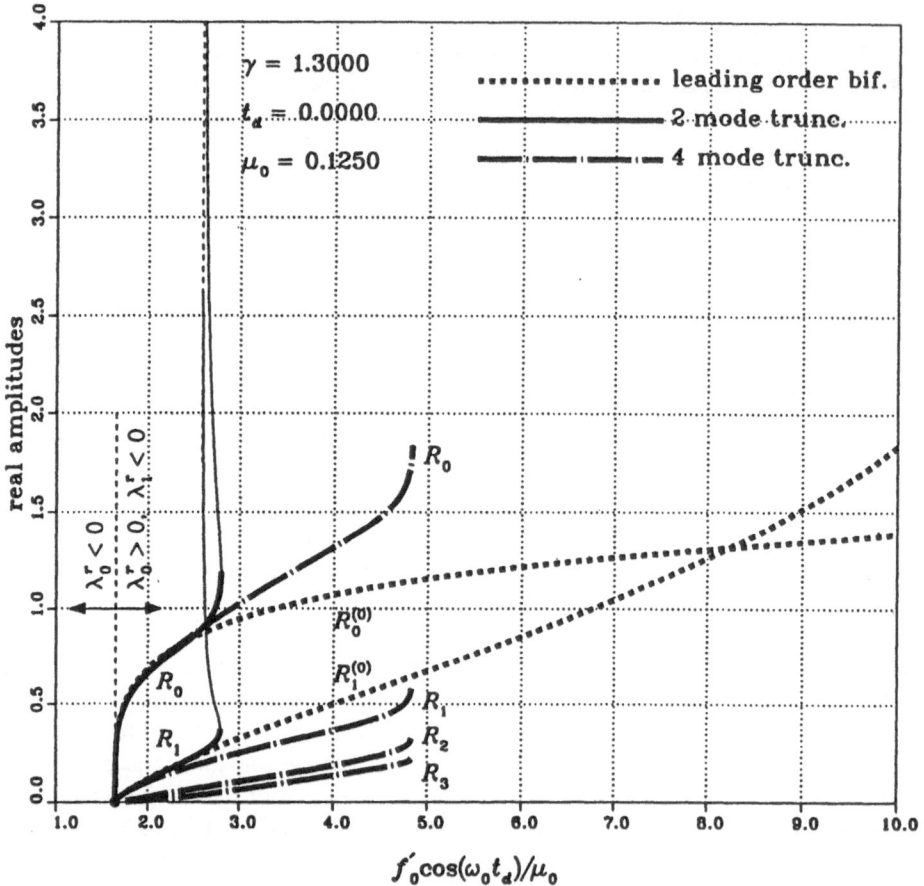

Fig. 2. *Real amplitudes for two- and four-mode approximations as functions of the parameter group* $\mathcal{F} = f_0' \cos(\omega_0 t_d)/\mu_0$. *Heavy (light) curves denote stable (unstable) branches. Parameter values are* $\gamma = 1.3$, $\mu_0 = 1/8$, $t_d = 0$, $g_0' = 0$, *and typical values of particle-related parameters which, in the regime considered here, enter into the expressions for the* λ_i^i [4].

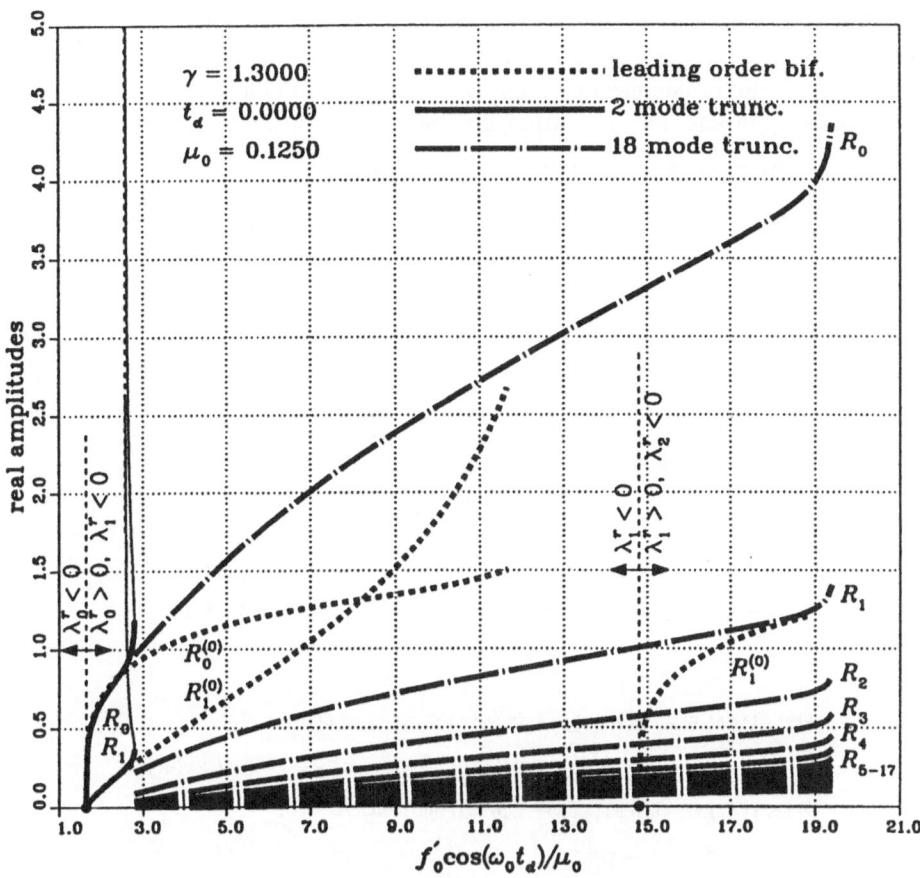

Fig. 3. *Same as Fig. 2, showing the acoustic response using eighteen modes. The second primary bifurcation, unlike the first, is unstable, and has no effect on the latter, which now extends past the point at which a second mode achieves a positive linear growth rate.*

5. Bifurcation Analysis

To obtain further insight into the nature of nonlinear acoustic interactions and to formalize the above description of the acoustic response, we now consider the full infinitely-coupled system in a parameter regime that corresponds to the onset of acoustic oscillations. As a guide, we first observe that the real amplitudes of the analytical limit-cycle solution obtained for $N = 2$ have, for $0 < \lambda_0^r \ll 1$ and $\lambda_1^r < 0$, the order-of-magnitude behavior $R_0 \sim (\lambda_0^r)^{1/4}$, $R_1 \sim (\lambda_0^r)^{3/4}$ (i.e., $R_1 \ll R_0 \ll 1$). More generally, we consider the primary acoustic bifurcation corresponding to the first passage of any one linear growth rate λ_j^r through zero, and seek a perturbation solution of the full system in that neighborhood.

We thus consider the regime $0 < \lambda_j^r \ll 1$, $\lambda_l^r < 0$ for all $l \neq j$, and define the small expansion parameter $\kappa \equiv (\lambda_j^r)^{1/4}$. By considering the manner in which successive modes of decreasing magnitude enter into (11) for arbitrarily large N, it becomes apparent that a solution for the real amplitudes can be sought as

$$R_0 = \cdots = R_{j-1} = 0, \qquad R_j \sim r_j^{(0)} \kappa + r_j^{(1)} \kappa^3 + \cdots ,$$

$$R_{j+1} = \cdots = R_{3j} = 0, \qquad R_{3j+1} \sim r_{3j+1}^{(0)} \kappa^3 + r_{3j+1}^{(1)} \kappa^5 + \cdots ,$$

$$R_{3j+2} = \cdots = R_{5j+1} = 0, \qquad R_{5j+2} \sim r_{5j+2}^{(0)} \kappa^5 + r_{5j+2}^{(1)} \kappa^7 + \cdots ,$$

$$\vdots$$

$$R_{(2l-1)j+l} = \cdots = R_{(2l+1)j+l-1} = 0,$$

$$R_{(2l+1)j+l} \sim r_{(2l+1)j+l}^{(0)} \kappa^{2l+1} + r_{(2l+1)j+l}^{(1)} \kappa^{2l+3} + \cdots ,$$

(18)

while solutions for *phase-differences* may be obtained as

$$\psi_{3j+1} \equiv \phi_{3j+1} - 3\phi_j \sim \psi_{3j+1}^{(0)} + \psi_{3j+1}^{(1)} \kappa^2 + \cdots ,$$

$$\psi_{5j+2} \equiv -\phi_{5j+2} + 2\phi_{3j+1} - \phi_j \sim \psi_{5j+2}^{(0)} + \psi_{5j+2}^{(1)} \kappa^2 + \cdots ,$$

$$\psi_{7j+3} \equiv \phi_{7j+3} - 2\phi_{3j+1} - \phi_j \sim \psi_{7j+3}^{(0)} + \psi_{7j+3}^{(1)} \kappa^2 + \cdots ,$$

$$\vdots$$

$$\psi_{(2l+1)j+l} \equiv -\phi_{(2l+1)j+l} + 2\phi_{(l+1)j+l/2} - \phi_j$$
$$\sim \psi_{(2l+1)j+l}^{(0)} + \psi_{(2l+1)j+l}^{(1)} \kappa^2 + \cdots \quad (l \text{ even}),$$

$$\psi_{(2l+1)j+l} \equiv \phi_{(2l+1)j+l} - 2\phi_{lj+|l/2|} - \phi_j$$
$$\sim \psi_{(2l+1)j+l}^{(0)} + \psi_{(2l+1)j+l}^{(1)} \kappa^2 + \cdots \quad (l \text{ odd}),$$

(19)

where the phase-differences $\psi_{(2l+1)j+l}$ are those that appear in terms that are proportional to $A_{(2k+1)j+k}^* A_{(k+1)j+[k/2]}^2$ (for k even) and $A_{(2k+1)j+k}(A_{kj+[k/2]}^*)^2$ (for k odd) in the amplitude equation for A_j. Thus, in the neighborhood of a simple acoustic bifurcation, energy transfer from the linearly unstable mode results in the excitation of only a resonant subset of the acoustic spectrum corresponding to frequencies that are odd integral multiples of the fundamental [i.e., $\omega_{(2l+1)j+l} = (2l+1)\omega_j$].

Substituting the expansions (18) - (19) into the infinitely-coupled system (11) written in terms of real amplitudes and phases, and collecting coefficients of like powers of κ, we obtain a decoupled subsystem for the leading-order coefficients $r_j^{(0)}$, $r_{3j+1}^{(0)}$ and $\psi_{3j+1}^{(0)}$ given by $\partial r_j^{(0)}/\partial \tau = 0$,

$$\frac{\partial r_j^{(0)}}{\partial \tau_4} = r_j^{(0)} - \lambda_{\bar{j},\bar{j},3j+1}^i \left(r_j^{(0)}\right)^2 r_{3j+1}^{(0)} \sin \psi_{3j+1}^{(0)} ,$$

$$\frac{\partial r_{3j+1}^{(0)}}{\partial \tau} = \lambda_{3j+1}^r r_{3j+1}^{(0)} + \lambda_{j,j,j}^i \left(r_j^{(0)}\right)^3 \sin \psi_{3j+1}^{(0)} ,$$

(20)

$$\frac{\partial \psi_{3j+1}^{(0)}}{\partial \tau} = \lambda_{d,1,j} + \lambda_{j,j,j}^i \left(r_j^{(0)}\right)^3 \left(r_{3j+1}^{(0)}\right)^{-1} \cos \psi_{3j+1}^{(0)} , \qquad \lambda_{d,1,j} \equiv \lambda_{3j+1}^i - 3\lambda_j^i ,$$

where $\tau_4 = \kappa^4 \tau$ is a new slow time. A nontrivial *time-independent* solution of (20) is analytically tractable and thus a formally correct first approximation of the steady limit-cycle solution is given by $\bar{r}_j^{(0)}$, $\bar{r}_{3j+1}^{(0)}$, $\bar{\psi}_{3j+1}^{(0)}$ [the last implies, from (19), that $d\phi_{3j+1}^{(0)}/d\tau = 3\, d\phi_j^{(0)}/d\tau$], where

$$\left(\bar{r}_j^{(0)}\right)^4 = -\lambda_{3j+1}^r \frac{\left[1 + (\lambda_{d,1,j}/\lambda_{3j+1}^r)^2\right]}{\lambda_{j,j,j}^i \lambda_{\bar{j},\bar{j},3j+1}^i} > 0,$$

$$\left(\bar{r}_{3j+1}^{(0)}\right)^2 = -\frac{\lambda_{j,j,j}^i}{\lambda_{3j+1}^r \lambda_{\bar{j},\bar{j},3j+1}^i} \left(\bar{r}_j^{(0)}\right)^2 > 0, \tag{21}$$

$$\bar{\psi}_{3j+1}^{(0)} = \frac{\pi}{2} \mp \frac{\pi}{2} \pm \mathrm{Sin}^{-1}\left[\left(\lambda_{\bar{j},\bar{j},3j+1}^i \bar{r}_j^{(0)} \bar{r}_{3j+1}^{(0)}\right)^{-1}\right], \quad \lambda_{d,1,j}\lambda_{j,j,j}^i \lessgtr 0, \tag{22}$$

where the inequalities in (21) are guaranteed by the fact that $\lambda_{\bar{j},\bar{j},3j+1}^i = \lambda_{j,j,j}^i$. Thus, if the jth mode is the only linearly unstable mode, energy is transferred primarily to the $(3j+1)$th decaying mode whose frequency ω_{3j+1} is three times the fundamental frequency ω_j. We thus refer to the latter mode as the first resonant harmonic of the former. At successively higher orders of approximation, only those modes whose frequencies are $(2l+1)\omega_j$ enter, with amplitudes which are $O(\kappa^{2l+1})$. The leading-order solution (21) - (22) is indicated by the heavy dashed lines in Figs. 2 and 3, where the initial bifurcation corresponds to $j = 0$.

It is clear that in the neighborhood of a simple bifurcation point, a two-mode approximation, whether obtained via direct truncation or from a formal bifurcation analysis, correctly represents a true solution of the full infinitely-coupled system. Demonstrating the linear stability of the first bifurcation given by (21) - (22), which is a steady solution of (20), requires a little care owing to the fact that $r_j^{(0)}$ evolves on a slower time scale than either $r_{3j+1}^{(0)}$ or $\psi_{3j+1}^{(0)}$. One approach is to suppress the slow time τ_4 by substituting its definition in terms of τ into the evolution equation for $r_j^{(0)}$. This explicitly introduces the small expansion parameter κ into (20), but the limit cycle solution is then shown to be stable in a standard fashion by calculating the eigenvalues of the stability matrix associated with the linearization of (20) about (21) - (22). The results are consistent with a strictly formal stability analysis, which is presented elsewhere [7].

6. Summary

The above analysis demonstrates the type of preferred coupling that exists in nonlinear acoustics, and serves as a basis for analyzing more complex mode interactions that arise when more than a single mode achieves a positive linear growth rate [7,8]. In particular, it has been shown that in a typical combustion-driven device, the solution can be represented as a superposition of classical acoustic eigenfunctions. The onset (bifurcation) of acoustic oscillations then occurs at a critical value of a driving parameter that corresponds to the first passage through zero of the linear growth rate of one of these acoustic modes. Beyond this value, a finite-amplitude limit cycle is established through a nonlinear resonant coupling between the linearly unstable (growing) mode and the linearly stable (decaying) mode whose frequency is three times that of the former.

Similar analyses can be applied to other combustion-driven acoustics problems, such as those associated with acoustic instabilities in rocket motors. One major distinction that has emerged, however, is the qualitative difference in the form of the amplitude equations corresponding to (11) [2,3,10,11]. This arises because of differences in the acoustic boundary conditions at leading order [4,11], which are closed/open (at the entrance/exit of the resonance tube) for the pulse combustor problem analyzed here, but are typically closed/closed for the rocket motor problem. This feature results in cubic nonlinearities in the former case, but quadratic nonlinearities in the latter, which in turn may lead to qualitatively different nonlinear dynamics for the two types of problems.

Acknowledgement. This work was supported by the U. S. Department of Energy under Contract DE-AC04-94AL85000.

Nomenclature:

a, b	transverse dimensions of resonance tube
A	complex amplitude of acoustic mode
f, g, h	pressure/velocity coupling functions
$\mathcal{F}$	reduced driving parameter
H	resonance tube length
L	heat of vaporization
M	Mach number
$\mathbf{n}$	normal direction
$p, \wp$	pressure, pressure perturbation
r	particle vaporization rate, real amplitude coefficient
R	gas constant, real amplitude
$t, t_{c,d,e}$	time, time delays
T, θ	temperature, temperature perturbation
$\mathbf{u}, \mathbf{v}$	velocity, velocity perturbation
x, y, z	transverse, longitudinal coordinates
δ	mass fraction of solid
ϵ, κ	small expansion parameters
γ	ratio of specific heats for the gas
λ	coefficient in amplitude equations
μ	viscosity parameter
ρ, ζ	density, density perturbation
τ	slow time
Φ	viscous dissipation term

ϕ, ψ phase, phase-difference variable

ω acoustic frequency

subscripts, superscripts:

C transverse boundary of resonance tube

j, k, l indices

π particle property

s steady quantity

r, i real, imaginary part

$-$ averaged value or steady solution

$\sim$ dimensional quantity

References:

1. Putnam, A. A., Belles, F. E. and Kentfield, J., *Pulse combustion*, Prog. Energy Combust. Sci. 12:43-79 (1986).

2. Culick, F. E. C. and Yang, V., *Prediction of the stability of unsteady motions in solid propellant rocket motors*, in Nonsteady Burning and Combustion Stability of Solid Propellants (L. DeLuca and M. Summerfield, Eds.), AIAA, New York, 1992, pp. 719-779.

3. Margolis, S. B., Stability of acoustic oscillations in a model Helmholtz-type pulse combustor, in Twenty-Fourth Symposium (International) on Combustion, The Combustion Institute, Pittsburgh, 1992, pp. 19-27.

4. Margolis, S. B., *Nonlinear stability of combustion-driven acoustic oscillations in resonance tubes*, J. Fluid Mech. 253:67-104 (1993).

5. Dec, J. E., Keller, J. O. and Hongo, I., *Time-resolved velocities and turbulence in the oscillating turbulent flow of a pulse combustor tailpipe*, Combust. Flame 83:271-292 (1991).

6. Barr, P. K., Dwyer, H. A. and Bramlette, T. T., *A one-dimensional model of a pulse combustor*, Combust. Sci. Tech. 58:315-336 (1988).

7. Margolis, S. B., *Resonant mode interactions and the bifurcation of combustion-driven acoustic oscillations in resonance tubes*, SIAM J. Appl. Math., to appear (1994).

8. Margolis, S. B., *The nonlinear dynamics of intrinsic acoustic oscillations in a model pulse combustor*, Combust. Flame, to appear (1994).

9. Grenda, J. M., Venkateswaran, S. and Merkle, C. L., *Three-dimensional analysis of combustion instabilities in liquid rocket engines*, AIAA 93-0235 (1993).

10. Jahnke, C. C. and Culick, F. E. C., *An application of dynamical systems theory to nonlinear combustion instabilities*, AIAA 93-0114 (1993).

11. Wang, M. and Kassoy, D. R., *Dynamic compression and weak shock formation in an inert gas due to fast piston acceleration*, J. Fluid Mech. 220:267-292 (1990).

5. Numerical Treatments

Modeling of Combustion of a Gaseous Sphere Using Mathematica

R.G. Rehm and H.R. Baum

National Institute of Standards and Technology, Gaithersburg, MD 20899, USA

Abstract. Transient combustion of a gaseous sphere of fuel is examined in the flame-sheet limit. The gas is considered thermally expandable, the Lewis numbers are taken as unity and a temperature-dependent thermal diffusivity is allowed. Evaluation of some approximations used in forming the model are evaluated by examining a spherical, nonlinear thermal conduction problem first. Mathematica is used to solve the PDEs arising from both problems by a Method of Lines. The solutions for the combustion problem show the expansion and subsequent collapse of the flame-sheet trajectory, the sharp initial temperature spike at the flame front and its later diffusive spreading, and the early peaked expansion velocity, followed by a double humped velocity profile.

Keywords. combustion, non-premixed; finite difference methods; flame-sheet analysis; fuel-pocket burning; mathematica; mathematical modeling; method of lines; transient combustion

1 Introduction

Traditional models of spherically symmetric burning [1], [2], [3] and [4] for example (and references therein) have focussed on the quasi-steady combustion of liquid (or solid) fuel droplets. They have guided understanding in the practical area of spray combustion and been widely accepted as the basis of most of the work in the field. While these models offer simplifications because the fuel sphere decreases slowly due to the large liquid-fuel, gas density ratio, they introduce complexity because of the two phases. A much simpler problem, which seems to have received much less attention in the literature, is that of purely gas phase combustion of an initial sphere of fuel in an oxidizing atmosphere. We address this problem.

The theme of this workshop was the use of both analytical and numerical methods in theoretical combustion. New computer software for combining these methods has become available over the past few years. These software packages, MATLAB, Maple. Mathematica and others, have introduced a computational environment in which expressions obtained from traditional analytical methods can be evaluated easily and quickly and can be combined with simple numerical methods to obtain and visualize solutions to a wide variety of problems. This software advance, coupled with the phenomenal growth in power of computer hardware, is providing a unique opportunity to re-examine

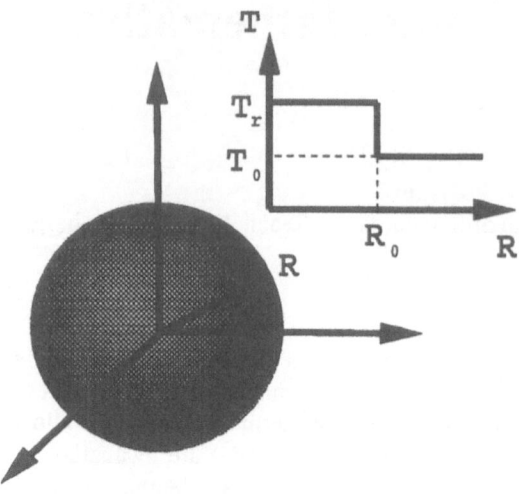

Fig. 1: Schematic diagram of spherical high-temperature region or spherical fuel parcel. Initial temperature profile for nonlinear conduction problem is shown as insert.

the relationship between analytical and numerical solution methods. It is the intention of this work to use the particular tool Mathematica to study the problem of spherical combustion noted above.

In the next section, we examine a simpler problem of nonlinear thermal conduction in a gas as a preamble to the combustion problem; the conduction problem requires the development of much of the methodology for combustion considerations. Then, we examine the combustion of a spherical pocket of fuel, first in the linear case and then for a nonlinear case under a number of assumptions stated in Section 3. Finally, some conclusions both about the fuel-pocket combustion and about Mathematica as a tool are summarized.

2 Transient, Nonlinear Conduction

As a preliminary example which both establishes the methodology and illustrates some interesting physical effects, we consider the transient, nonlinear conduction of a spherical region of high temperature in a thermally-expandable gas at ambient temperature in which the thermal diffusivity is a function of temperature. See Figure 1 where a schematic diagram is shown.

By a thermally-expandable gas, we mean one for which the pressure remains approximately constant [5], a nearly isobaric process:

$$\rho T = \rho_\infty T_\infty \qquad (1)$$

For the dependence of the thermal diffusivity upon temperature, we take

$$\rho^2 \kappa = \rho_\infty^2 \kappa_\infty \qquad (2)$$

The governing equations are the continuity and the energy equations:

$$\frac{\partial \rho}{\partial t} + \frac{1}{r^2}\frac{\partial}{\partial r}\left(\rho r^2 u\right) \doteq 0 \tag{3}$$

$$\rho r^2 \left(\frac{\partial T}{\partial t} + u\frac{\partial T}{\partial r}\right) = \frac{\partial}{\partial r}\left(\rho r^2 \kappa \frac{\partial T}{\partial r}\right) \tag{4}$$

Here, all symbols have their usual meanings: ρ is density, u is the velocity, T is temperature, r is radial distance, t is time and κ is the thermal diffusivity.

Combination of these equations [5] yields a relation for the velocity:

$$u = \frac{\kappa_\infty T}{T_\infty^2}\frac{\partial T}{\partial r} \tag{5}$$

and this relation can be used to eliminate the velocity in the continuity equation. Also, the density and the thermal diffusivity can be eliminated in favor of the temperature. We make the equations dimensionless using the ambient temperature T_∞, the diffusivity κ_∞ and the initial radius of the spherical blob r_0; then the the appropriate time scale is $\tau = r_0^2/\kappa_\infty$ and T_{init} is the initial dimensionless temperature of this blob. Then, the equation in dimensionless form is

$$\frac{1}{T^2}\frac{\partial T}{\partial t} = \frac{1}{r^2}\frac{\partial}{\partial r}\left(r^2\frac{\partial T}{\partial r}\right) \tag{6}$$

where we regard all variables to be dimensionless. The initial conditions are

$$T = T_{init}, \quad 0 \le r \le 1; \qquad\qquad T = 1, \quad 1 \le r \le \infty$$

with boundary conditions that

$$\frac{\partial T}{\partial r} = 0, \quad r = 0; \qquad\qquad T \to 1, \quad r \to \infty$$

To solve this nonlinear equation, we use the method-of-lines. Let $L =$ the number of cells inside the sphere ($0 \le r \le 1$); $\delta \equiv 1/L$; $r_i = (i - 1/2)\delta = (i - 1/2)/L$ for $1 \le i \le M$ (where $M\delta$ is the outer cutoff radius for the truncated problem over the infinite domain); $T_i = T(r_i) = T((\delta(i-1/2))$. Then the ordinary differential equation (ODE) system obtained by uniformly discretizing the partial differential equation (PDE) for the temperature in the radial coordinate becomes:

$$\left(\frac{i - 1/2}{LT_i}\right)^2 \frac{dT_i}{dt} = i^2(T_{i+1} - T_i) - (i - 1)^2(T_i - T_{i-1}) \tag{7}$$

for $2 \le i \le (M - 1)$ while the ODEs arizing from the boundary conditions become, at the inner and outer boundaries,

$$\left(\frac{1}{LT_i}\right)^2 \frac{dT_1}{dt} = T_2 - T_1; \qquad\qquad \frac{dT_M}{dt} = 0$$

We note that there are only three parameters which describe this problem: first, the physical parameter T_{init}, which is the ratio of the initial temperature in the blob to the ambient temperature. Then, there are two numerical parameters, L, the number of nodes in the initial blob and M, the total number of nodes used in the truncated region.

The velocity induced by the thermal conduction in the thermally expandable fluid is, for this example, evaluated at nodes midway between those used to evaluate the temperature; hence $u_i \equiv u(\delta(i-1))$,

$$u_i = (L/2)(T_i + T_{i-1})(T_i - T_{i-1}) \tag{8}$$

Mathematica [7] has been used to evaluate both the analytical expressions derived for the linear problem above and to solve the ODE system for the nonlinear problem. From our vantage point, there are two important advantages for using a tool such as Mathematica. First, the programming effort and the number of lines of "code" for either task, evaluation of the analytical expressions or numerical integration of the ODE system, is rather small. Second, and maybe more important, is that the graphics is integrated into the Mathematica system and, therefore, also is rather straightforward. Provided that the computational task is not too intensive, results can be obtained quickly, viewed graphically and revised or rerun for different conditions in a rather short time; insight is rapidly obtained.

To illustrate, we compare the solutions between the linear and the nonlinear problems. The linear problem is that in which the gas density and thermal diffusivity remain constant; it is Eq. (6) in which the undifferentiated temperature on the left of the equation is taken as unity. First, we note that the initial temperature ratio was chosen as $T_{init} = 1.5, 3$ and 5, reasonable values for a pocket of fuel that has been burned. Also, $L = 12$ and $M = 49$, which is quite adequate to smoothly represent the solutions. The temperature as a function of radius starts off as T_{init} inside the sphere and unity outside. As time increases, the high temperature diffuses until, at large times, the temperature approaches unity throughout space. The temperature decays much more quickly with time in the nonlinear case.

In Figure 2 is shown a comparison between the linear and the nonlinear solutions for the temperature for $T_{init} = 1.5, 3$ and 5. This comparison is made by showing the normalized temperature, $f = (T(r, t) - 1)/T_{init} + 1$, near the origin as a function of time for each of these initial temperature ratios. Since the "effective" diffusivity is proportional to T^2 in the nonlinear problem and since the fluid is collapsing inward as described below, the temperature in the hot pocket is eroded much more quickly than in the linear problem.

The velocity induced by the thermally expandable gas can be determined, i.e., combining the equations of mass and energy conservation and the equation of state in a consistent fashion. It is found that the flow is all inward. Initially, the hot blob is at the ambient pressure, which means that the density in the blob is low. Therefore, while the heat is being conducted outward, the fluid is moving inward to "fill" this low density region.

Buckmaster et al. [2] observed this effect in connection with the collapse of a diffusion flame after extinction. Using asymptotic methods, they examined the rapid adjustment in the temperature (and other fields) as extinction occurs and then the slower collapse of the high temperature region similar to the model discussed above. Their Eqs. (4.1) and (4.2) yield a model similar to that described above. However, the temperature-dependent thermal diffusivity included here shortens the effective time scale of collapse.

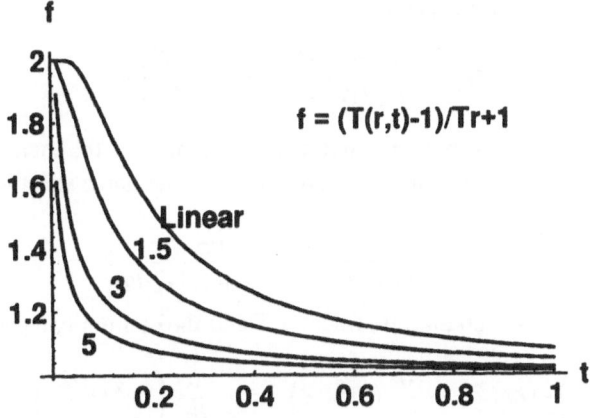

Fig. 2: A comparison of the linear and nonlinear solutions for the nonlinear conduction problem. The ordinate $f = (T(r_1, t) - 1)/T_{init} + 1$ shows that the high-temperature-region collapse occurs much more rapidly with increasing nonlinear effects.

3 Combustion of a Spherical Fuel Parcel

Having studied nonlinear conduction, we now turn to the spherically symmetric combustion of a gaseous fuel parcel, the real problem of interest here. We consider a thermally expandable fluid with a temperature-dependent diffusivity as in the previous example. The equations of conservation for mass, fuel, oxidizer and energy, together with the equation of state for a thermally expandable gas, are

$$r^2 \frac{\partial \rho}{\partial t} + \frac{\partial}{\partial r}\left(\rho r^2 u\right) = 0$$

$$\left(\frac{\partial \rho r^2 Y_f}{\partial t} + \frac{\partial \rho r^2 u Y_f}{\partial r}\right) - \frac{\partial}{\partial r}\left(\rho r^2 D \frac{\partial Y_f}{\partial r}\right) = -\dot{m}/W_f \nu_f \qquad (9)$$

$$\left(\frac{\partial \rho r^2 Y_O}{\partial t} + \frac{\partial \rho r^2 u Y_O}{\partial r}\right) - \frac{\partial}{\partial r}\left(\rho r^2 D \frac{\partial Y_O}{\partial r}\right) = -\dot{m}/W_O \nu_O$$

$$\rho C_p \left(\frac{\partial \rho r^2 T}{\partial t} + \frac{\partial \rho r^2 u T}{\partial r}\right) - \frac{\partial}{\partial r}\left(\rho r^2 K \frac{\partial T}{\partial r}\right) = \Delta H \dot{m}$$

$$\rho T = \rho_\infty T_\infty$$

where Y_f, Y_O are the mass fractions of fuel and oxidizer respectively, W_f, W_O are the molecular weights of fuel and oxidizer, ν_O, ν_f are the stoichiometric coefficients, T is the temperature, D is the diffusion coefficient (assumed to be the same for fuel and oxidizer), K is the thermal conduction coefficient ($K = \kappa \rho C_p$), $\dot{m}$ is the chemical reaction rate, assumed to be large, and all other quantities have their usual meanings.

As is common for problems with rapid reaction rates, it is assumed that the Lewis numbers are unity and that the reaction takes place at a flame sheet so that a mixture-fraction variable can be used. Define

$$Z \equiv \frac{Y_f/W_f\nu_f - (Y_O - Y_{O\infty})/W_O\nu_O}{Y_{f\infty}/W_f\nu_f + Y_{O\infty}/W_O\nu_O}$$

Here it has been assumed, as is common in such problems, that the chemistry is infinitely fast so that a flame sheet separates fuel and oxidizer. The flame sheet location is given by

$$Z_* = \frac{Y_{O\infty}/W_O\nu_O}{Y_{f\infty}/W_f\nu_f + Y_{O\infty}/W_O\nu_O}$$

Then the equations for the mixture fraction Z and the continuity equation become:

$$\frac{\partial}{\partial t}\left(\rho r^2 Z\right) + \frac{\partial}{\partial r}\left(\rho r^2 u Z\right) = \frac{\partial}{\partial r}\left(\rho r^2 D \frac{\partial Z}{\partial r}\right)$$

$$r^2\frac{\partial \rho}{\partial t} + \frac{\partial}{\partial r}\left(\rho r^2 u\right) = 0 \tag{10}$$

where, as in the nonlinear conduction problem, we assume the subsidiary relations for a thermally expandable gas and a variable diffusivity:

$$\rho T = \rho_\infty T_\infty; \qquad\qquad \rho^2 D = \rho_\infty^2 D_\infty$$

We assume that there are "state relations" between the temperature and the mixture fraction:

$$T = 1 + h\frac{1 - Z_*}{Z_*}Z \qquad 0 \le Z \le Z_*$$
$$T = 1 + h(1 - Z) \qquad Z_* \le Z \le 1 \tag{11}$$

where

$$h = \frac{\Delta H}{C_p T_\infty}\frac{Y_{O\infty}}{W_O\nu_O}$$

and where the subscript $*$ generally denotes the flame-sheet values.

For this problem, we take as initial conditions that $Z = 1$ for $0 \le r \le r_0$ and $Z = 0$ for $r_0 \le r < \infty$ at $t = 0$ while the boundary conditions are $\partial Z/\partial r$ at $r = 0$ and $Z = 0$ as $r \to \infty$.

To make the equations dimensionless, we assume that temperature, density and diffusivity are referenced to their ambient conditions, length is referenced to the initial fuel-sphere radius and time is referenced to the time scale for diffusion across the initial fuel sphere. From here on, all quantities will be considered to be dimensionless.

3.1 The Linear Problem

When we ignore the thermal expandability of the gas and the velocity induced, the problem becomes linear. The linear problem can be written in dimensionless form as

$$\frac{\partial Z}{\partial t} = \frac{1}{r^2}\frac{\partial}{\partial r}\left(r^2\frac{\partial Z}{\partial r}\right) \tag{12}$$

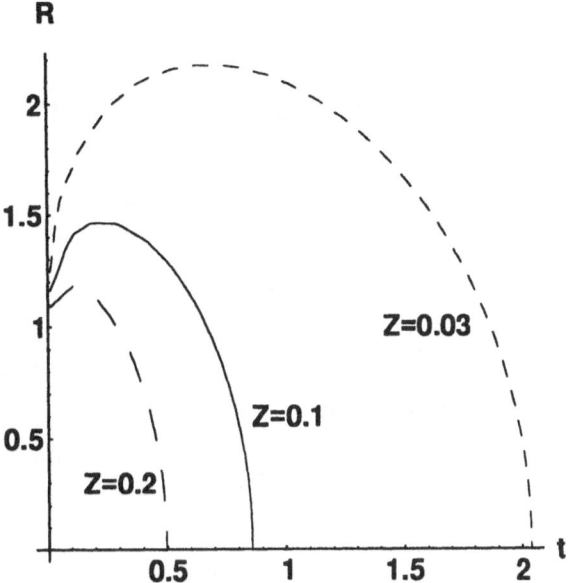

Fig. 3: Trajectories of the flame front calculated from solution to the linear mixture-fraction problem for three values of Z_*. Here R is the radial location of the flame front and t is time.

with the initial conditions and boundary conditions specified above.

The solution to this linear problem was determined analytically [6] and was used as a comparison both for this problem and for the heat conduction problem described above. In Figure 3, we plot the location of the flame front as a function of time, the flame-front trajectory, for three values of mixture-fraction at the flame front Z_*.

3.2 Further Reformulation

As in the first example for nonlinear conduction, we determine the velocity by combining the mass, energy and state equations; the expressions for the velocity differ by a function of time on the two sides of the flame sheet. To evaluate these functions of time, we use the facts that $u \to 0$ as $r \to 0$, and $T \to 1$ as $r \to \infty$. Also, at the flame sheet, we require that $Z_{*-} = Z_{*+}$, and this implies that $T_{*-} = T_{*+}$. Also, we require that $\left(\frac{\partial Z}{\partial r}\right)_{*-} = \left(\frac{\partial Z}{\partial r}\right)_{*+}$, and this relation implies that $\left(\frac{\partial T}{\partial r}\right)_{*+} = -\frac{1-Z_*}{Z_*}\left(\frac{\partial T}{\partial r}\right)_{*-}$. Finally, we require that $u_{*-} = u_{*+}$.

Therefore, we can state the problem as

$$r^2\left(\frac{\partial Z}{\partial t} + u\frac{\partial Z}{\partial r}\right) = T(Z)\frac{\partial}{\partial r}\left(r^2 T(Z)\frac{\partial Z}{\partial r}\right) \tag{13}$$

where the velocity is given by the relations: for $0 \le r \le r_*(t)$ or $1 \ge Z \ge Z_*$,

$$u = -hT(Z)\frac{\partial Z}{\partial r} \tag{14}$$

and for $r_*(t) \leq r < \infty$ or $Z_* \geq Z \geq 0$,

$$u = -h \frac{T_*}{Z_*} \frac{r_*^2}{r^2} \left(\frac{\partial Z}{\partial r} \right)_* + h \frac{1 - Z_*}{Z_*} T(Z) \frac{\partial Z}{\partial r} \tag{15}$$

and where $T(Z)$ is given above.

To use the method of lines in its simplest form in subsequent numerical integration, we make one more transformation so that the flame front remains at a fixed location in the new coordinates. Let $y \equiv \frac{r}{r_*}; t' = t$ so that

$$\frac{\partial}{\partial t} = \frac{\partial}{\partial t'} - \frac{\dot{r}_*}{(r_*(t))^2} y \frac{\partial}{\partial y}; \qquad\qquad \frac{\partial}{\partial r} = \frac{1}{r_*(t)} \frac{\partial}{\partial y}$$

Then, the integration problem becomes:

$$\frac{\partial Z}{\partial t} + \frac{u - \dot{r}_* y}{r_*(t)} \frac{\partial Z}{\partial y} = \frac{T(Z)}{(y r_*(t))^2} \frac{\partial}{\partial y} \left(y^2 T(Z) \frac{\partial Z}{\partial y} \right) \tag{16}$$

where

$$\dot{r}_* = \frac{dr_*}{dt} = -\frac{1}{r_*(t)} \frac{T_*^2}{\left(\frac{\partial Z}{\partial y} \right)_*} \left[\frac{\partial}{\partial y} \left(y^2 \frac{\partial Z}{\partial y} \right) \right]_* \tag{17}$$

and where, $0 \leq y \leq 1$ or $1 \geq Z \geq Z_*$,

$$u = -\frac{h T(Z)}{r_*(t)} \frac{\partial Z}{\partial y} \qquad\qquad T(Z) = 1 + h(1 - Z) \tag{18}$$

and for $1 \leq y < \infty$ or $Z_* \geq Z \geq 0$,

$$u = \frac{h}{r_*(t)} \left[-\frac{T_*}{Z_*} \frac{1}{y^2} \left(\frac{\partial Z}{\partial y} \right)_{y=1} + \frac{1 - Z_*}{Z_*} T(Z) \frac{\partial Z}{\partial y} \right]$$

$$T(Z) = 1 + h \frac{1 - Z_*}{Z_*} Z \tag{19}$$

3.3 The Numerical Solution

Numerical computations with Mathematica were performed using a numerical discretization and a method-of-lines computation similar to that described above for the high-temperature sphere. The numerical method was tested by comparing with an analytical solution to the 1D nonlinear problem in which $Z = 1$ at $x = 0$ for all time and $Z = 0$ for $x \geq 0$ initially. This 1D problem can be solved using a Howarth transformation and yields a analytical solution dependent upon a similarity variable $\eta = x/2\sqrt{t}$.

Let there be L panels in $0 \leq y \leq 1$, so that $y_i = (i - 1)/L$ and the flame sheet is at $i = L + 1$. As before, let $Z_i = Z(y_i) = Z((i - 1)/L)$, and $Z_{L+1} = Z_*$.

The mixture-fraction equation becomes, for $2 \leq i \leq L$ and for $L + 2 \leq i \leq M$,

$$
\begin{aligned}
\frac{dZ_i}{dt} &= -\frac{Lu_i - \dot{r}_*(i-1)}{2r_*(t)}(Z_{i+1} - Z_{i-1}) \\
&+ \frac{T_i}{2(r_*(t))^2}\frac{L^2}{(i-1)^2}[(i-1/2)^2(T_{i+1} + T_i)(Z_{i+1} - Z_i) \\
&- (i-3/2)^2(T_i + T_{i-1})(Z_i - Z_{i-1})]
\end{aligned} \tag{20}
$$

where, for $2 \leq i \leq L$,

$$
u_i = -\frac{h}{r_*(t)}\frac{L}{2}T_i(Z_{i+1} - Z_{i-1}) \qquad\qquad T_i = 1 + h(1 - Z_i) \tag{21}
$$

and, for $L + 2 \leq i \leq M$,

$$
\begin{aligned}
u_i &= \frac{h}{r_*(t)}\left[-\frac{L^2}{(i-1)^2}\frac{T_*}{Z_*}(Z_{L+2} - Z_L)\frac{L}{2} + \frac{1 - Z_*}{Z_*}T_i\frac{(Z_{i+1} - Z_{i-1})L}{2}\right] \\
T_i &= 1 + h\frac{1 - Z_*}{Z_*}Z_i
\end{aligned} \tag{22}
$$

At the boundaries, including the flame sheet, we have

$$
\begin{aligned}
\frac{dZ_1}{dt} &= \frac{6L^2}{r_*^2}T_1^2(Z_2 - Z_1) \\
\frac{dZ_{L+1}}{dt} &= 0 \qquad\qquad \frac{dZ_{M+1}}{dt} = 0 \\
r_*(t)\frac{dr_*}{dt} &= -\frac{2T_*^2}{L}\frac{[(L + 1/2)^2(Z_{L+2} - Z_*) - (L - 1/2)^2(Z_* - Z_L)]}{(Z_{L+2} - Z_L)}
\end{aligned} \tag{23}
$$

where we have taken $Z_0 = Z_2$ and $T_0 = T_2$.

Finally, for initial conditions, we take $Z_i(0) \approx 0.5\, erf\,[0.5(1 - x_*(i-1)/L)/\sqrt{t_{init}}] + 0.5\, erf\,[0.5(1 + x_*(i - 1)/L)/\sqrt{t_{init}}]$ for $1 \leq i \leq M + 1$, where x_* is the solution to the equation $Z_* \equiv 0.5\, erf\,[0.5(1 - x_*)/\sqrt{t_{init}}] + 0.5\, erf\,[0.5(1 + x_*)/\sqrt{t_{init}}]$ for some small value of t_{init} (here taken as 0.01).

Once again, the built-in capabilities of Mathematica, including the ODE solver and the graphics, made these computations relatively simple. The computational time required on an SGI Indigo 2 workstation with an R4000 100MHZ MIPS microprocessor and 64 MB of memory was no more than a few minutes of clock time for the maximum computation reported here.

In Figure 4 are shown two trajectory plots for the parameters $h = 1$, $z_* = 0.3$, one for the numerical parameters $L = 6$, $M = 24$ and the other for the numerical parameters $L = 8$, $M = 32$. (A value of $h = 1$ represents a rather mild heat release.) The fact that these two trajectories are close demonstrates that the computations are nearly converged. As in the linear case, the flame-front trajectory at first increases rapidly from near $r = 1$ and then collapses as the fuel is consumed. The whole process occurs more rapidly than in the linear case, however, due primarily to the nonlinear conduction.

Temperature as a function of radius for the burning sphere is shown at three times in Figure 5. The temperature profile is very sharp initially, the peak moving radially out

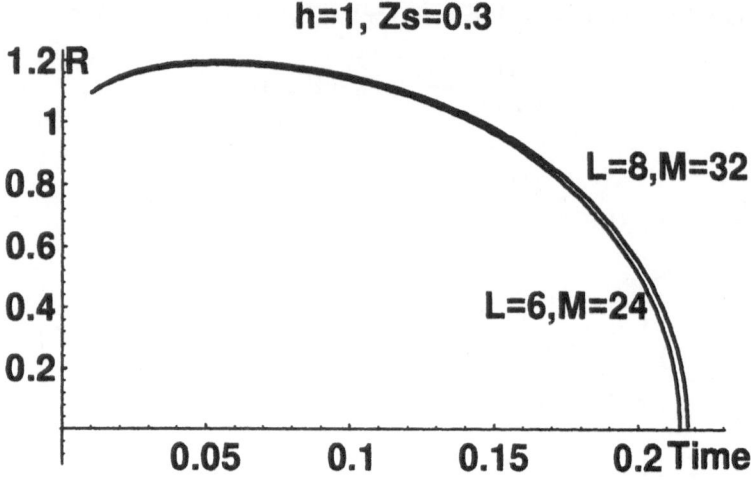

Fig. 4: Plots of the trajectory of the flame front for $h = 1$, $z_* = 0.3$ for $L = 6$, $M = 24$ and also for $L = 8$, $M = 32$, showing that the solutions do not depend sensitively on the number of nodes for $M = 32$.

from near $r = 1$. Later, the temperature peak broadens and the peak moves toward the origin as the fuel is consumed.

Three velocity profiles are shown in Figure 6. Initially, an outer shell of heated gas is ejected radially outward, but as time increases, the temperature gradients induce a double-peaked velocity profile.

4 Conclusions

Transient combustion of a spherical pocket of fuel has been examined using a simple mixture-fraction, flame-sheet model and Mathematica as the solution, evaluation and visualization tool. The model assumed a nearly isobaric process and included temperature-dependent diffusion, both processes introducing important nonlinear effects. First, a simple, nonlinear thermal conduction problem was solved using Mathematica to introduce a method-of-lines (MOL) methodology and to illustrate the importance of including the isobaric process and a temperature-dependent conductivity. Then the spherical combustion problem was similarly solved. The solutions show the expansion and subsequent collapse of the flame-sheet trajectory, the sharp initial temperature spike at the flame front and its later diffusive spreading, and the early peaked expansion velocity, followed by a double humped velocity profile. Mathematica is an excellent tool for combining the most useful aspects of analytical techniques and numerical methods, for the visualization of results and for gaining understanding of phenomena. However, at present, Mathematica still has limitations for computationally intensive tasks.

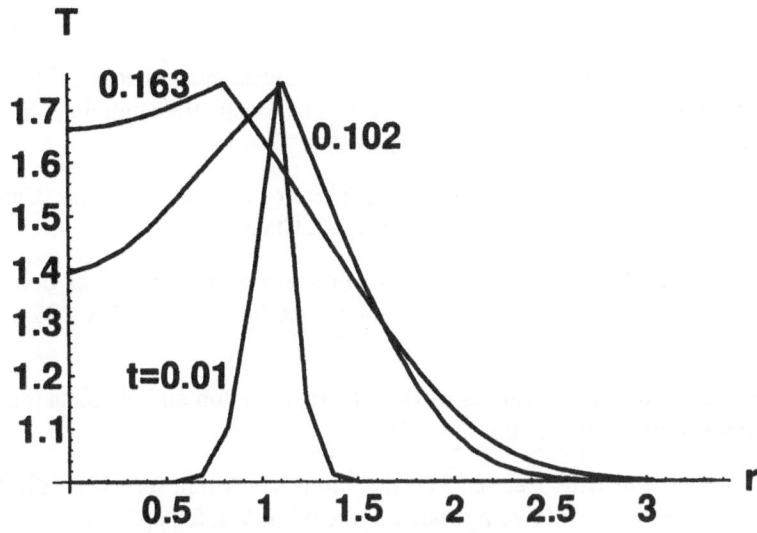

Fig. 5: Plots of temperature as functions of radius for three times during combustion for $h = 1$, $z_* = 0.3$ and $L = 8$, $M = 32$. A temperature spike rapidly develops, moving outward and spreading before collapsing as the fuel is consumed.

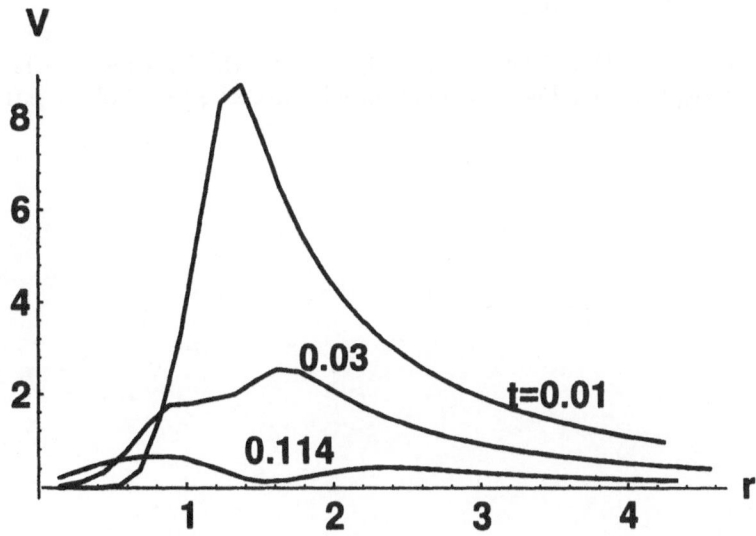

Fig. 6: Plots of velocity as functions of radius for three times during combustion for $h = 1$, $z_* = 0.3$ and $L = 8$, $M = 32$. Initially, heated gas is ejected radially outward. Later, temperature gradients induce a double-peaked velocity profile.

References

[1] Williams, F.A., *Combustion Theory, The Fundamental Theory of Chemically Reacting Flow Systems*, Second Edition, The Benjamin/Cummings Publishing Company, Inc., Reading, MA, 1985, p. 492.

[2] Buckmaster, J.D. and Ludford, G.S.S., *Theory of Laminar Flames*, Cambridge University Press, Cambridge, England, 1982, Chapter 6.

[3] Buckmaster, J.D. and Ludford, G.S.S., *Lectures on Mathematical Combustion*, Society of Industrial and Applied Mathematics, Philadelphia, PA, 1983, Lecture 9.

[4] Law, C.K., "Recent Advances in Droplet Vaporization and Combustion," Prog. Energy Combust. Sci., **8**, pp. 171-201.

[5] Rehm, R.G. and H.R. Baum, "The Equations of Motion for Thermally Driven, Buoyant Flows", *Journal of Research of the NBS*, Vol. 83, pp 297-308, May-June 1978.

[6] Carslaw, H.S. and Jaeger, J.C., *Conduction of Heat in Solids*, Second Edition, Clarendon Press, Oxford 1986, p. 230 and p. 54.

[7] Wolfram, Stephen, *Mathematica, A System for Doing Mathematics by Computer*, Second Edition, Addison-Wesley Publishing Company, Inc., The Advanced Book Program, New York, 1991.

[8] Buckmaster, J., D.S. Stewart and A. Ignatiadis, "On the Wind Generated by a Collapsing Diffusion Flame," Combust. Sci. and Tech., 1986 Vol. 46, pp145-165.

Wavelets as a Numerical Tool

Leland Jameson[1], T.L. Jackson[1] and D. Glenn Lasseigne[2]

[1]ICASE, NASA Langley Research Center, Hampton, VA 23681-0001, USA
[2]Department of Mathematics and Statistics, Old Dominion University, Norfolk, VA 23529, USA

Abstract. The stability of a plane, premixed, adiabatic flame is re-examined for finite activation energies. Preliminary results are given for one-dimensional disturbances. The full governing equations for an unsteady propagating flame are then solved using a novel numerical approach called WOFD, a wavelet-optimized finite difference method. This method is briefly described here and preliminary results are reported. As a result of this work, we strongly recommend that new and old algorithms should be tested at activation energies of 40 and 60 and compared to the results presented here as a gauge to the accuracy of the numerical solution.

Keywords. Finite activation energy, stability, wavelets

1   Introduction

One of the main purposes of the 1981 GAMM-Workshop on Numerical Methods in Laminar Flame Propagation was to bring together scientists working in the field of numerical methods in flame propagation and confront them with recent results obtained by large activation energy asymptotics and to check these numerically [1]. One of the test problems considered was the unsteady propagation of a plane, premixed, adiabatic flame with one-step chemistry and Lewis number different from unity. For Lewis number equal to two and nondimensional activation energy of 20, it was observed that the flame oscillates with a period of about 10 nondimensional time units. The goal was to calculate the flame position accurately as a function of time. It was concluded that extreme care should be taken when calculating unsteady flames with non-unity Lewis numbers. At the time this was considered a challenging test problem, and remains so even today. Thus, the unsteady propagating flat flame has become a standard test problem when developing new adaptive algorithms.

As a result of the GAMM meeting, Rogg [2] computed the neutral stability boundary for the pulsating branch. The approach taken was to solve the unsteady equations numerically at a fixed value of the activation energy, and then decrease the Lewis number until reaching the neutral value where the flame no longer oscillates. In this way, the neutral branch in the activation energy-Lewis number plane was determined. The numerical scheme was based on the method of lines, but it is not clear from the text if the mesh was uniform or not. Rogg obtained the interesting result that as the activation energy increased, the neutral stability boundary reversed itself (cf., figure 1). This is almost certainly unphysical, since large activation energy asymptotics shows that the neutral curve should approach unit Lewis number as the activation energy is increased. The probable explanation is that the numerical simulation at large but finite activation energy did not resolve the reaction zone adequately, resulting in this unphysical phenomena.

More recently, Denet and Haldenwang [3] computed growth rates from a numerical simulation and showed that as the activation energy increased from 10 to 20, the growth rates tended towards those obtained from large activation energy asymptotics. It should be emphasized that actual convergence tests were not performed since the activation energy used in the numerical computations never exceeded 20. This brings up two interesting questions: (i) what happens to the numerical solution as the activation energy is increased further, and (ii) how do the computed growth rates compare to those obtained from a stability analysis using finite activation energy?

It is apparent from the above discussions that stability results for finite activation energy are needed and this has not been performed to date. Peters [4] notes, "A stability analysis of the problem based on finite activation energy should - if it could be done correctly - predict the oscillations in the numerical calculations". Therefore the central issues of this work is two-fold: (i) to provide the neutral stability boundary and growth rates at finite activation energy as a test-bed for numerical simulations; and (ii) to discuss a new numerical approach, based on wavelets, which can resolve the reaction zone at the larger activation energies.

2 Steady State Analysis

The high activation energy analysis of a plane, premixed, steadily propagating adiabatic flame with weak heat release is well established. Under these conditions, there exists a unique flame speed determined analytically by treating the reaction zone as a discontinuity with jump conditions derived from an integration of the source term. The leading order flame speed value thusly determined is used to non-dimensionalize velocities and the diffusion length scale non-dimensionalizes the spatial variable. The steady equations for the normalized temperature T^s and mass fraction Y^s in a reference frame moving with the flame at speed V_f are

$$T^s_{\xi\xi} - T^s_\xi + R = 0, \qquad Le^{-1}Y^s_{\xi\xi} - Y^s_\xi - R = 0, \tag{1}$$

$$R = \frac{\beta^2}{2LeV_f^2} Y^s exp \left[\frac{-\beta(1 - T^s)}{1 - \alpha(1 - T^s)} \right], \tag{2}$$

where Le is the Lewis number, β the activation energy, α the heat release and V_f the flame speed which is determined below as a function of β, Le, and α. The appropriate boundary conditions in terms of these variables are

$$T^s = 0, \quad Y^s = 1 \quad as \quad \xi \to -\infty, \tag{3}$$

$$\frac{dT^s}{d\xi} = \frac{dY^s}{d\xi} = 0, \quad as \quad \xi \to +\infty. \tag{4}$$

For finite activation energy, the system (1)-(4) must be solved numerically as an eigenvalue problem for the flame speed V_f. The condition (5) is replaced with $T^s = 1$ (consistent with complete burning of the fuel), then a shooting method determines the value of V_f for which the original boundary condition (5) is satisfied. For $\alpha = 0.8$, the flame speed as a function of activation energy is shown in figure 1. For this value of the heat release, the cold boundary difficulty is circumvented but the numerical procedure is still sensitive to other factors. Two remarks concerning figure 1 are in order: first, consistent with the non-dimensionalization, the flame speed asymptotes to unity independent of the Lewis number; second, this approach allows for O(1) variations in the Lewis number whereas the large activation energy analysis is restricted to near unity Lewis numbers.

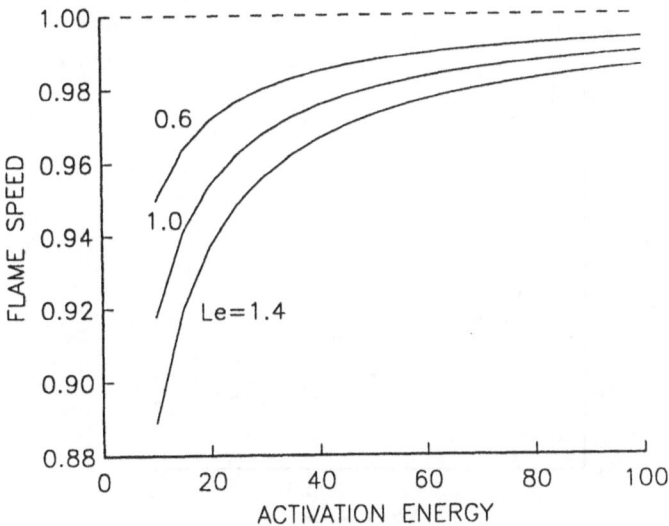

Figure 1. Flame speed as a function of activation energy for various Lewis numbers. The dashed line corresponds to the flame speed obtained by large activation energy asymptotics.

3 Stability Analysis

We seek infinitesimal perturbations of the steady solution by writing

$$T = T^s(\xi) + \epsilon T_0(\xi)e^{\omega t}, \qquad Y = Y^s(\xi) + \epsilon Y_0(\xi)e^{\omega t}, \tag{5}$$

where ω is the (complex) growth rate and is a function of the parameters α, β, and Le. Substitution into the full governing equations and appropriate boundary conditions, and taking the limit $\epsilon \to 0$, results in a linear homogeneous system with non-trivial solutions for certain values of the eigenvalue ω. The resulting system was solved using the compound matrix method and shooting on the eigenvalue ω. Preliminary results are given below for one-dimensional disturbances. A complete discussion, including multi-dimensional disturbances and effect of heat loss, will be given at a later date [5].

Only the pulsating branch of the neutral stability boundary can be computed from a one-dimensional analysis. Shown in figure 2 is a plot of the neutral stability boundary ($\text{Re}(\omega) = 0$, $\text{Im}(\omega) \neq 0$) in the activation energy-Lewis number plane. The solid curve corresponds to the results of large activation energy asymptotics, and the boxes to those of the finite activation energy calculations. At a fixed activation energy of $\beta = 30$, there is only a 6% error between the respective neutral Lewis numbers, but at $\beta = 10$, the relative error grows to about 41%. Thus, all future numerical simulations should be compared to the results obtained here for finite activation energy.

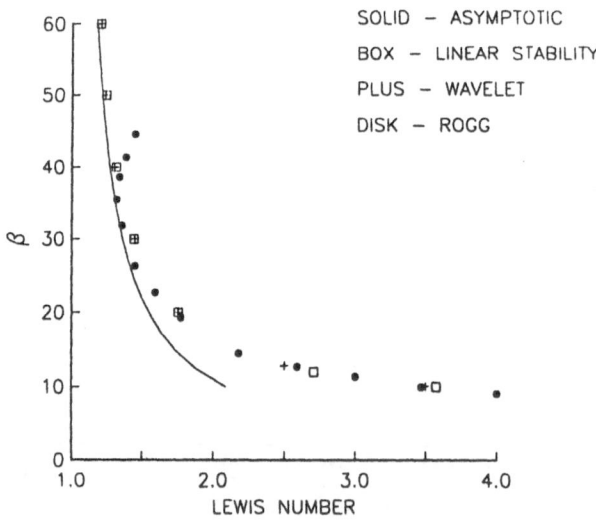

Figure 2. Neutral stability boundaries in the β-Le plane obtained from (i) large activation energy asymptotics (solid), (ii) finite activation energy calculations (box), (iii) numerically computed points obtained from Rogg's [2] calculations (disk), and (iv) numerically computed points using WOFD (plus).

4 Numerical Simulation using WOFD

Wavelet based numerical methods as applied to physical problems of interest have recently become quite popular because of their potential to resolve small scale features that might develop within a given situation. Exploiting this strength is appealing: costs associated with numerical simulation can be reduced allowing more complex problems to be solved, by using dense grids where small scale features exists, and sparse grids where large scale features exist. Wavelets provide a natural mechanism for decomposing a solution into a set of coefficients which depend on scale and location which can be used to determine a near optimal grid. However, major obstacles must be overcome before wavelets, as a general numerical tool, can be applied robustly. The difficulties with using standard Daubechies-based wavelet methods are two-fold: (i) proper numerical boundary conditions cannot currently be constructed for non-periodic boundary conditions; and (ii) nonlinear terms require a transformation back to physical space for evaluation. Recently, Jameson [6] proposed a new numerical method which utilizes the strength of wavelets while avoiding the difficulties just mentioned. The interested reader is referred to this work for a general discussion and details. Here, we only give a very brief outline of wavelet optimized finite-difference (WOFD), and its application to the unsteady propagating flame problem.

WOFD uses Daubechies-based wavelets in their finite-difference form, which is equivalent to explicit finite difference methods with grid refinement. In this way, wavelets are used to define the (arbitrary) grid for finite difference methods. The advantages of this approach is summarized as follows: (i) a full multi-resolution numerical method is achieved such that degrees of freedom are added and removed depending on the local behaviour of the function; (ii) all calculations are performed on the pointvalues of the function in physical space removing the difficulties associated with the nonlinear terms; (iii) any boundary condition applied to an ordinary finite difference method can be applied to WOFD, alleviating the greatest difficulty with current wavelet methods; and (iv) grid adaptation is automatic and depends only on the function or signal being analyzed.

The WOFD numerical algorithm used in this study is constructed as follows:

1. Given the initial condition on a uniform grid, use a wavelet decomposition on the point values to define a new and optimal grid;

2. Given this new grid, build the 4th-order spatial differentiation matrices by fitting a polynomial of degree 4 through five neighboring points and differentiating in the middle of the domain;

3. Advance N time units using a 4th-order Runge-Kutta scheme, where N is the number of time steps between grid updates; and

4. Reconstruct the function on a uniform grid and return to step 1.

Preliminary results are given below for the one-dimensional problem. A complete discussion, including the extension in two space dimensions, will be given at a later date [7]. Numerical experiments were conducted to determine the neutral stability boundary. In all cases reported here, the number of grid points never exceeded 190, with this value occurring at $\beta = 60$. The approach for determining the neutral boundary is identical to that of Rogg: solve the unsteady equations numerically and, at a fixed value of the activation energy, decrease the Lewis number until the flame no longer oscillates. The largest value of the Lewis number for which no oscillations is observed is called the neutral value. Shown in figure 2 are the neutral stability boundaries in the β-Le plane obtained using WOFD (denoted by the plus symbol). Several remarks are in order. First, there is excellent agreement between the results from the linear stability theory developed here and those obtained using WOFD. Second, note that Rogg's calculations are quite good for activation energies below 22, but deteriorate as the activation energy is increased further. This is almost certainly due to the unresolved thin reaction zone at the high activation energies.

Finally, our conclusion is identical to that first noted by the GAMM meeting: extreme care should be taken when calculating unsteady flames with non-unity Lewis numbers. We strongly recommend that new and old algorithms should be tested at activation energies of 40 and 60 and compared to the results presented here as a gauge to the accuracy of the numerical solution.

5 References

1. Peters, N. and Warnatz, J., eds., *Numerical Methods in Laminar Flame Propagation*, Vieweg, Braunschweig/Wiesbaden, (1982).

2. Rogg, B., "The effect of Lewis number greater than unity on an unsteady propagating flame with one-step chemistry" in *Numerical Methods in Laminar Flame Propagation*, N. Peters and J. Warnatz (eds.), Vieweg, Braunschweig/Wiesbaden, pp. 38-48, (1982).

3. Denet, B. and Haldenwang, P., "Numerical study of thermal-diffusive instability of premixed flames", *Combust. Sci. Technol.* 86:199-221 (1992).

4. Peters, N., "Theoretical implications of nonequal diffusivities of heat and matter on the stability of a plane premixed flame", in *Numerical Methods in Laminar Flame Propagation*, N. Peters and J. Warnatz (eds.), Vieweg, Braunschweig/Wiesbaden, pp. 29-37, (1982).

5. Lasseigne, D.G., Jackson, T.L. and Grosch, C.E., "Stability of premixed, nonadiabatic flames with finite activation energy", *in preparation*, (1994).

6. Jameson, L., "On the wavelet optimized finite difference method", *ICASE Report No. 94-9*, (1994).

7. Jameson, L., Jackson, T.L. and Hussaini, M.Y., "WOFD with applications to unstable problems in combustion", *in preparation*, (1994).

Numerical Simulation of Dynamic Interaction of a Droplet with a Vortex using a CIP-Combined Unified Procedure Method

A. Koichi Hayashi, Yasushi Nagumo, and Toshi Fujiwara

Department of Aerospace Engineering
Nagoya University, Nagoya 464-01, Japan

Abstract. A droplet colliding with a vortex is studied numerically using two-dimensional incompressible Euler equations and including a surface tension term. A new numerical scheme is applied to this equation system to discretize the two-phase problem without establishing boundary conditions between two phases. The nondimensional values for this problem include Weber numbers and relative Reynolds numbers of 100 and about 320, respectively. The density ratios between the droplet and gas used in the present problem are 100 and 200. The results show the detail structure of the dynamic interaction between the two phases; and the deformation and initial fragmentation of both the droplet and the vortex.

Keywords. Numerical simulation, new scheme, droplet, vortex, droplet fragmentation

1. Introduction

Automobile engines, boilers, and many other engineering applications use spray combustion for their energy sources. The problem of spray combustion involves two-phase interaction, turbulent mixing, and droplet deformation, evaporation and combustion. The problem of non-deformable droplets has been studied by a number of researchers. The real cases, however, shows that the droplet breaks up as well as deforms by either shear flows or turbulent flows.

The present study shows the numerical simulation of a single cylindrical droplet colliding with a single cylindrical vortex, which idea follows recent numerical work on droplet deformation [1,2]. The simulation is performed using the two-dimensional Euler equations which, at present, do not include the energy conservation equations. The CIP (Cubic-Interpolated-Pseudoparticle) combined Unified Procedure (C-CUP) method is developed and used to descretize the Euler equation system together with the surface tracking function and the van der Waals equation of state. This paper shows new results about the interaction between the droplet and vortex, which includes the details of pressure, density, and velocity vector profiles for a Weber number of 100 and for density ratios between the droplet and the gas of 100 and 200.

2. Governing Equations

The fundamental equations for the present problem are the two-dimensional incompressible Euler equations with a surface tension term, but without expressing the energy terms. The equations of continuity and momentum, with the surface tension γ, are as follows[3]:

$$\frac{\partial u}{\partial x} + \frac{\partial v}{\partial y} = 0 \ , \tag{1}$$

$$\frac{\partial u}{\partial t} + u\frac{\partial u}{\partial x} + v\frac{\partial u}{\partial y} = -\frac{1}{\rho}\frac{\partial p}{\partial x} + \frac{\gamma\kappa}{\rho}\frac{\partial S_f}{\partial x} \ , \tag{2}$$

$$\frac{\partial v}{\partial t} + u\frac{\partial v}{\partial x} + v\frac{\partial v}{\partial y} = -\frac{1}{\rho}\frac{\partial p}{\partial y} + \frac{\gamma\kappa}{\rho}\frac{\partial S_f}{\partial y} \ . \tag{3}$$

where we introduce the surface tracking function S_f, curvature κ, and the normal vector to the droplet surface $\mathbf{n}$, as follows:

$$S_f = \begin{cases} 0: & gas - phase \\ 1: & liquid - phase \end{cases}, \tag{4}$$

$$\kappa = \nabla \cdot n \ , \tag{5}$$

$$n = -\frac{\nabla S_f}{|\nabla S_f|}. \tag{6}$$

The philosophy of this problem is to simulate an arbitrary configuration of the droplet surface while the vortex interacts with it: the numerical calculation is performed without the boundary condition between two phases. A tracking function S_f[4] is introduced by a transport equation, as follows:

$$\frac{\partial F_S}{\partial t} + u\frac{\partial F_S}{\partial x} + v\frac{\partial F_S}{\partial y} = 0 \ ; \ F_s = \tan[\,0.99\pi(S_f - 0.5)\,]. \tag{7}$$

This transport equation provides a smooth density profile between two phases.

3. Ccomputational Method

A CIP Combined Unified Procedure (C-CUP) method is developed for problems with flexible boundaries and phase transitions using the van der Waals equation of state. A non-conservative form of the Euler equation system is integrated by the Cubic-Interpolated-Pseudoparticle (CIP) method originally proposed by Yabe[5], which can even solve a compressible flow problem with shock waves accurately. The detail of the CIP method may be found elsewhere [5]. Since the C-CUP method uses a non-conservative description, it can be applied to the multi-phase flow problems of the present kind uniformly from the liquid-phase to the gas-phase.

The numerical problem associated with the phase transition is that the small volume change triggers a large pressure change in the van der Waals equation. The standard CFD analysis is to start with calculating the density and then the pressure from the equation of state. This approach may yield a larger numerical error in the pressure even when a small numerical error occurs in the density. The present method calculates the pressure first and follows with the density calculation. The following steps are taken using the Euler equation in the present calculation:

$$\frac{\partial \rho}{\partial t} + u \cdot \nabla \rho = -\rho \nabla \cdot u \quad , \tag{8}$$

$$\frac{\partial u}{\partial t} + u \cdot \nabla u = -\frac{1}{\rho} \nabla \cdot p \quad , \tag{9}$$

The CIP method is appropriate to calculate the pressure first since it treats the non-advection terms and advection terms separately. Especially the phase transition (not for the present calculation) is in the non-advection terms. The non-advection terms are integrated using the following finite difference method:

$$\frac{\rho^* - \rho^n}{\Delta t} = -\rho^n \nabla \cdot u^* \quad , \tag{10}$$

$$\frac{u^* - u^n}{\Delta t} = -\frac{1}{\rho^n} \nabla \cdot p^* \quad , \tag{11}$$

$$\frac{p^* - p^n}{\Delta t} = -\rho^n C_S^2 \nabla \cdot u^* \quad , \tag{12}$$

where Eq. 12 is derived from the equation of sound (C_S is the speed of sound) and is obtained from the thermal and caloric equations of state under the isentropic condition. Equation 12 is described using Eq. 10 as follows:

$$\frac{p^* - p^n}{\Delta t} = C_S^2 \frac{\rho^* - \rho^n}{\Delta t} \tag{13}$$

which shows that the density change is included in the pressure change through the speed of sound, hence the density at the next time step is not necessarily calculated at this point. Since Eqs. 11 and 12 have to be coupled compatibly, the following relation is derived after taking a divergence of Eq. 11:

$$\nabla \cdot \left(\frac{\nabla p^*}{\rho^n}\right) = \frac{p^* - p^n}{\rho^n C_S^2 (\Delta t)^2} + \frac{\nabla \cdot u^n}{\Delta t} \tag{14}$$

which is described by the following pressure diffusion equation:

$$\frac{\partial p}{\partial t} = (\rho^n C_S^2 \Delta t) \nabla \cdot \left(\frac{\nabla p}{\rho^n}\right) + \rho^n C_S^2 \nabla \cdot u \quad . \tag{15}$$

Eq. 14 indicates that:

(i) when the speed of sound is infinite, the first term of the right-hand-side of Eq. 14 disappears, then the equation becomes the incompressible fluid equation;

(ii) when $\Delta t \ll (\nabla^2 p/\rho^*)^{-1/2}$, the left-hand-side term is much smaller than the first term of the right-hand-side of Eq. 14, then the equation becomes the compressible fluid equation.

From the above features, Eq. 14 can be used for both compressible and the incompressible flow to determine the speed of sound. Once the pressure p^* is obtained, u^* is calculated from Eq. 11 and then ρ^* from the continuity equation. The next time step is calculated using the CIP scheme.

4. Numerical Conditions

In order to simulate the problem of the interaction between the two-dimensional cylindrical droplet and two-dimensional cylindrical vortex, we initially set the droplet to stay in the gas-phase flow stream and the vortex to flow toward the droplet with some rotational and convective velocities. The initial conditions for this problem are listed in Table 1, where the rotational velocity distribution of the vortex is given as follows [6]:

$$u_\theta = u_{\theta\,max} \sin\left(\frac{\pi r}{R_0}\right). \tag{16}$$

where $u_{\theta max}$ is the maximum rotational velocity and R_0 the radius of the vortex. This velocity distribution is superimposed on a uniform flow to describe the vortex motion. The direction of rotation is originally clockwise and the center axis of vortex motion is slightly shifted upward by about two ninths of the droplet diameter.

Table 1 Numerical initial conditions

	Non-dimensional value	Dimensional value	Dimension
Weber number[1]	100.0	-	-
Density (gas)	1.0	1.0	kg/m^3
Density (liquid)	100.0 200.0	100.0 200.0	kg/m^3
Pressure (gas)	10.0	10^5	Pa
Pressure (liquid)	10.0	10^5	Pa
Rotational Velocity	1.0	100.0	m/sec
Convective velocity	0.5	50.0	m/sec
Droplet radius	1.0	100.0	μm

The boundary conditions for the present problem are that the initial values are provided at the inflow condition and a zero-th order extrapolation is applied at the outflow condition. The grid system applied for the present modeling is 150x100 in the two-dimensional coordinates, with $\Delta x = \Delta y = 0.06$ mm.

---

[1]The Weber number is a ratio of the relative inertia between two phases to the surface tension: $We = 2r\rho_g |u_l - u_g|^2/\gamma$.

5. Results and Discussion

The numerical simulation of the interaction between the droplet and vortex was performed using the C-CUP method. This method uses no boundary conditions between the liquid and gas phase. The Weber number for the present problem is 100, the relative Reynolds number is about 320, and the density ratios between liquid and gas are 100 and 200. When the density ratio between two phases increases, we have encountered some problems in the convergence of computation in the present calculation situation.

The interaction between the droplet and vortex becomes less active with increasing density ratio. This can be seen by comparing Figs. 1 and 2. Figures 1 and 2 are for density ratios between two phases of 100 and 200. Larger droplet deformation occurs in the smaller density ratio, where the vortex initially rotates clockwise and its center axis is shifted upward by about two nineths of the droplet diameter. For both density ratios, however, the vortex separates in two or more parts early during the collision. When the strong momentum of the vortex hits the droplet surface, the pressure waves go through the inside of the droplet to show their diffractive patterns (Figs. 1-(b)~(h)). The reflected waves from the droplet surface affect the vortex configuration as seen in Fig. 1-(d), where, by this moment, the vortex is separated into several parts (this separation is also seen in Fig. 4-(d)). Figure 1-(e) shows a quite different pressure wave pattern from the other figures, which type of pattern does not appear in the case of the density ratio of 200 (Fig. 2). This sudden appearance of dense pressure profiles may come from the phenomena that the characteristic scales of wave propagation between the droplet and vortex are matched, or for some other reasons. At the same time or just after the vortex collides with the droplet, it deforms quickly. Since the droplet surface tension is not strong (We=100), the droplet surface deforms irregularly. Figures 1-(g) and (h) show the beginning of the droplet fragmentation at the tip of the droplet peninsula.

In the case of the density ratio of 200 (Fig. 2), the figures show that the droplet surface is just flattened at the side of collision and is not deformed much for this period studied. Figure 3 shows the density profiles of gas and liquid in the case of We=100. Since the droplet surface is described by Eq. 7, its surface configuration is rather distributed. The vortex rotational motion scrapes off the droplet surface and carries away the surface fluid with its motion as shown in Figs. 3-(d)~(h), since the droplet fluid is not dense enough not to be carried away or the surface has a skirt-like density profile where its foot is easily scraped off. The gas velocity vector profiles (Fig. 4) show the dynamics of the vortex motion, where the vortex is broken into many vortices due to the interaction with the droplet. After the vortex collides with the droplet, some of the broken-up vortices bounce back upstream probably due to their strong inertia forces.

6. Conclusion

The interaction between a cylindrical vortex and a cylindrical droplet is simulated using a new numerical method called the C-CUP method. The philosophy of calculating the droplet-gas two-phase flow without setting its boundary worked well for the present problem up to the certain Weber number. The results show the detail dynamics of the interaction between the droplet and the vortex and the droplet surface deformation including its fragmentation. Future work should include calculation of cases of evaporation and combustion as well as smaller Weber number and larger density ratio cases. Improvement of the treatment of the droplet surface function, F_S, should also be considered.

232

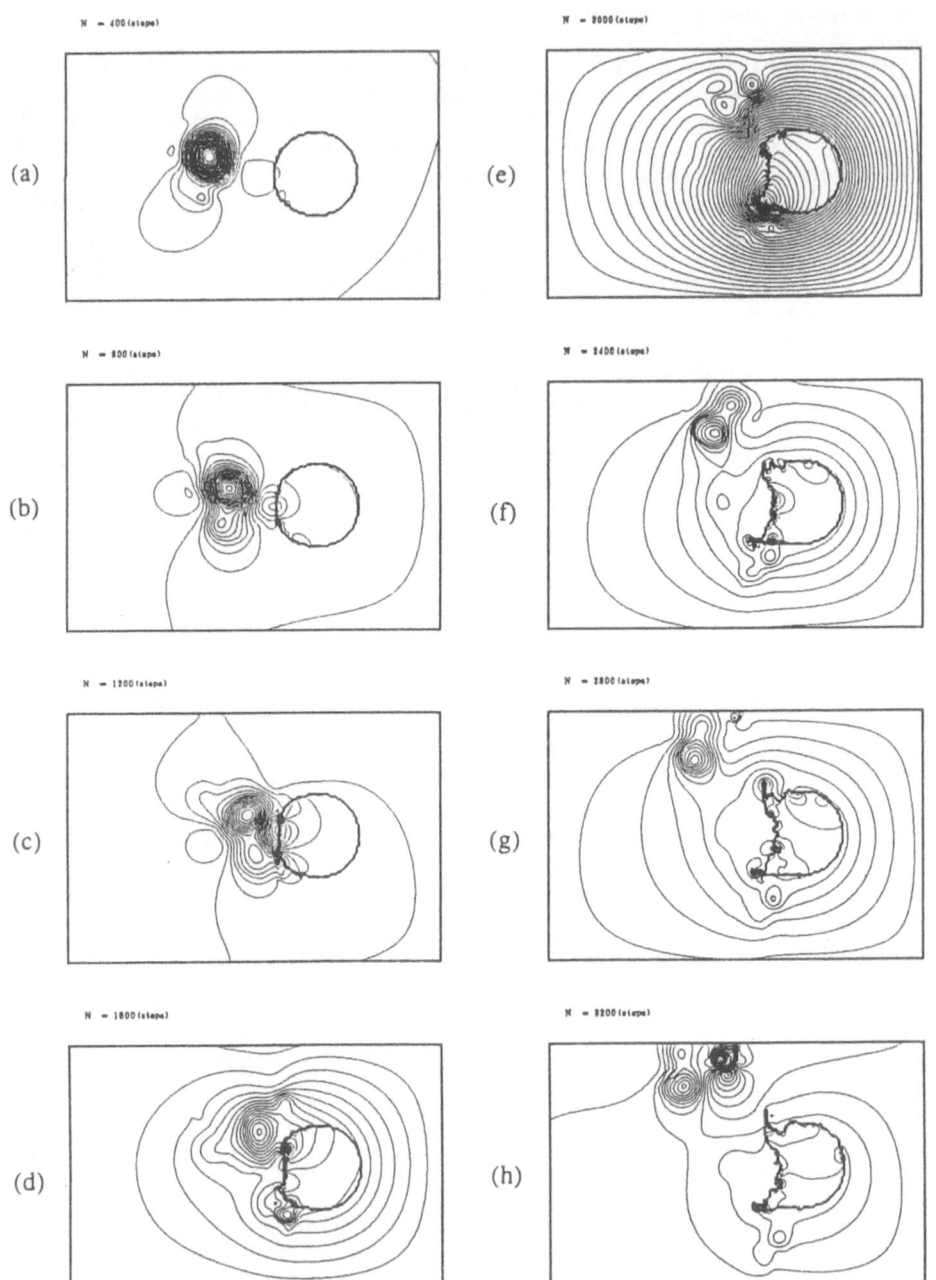

Fig.1 Pressure profiles in the flowfield of the droplet-vortex interaction for the case of a density ratio of 100 and a Weber number of 100: (a)t=1.95 μs, (b)t=3.76 μs, (c)t=5.63 μs, (d)t=7.92 μs, (e)t=10.42 μs, (f)t=13.30 μs, (g)t=15.77 μs, (h)t=17.38 μs.

233

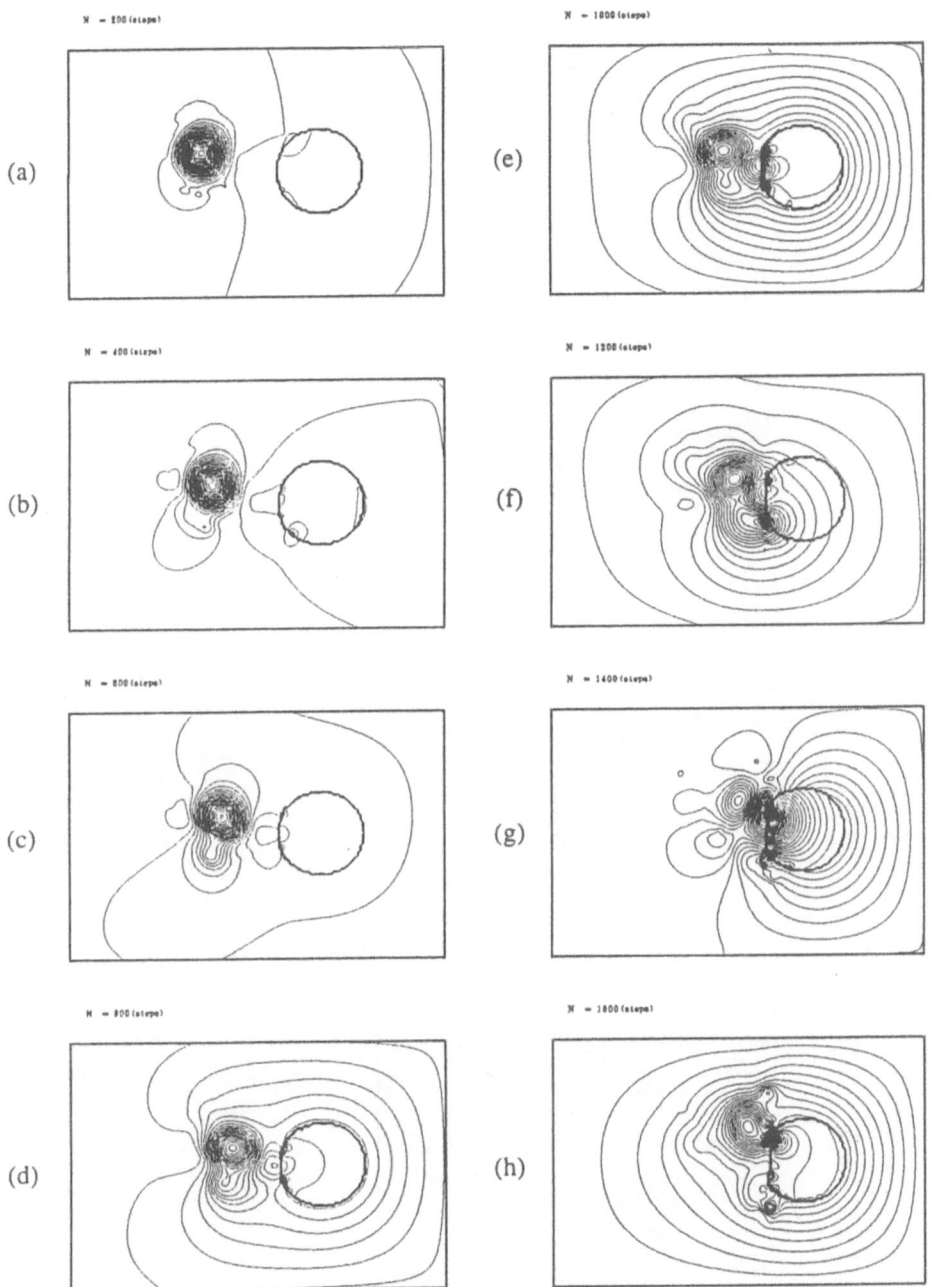

Fig.2 Pressure profiles in the flowfield of the droplet-vortex interaction for the case of a density ratio of 200 and a Weber number of 100: (a)t=0.97 μs, (b)t=1.95 μs, (c)t=2.88 μs, (d)t=3.76 μs, (e)t=4.65 μs, (f)t=5.63 μs, (g)t=6.77 μs, (h)t=7.97 μs.

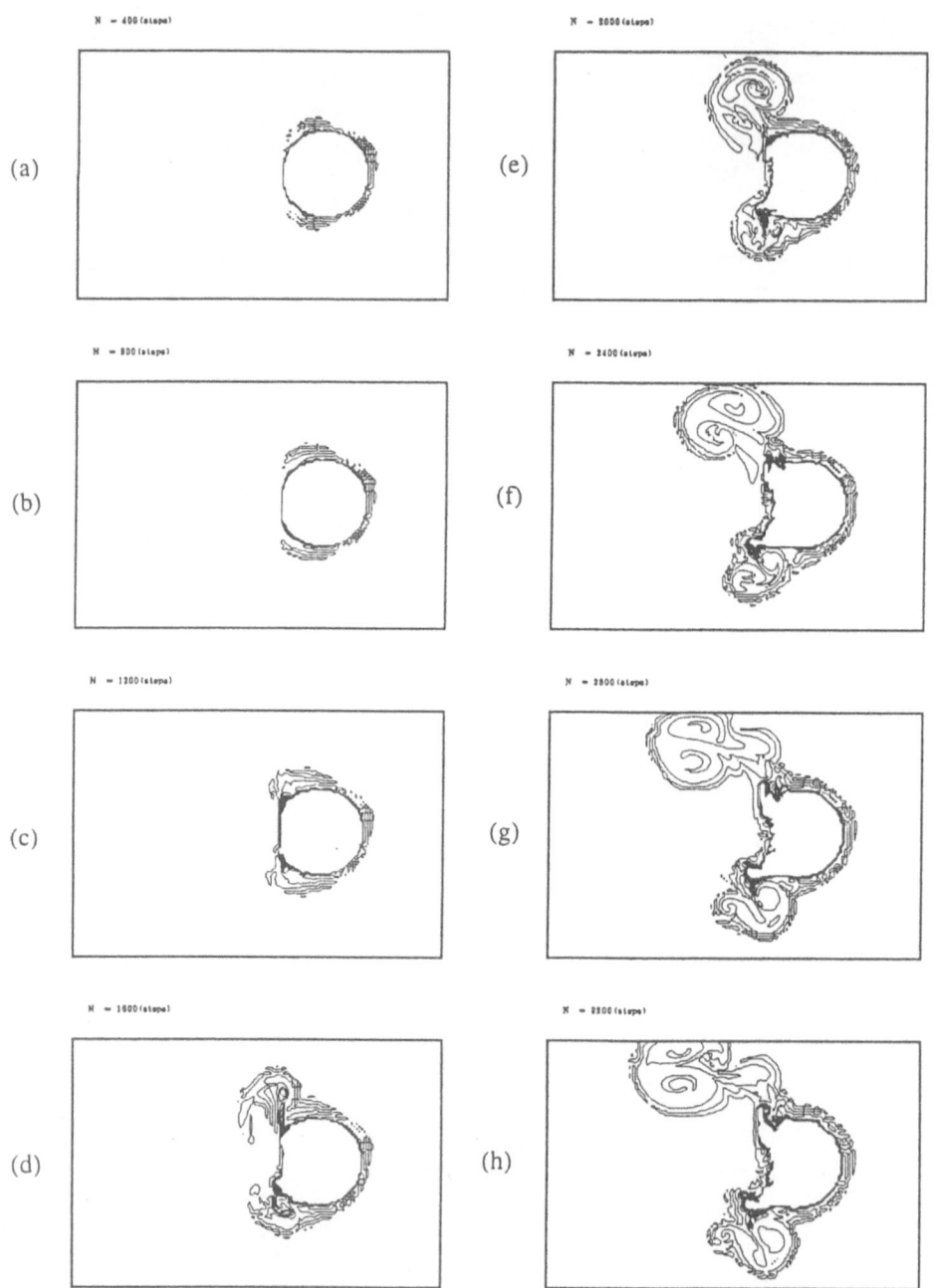

Fig.3 Density profiles in the flowfield of the droplet-vortex interaction for the case of a density ratio of 100 and a Weber number of 100: (a)t=1.95 μs, (b)t=3.76 μs, (c)t=5.63 μs, (d)t=7.92 μs, (e)t=10.42 μs, (f)t=13.30 μs, (g)t=15.77 μs, (h)t=17.38 μs.

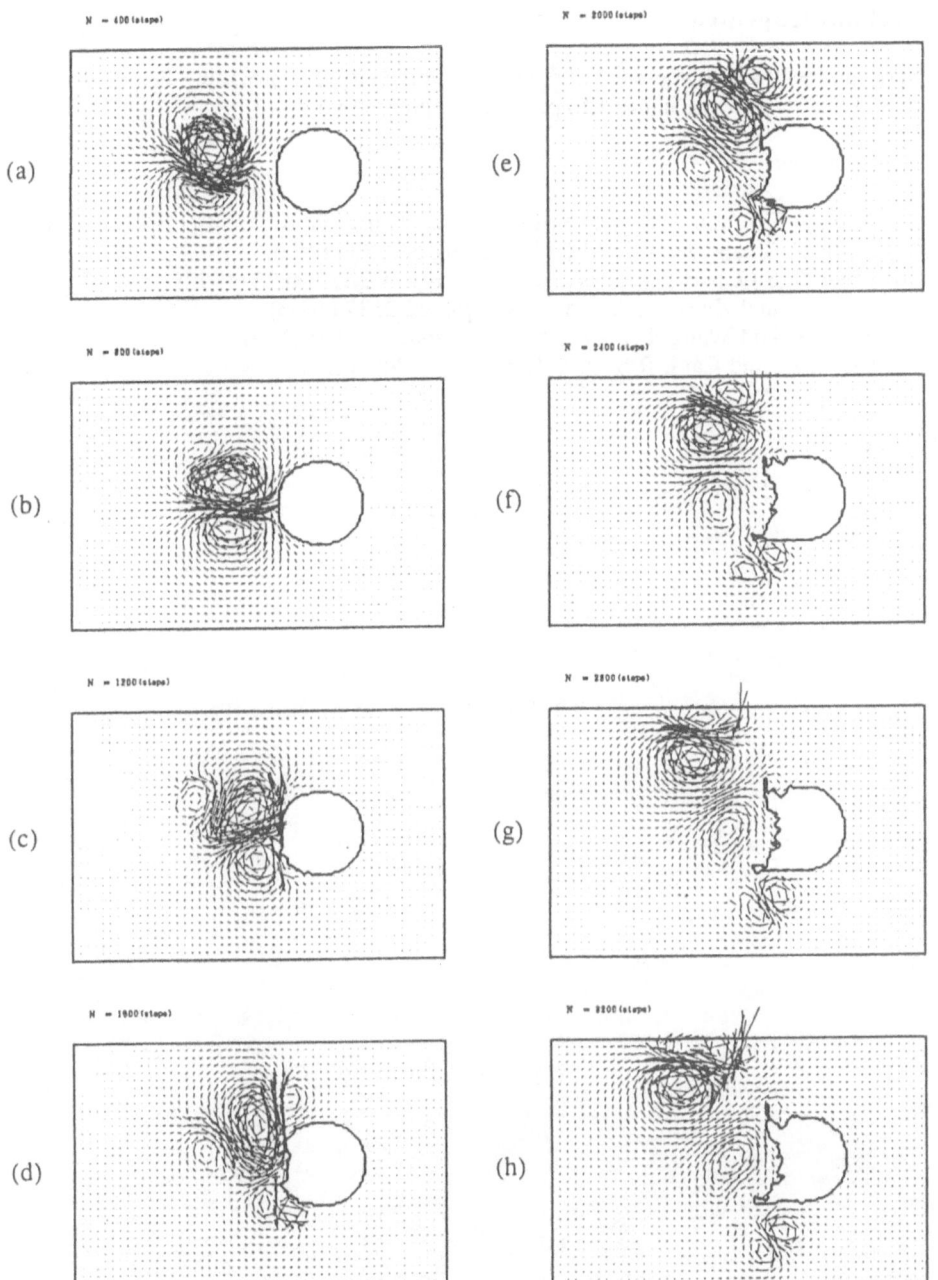

Fig.4 Velocity vector profiles in the flowfield of the droplet-vortex interaction for the case of a density ratio of 100 and a Weber number of 100: (a)t=1.95 μs, (b)t=3.76 μs, (c)t=5.63 μs, (d)t=7.92 μs, (e)t=10.42 μs, (f)t=13.30 μs, (g)t=15.77 μs, (h)t=17.38 μs.

Acknowledgement

The authors thank Professor Yabe, Gunma University, for providing the information about his CIP scheme in order to perform the present work.

References

1. Fyfe, D.E., Oran, E.S., and Fritts, M.J., *J. Comp. Phys.* 76:349 (1988).
2. Deng, Z.-T. and Jeng, S.-M., *AIAA J.* 30, 5:1290 (1992).
3. Unverdi, S.O. and Tryggvason, G., *J. Comp. Phys.* 100:25 (1992).
4. Yabe, T. and Xiao, F., *J. Phys. Soc. Jpn.* 62:2537 (1993).
5. Yabe, T. and Wang, P.Y., *J. Phys. Soc. Jpn.* 60:2105 (1991).
6. Chate, H. and Cant, R.S., *Combust. Flame* 74:1 (1988).

Gasification and Combustion of a Moving Droplet Undergoing Continuous Phase Change in Supercritical Ambience

Akira Umemura and Wei Jia

Department of Mechanical Engineering, Yamagata University

Yonezawa 992, Japan

Abstract

Numerical calculations were performed of the gasification and combustion of a liquid fuel droplet which was injected into a quiescent gas at a thermodynamic state far above the critical point of the fuel. Since spatially continuous phase change occurs from the beginning, the droplet is free from surface tension and latent heat of gasification. The calculation was conducted for the two-dimensional version of the problem.

The effect of initial droplet speed emerges through the deformation of droplet shape, yielding different gasification/combustion characteristics from the subcritical case. It is found that, at large Reynolds numbers, the gasification rate is almost invariant with time and the gasification lifetime shortens inversely proportional to the initial droplet speed. The flame sheet calculation conducted for a small fuel-to-oxygen stoichiometric ratio shows that the gasified fuel is accumulated in a region enclosed by the flame and then burns gradually at a rate controlled by the diffusion of oxygen in the quiescent ambience. The combustion lifetime is, therefore, independent of initial droplet speed. The underlying physics were also explored.

1. INTRODUCTION

Recent progress in both computer ability and computational techniques has made it possible to numerically simulate practical combustion problems such as occurring in a diesel engine and others. Nevertheless, due to yet limited computer capacities, a liquid fuel spray is modeled as a continuum or discrete sources in most of existing simulators. This situation seems unchanged in the near future. The simulators will provide reasonable results if the fundamental physics underlying individual droplet gasification and combustion are well understood and their overall effects are properly integrated into the governing equations (in particular, spray equation).

The gasification/combustion of a single fuel droplet have been studied for a long time. In recent years, the development of higher powered diesel engines and rocket motors require detailed knowledge on droplet gasification and combustion characteristics for the condition that ambience pressure and temperature exceed the critical point of fuel, stimulating recent active researches from experimental[1], theoretical[2] and numerical[3, 4] points of view. It has been made clear by Umemura[5] that, for supercritical ambiences, there are three types of gasification regimes, named subcritical, transitional, and supercritical regimes. Which regime is realized depends on the magnitudes of ambience pressure and temperature as well as initial droplet temperature. In the supercritical regime which is attainable for ambient conditions far above the critical point of fuel, the droplet no longer has a liquid-gas interface characterized by a discontinuity in fuel concentration. Correspondingly, the droplet is freed from surface tension and latent heat of gasification. The phase change takes place spatially continuously, and the droplet temperature can rise close to the ambience or flame temperature by conduction.

Numerical calculations of spherically symmetric droplet gasification and combustion in supercritical ambiences have been performed by one of the authors[6, 7]. The present paper extends the numerical method to a case where the droplet possesses a translational velocity as is encountered in practical spray combustion. Vorticous flows are induced both inside and outside of the droplet due to forced convection, and the droplet may be largely deformed from its initial shape. The objective of the present study is to clarify the effects of forced convection on gasification and combustion of a liquid fuel droplet undergoing continuous phase change. As far as gasification is concerned, the gasification rate and lifetime are of primary interest. When combustion is included, the prediction of flame size and temperature as well as combustion lifetime become important issues. Simple relationships between these characteristics will be derived with physical interpretations on the basis of numerical simulation results.

2. PHYSICAL MODEL

An idealized situation is considered as a prototype to investigate forced convection effects on supercritical droplet gasification and combustion; A liquid fuel droplet of radius $\bar{a}_0$ and density $\bar{\rho}_*$ is initially injected with a translational speed $\bar{U}$ into an unbounded quiescent gas of density $\bar{\rho}_\infty$. Since the thermodynamic state of the ambient gas is far above the critical point of the fuel, the droplet undergoes continuous phase change from the beginning. (The term "droplet" is used to imply that the continuous phase change occurring within a thin film gives an optical image as if there were a droplet. Later, another term "liquid core" will be introduced to indicate the part where the fuel has the same property as liquid.) Two problems are treated of (i) pure gasification and (ii) combustion. In the pure gasification problem, the system consists of fuel and inert gas while oxygen and combustion products add to them in the combustion problem. In either problem, the system is regarded as a pseudo-binary component system by postulating that chemical species other than fuel have the same physical properties. A one-step chemical reaction model along with the flame sheet approximation is adopted in the combustion problem. The Redlich-Kwong equation of state with adequate mixing rules is employed to describe the local equilibrium state of the mixture. The thermodynamic properties (density $\bar{\rho}$, specific heat at constant pressure $\bar{c}_p$ etc.) as well as the mixture thermal conductivity $\bar{\lambda}$ and binary diffusion coefficient $\bar{D}$ are all given as functions of pressure $\bar{P}$, temperature $\bar{T}$ and fuel mass fraction Y_F or oxygen mass fraction Y_O. Regarding the methods of estimating these physical properties, refer to Ref. [2, 8, 9].

The study is limited to a low mach number flow which is the most case in practical situations. The thermodynamic pressure $\bar{P}$ is fixed to the level of ambience pressure. The hydrodynamic counterpart $\bar{p}$ which is defined by the fluctuation over the thermodynamic one is retained only in the momentum equations.

In the combustion problem, the fuel and oxidizer are completely separated by the flame sheet. The Shvab-Zeldovich coupling function[10] $\beta_Y = Y_F - \sigma_Y Y_O$ is introduced to reduce the conservation equations for fuel and oxygen to a single homogeneous equation for β_Y. Similarly, the reaction terms in the energy and oxygen conservation equations may be eliminated by introducing another coupling function $\beta_T = T + \sigma_T Y_O$ but the resulting equation contains additional terms involving the oxygen mass fraction. In the case when the physical properties of the mixture are constant and the Lewis number is equal to unity, they vanishes identically, resulting in a homogeneous equation for β_T. Such simplification which is valid at low pressures, however, seemingly is not available for the present combustion problem where each physical property and local Lewis number vary significantly from point to point. The following considerations, nevertheless, indicate that the additional terms drop out for the present problem as well. In the fuel region where the Lewis number changes from a relatively large value to unity, there exists no oxygen so that the additional terms vanish. On the other hand, the non-condensable components in the oxygen region have a Lewis number around unity. Hence, the Shvab-Zeldovich coupling functions governed by homogenous equations can be solved in a similar way to the pure gasification problem. The location of flame sheet is given by the $\beta_Y = 0$ contour. The temperature and fuel and oxygen mass fractions are determined from the Shvab-Zeldovich functions to estimate the variable physical properties.

3. NUMERICAL RESULTS AND DISCUSSION

Computations were conducted for the two-dimensional version of the problem; by replacing an initial, spherical droplet by a circular cylindrical form of liquid fuel. The thermodynamic pressure, initial droplet and gas temperatures nondimensionalized by the fuel critical values were prescribed as $P = 4.0$, $T_{fuel} = 1.176$, $T_{gas} = 1.882$. These conditions correspond to an initial fuel-to-gas density ratio of 5.27. However, the maximum density ratio can rise up to about 7 in the combustion problem. The concentration of oxygen in the ambient gas was set $Y_{O\infty} = 0.1$ and the fuel-to-oxygen stoichimetric ratio, $\sigma_Y = 1/3$ for a hydrocarbon fuel.

A finite difference numerical scheme which account for the compressible effects due to temperature and mass fraction distributions at zero Much number limit was employed. The spatial derivatives in convective terms were calculated by the QUICK[11] method and the others were approximated by central differences. The time integration was performed by an explicit method. The hydrodynamic pressure was solved from the elliptic equation which was derived from the momentum and state equations.

3.1 Deformation of Material Interface

In a limit of infinite Peclet number, effects of heat conduction and mass diffusion are all negligible that the phenomenon is governed by the incompressible continuity equation and Navier-Stokes equations only. The fuel concentration keeps the value of unity inside the droplet and zero outside. The jump in fuel mass fraction and thereby density at the droplet boundary is an asymptotic expression for the continuously changing phase in the vanishingly thin film[12]. In the calculation, the dynamic viscosity of both liquid and gas phases was assumed to be a constant for simplicity. The motion of the droplet with large inertia induces hydrodynamic pressure and viscous stress distributions which feed back to deform the droplet shape as seen in Fig. 1.

3.2 Pure Gasification

Computations were conducted for various values of dimensionless parameters involved. The following discussion is concentrated on a typical case of Reynolds number $Re = \bar{U}\bar{a}_0/\bar{\nu} = 120$, Peclet number $Pe = \bar{U}\bar{a}_0/\bar{\kappa} = 24$ and Lewis number $Le = \bar{\lambda}/\bar{\rho}\bar{D}\bar{c}_p = 5$

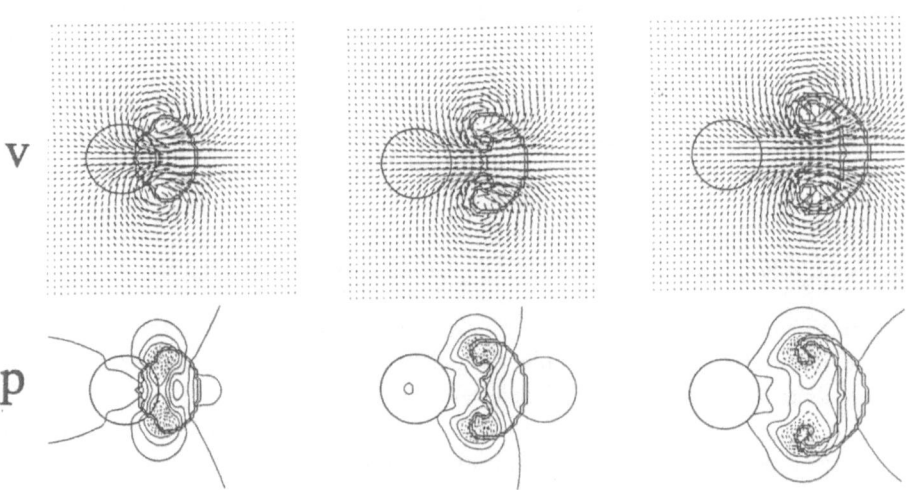

Fig. 1 Flow induced by liquid fuel droplet injected into quiescent gas and deformation of fuel-gas interface shape: $Re = 120$ and $Pe = \infty$. N.D.time $t = 1, 2, 3$ from left to right. Heavy lined circles denote initial droplet location.

unless otherwise stated. The computed flow fields are shown in Fig. 2, which displays the time sequences of velocity vectors, pressure, temperature and fuel mass fraction contours. The dashed and solid lines, respectively, correspond to values lower and higher than the arithmetic average of minimum and maximum values of each quantity.

The liquid core which has a similar property to liquid is identified with a region enclosed by the contour $Y_F = 0.9$ of heavy line. The displacements of the leading, rear, upper and lower edges of the liquid core and of its mass center are plotted in

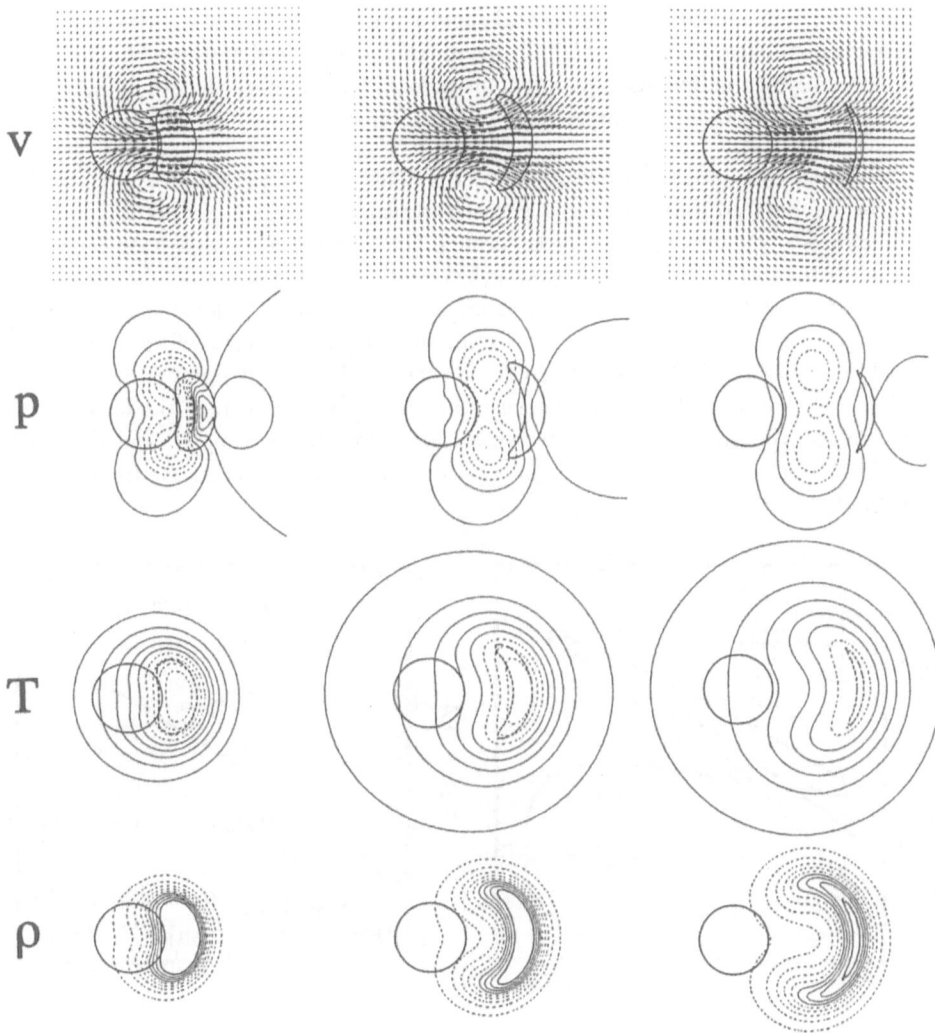

Fig. 2 Gasification of fuel droplet travelling through quiescent gas: $Re = 120$, $Pe = 24$ and $Le = 5$. N.D. $t = 1, 2, 3$ from left to right. Liquid core region ($Y_F > 0.9$) is indicated by heavy line.

Fig.3. The patterns are similar to those for the previous non-gasifying case. At the relatively low Peclet number and high Lewis number, heat conduction overwhelms the effects of convection and mass diffusion so that the temperature contours becomes concentric circles with small distortions due to convection. Therefore, it may be considered that the droplet is heated from all directions equally.

Figure 4 shows the variations with time of the areas of regions enclosed by the $Y_F = 0.9$ and 0.1 contours. It is found that the liquid core region decreases its area linearly with time. Furthermore, the dimensionless gasification lifetime, defined as the time elapsed until the core disappears, is approximately independent of the Reynolds number ranging from 100 to 500. Since the initial droplet speed has been chosen as the velocity scale in the definition of Reynolds number, this implies that the gasification lifetime decreases in inverse proportion to initial droplet speed.

To estimate the effect of convection on gasification rate, a square droplet is considered to deform into a distorted rectangular droplet under the action of self-induced hydrodynamic pressures as it travels through the ambient gas. Mass transfer is enhanced at the front side of the droplet and suppressed at the rear side. In a reference frame moving with the droplet, the maximum pressure must be located at the leading stagnation point. If the droplet is still, the stagnation point will appear at an infinite distance from the droplet. In the present problem, this point is detached at a short distance from the droplet because the Stefan flow must meet the far uniform flow. Hence, the undisturbed ambient gas comes close to the droplet, and the mass transfer associated with gasification is enhanced to a great degree at the front side of the droplet.

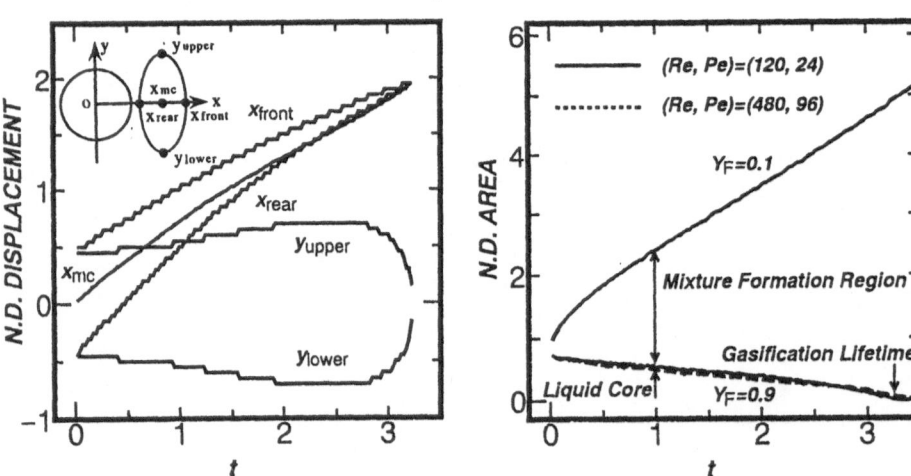

Fig. 3 Displacements of five points characterizing liquid core deformation.

Fig. 4 Time evolutions of droplet core ($Y_F > 0.9$) and mixture formation region ($0.1 < Y_F < 0.9$).

On the other hand, since the twin vortices are generated behind both sides of the droplet, the fuel mass transfer at the rear part of the droplet has a similar property to the case of non-translating droplet except a minor effect of vortex-induced velocity impinging on the rear edge. As a result, the front edge of liquid core is cut out by rapid gasification and the droplet becomes a distorted rectangular shape with decreasing thickness. Such a feature is clearly seen in the mass fraction contours.

The thickness of concentration boundary layer developed in front of the moving droplet is estimated to be of $O(\bar{D}/\bar{u}_{mc})$, where $\bar{u}_{mc}$ is the velocity of center of mass of liquid core. Across this layer, the fuel mass fraction varies from unity to zero. The resulting fuel mass flux, therefore, is of the order of $\bar{\rho}\bar{D}/(\bar{D}/\bar{u}_{mc}) = \bar{\rho}\bar{u}_{mc}$. This indicates that the liquid fuel gasified per unit time at the leading edge of liquid core is proportional to the mass center velocity. On the other hand, since the fuel vapor advected from the front edge is entrained into the recirculation region behind the droplet, the fuel mass flux at the rear edge is even lower than the spherically symmetric gasification case in which the gasification rate takes a value of $O(\bar{\rho}\bar{D}/\bar{a}_0)$. The ratio of the fuel mass fluxes at the front and rear edges is, thus, considered to be proportional to $\bar{u}_{mc}\bar{a}_0/\bar{D} \sim LePe$.

Again, suppose a rectangular droplet with dimensionless width a and thickness b scaled by the initial droplet radius $\bar{a}_0$. If only the gasification on the front edge is considered, the dimensionless area of droplet, ab, reduces according to

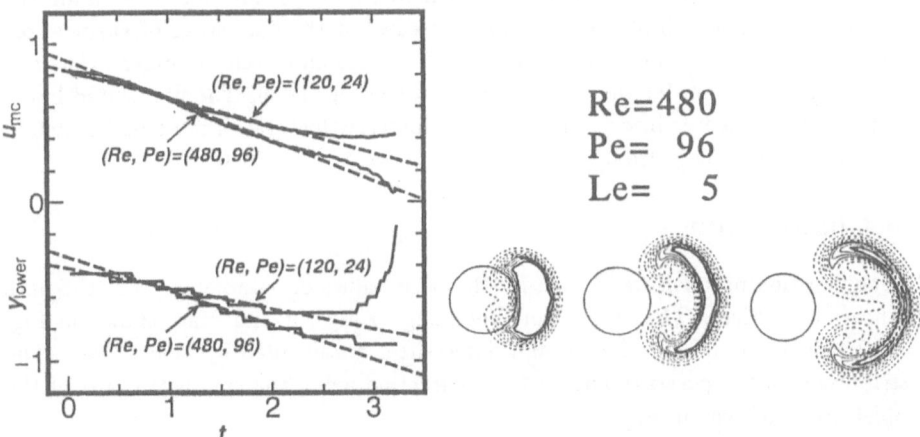

Fig. 5 Time evolution of mass center velocity compared with transverse displacement of lower side edge of liquid core for different Reynolds and Peclet numbers. Dashed lines denote least square fit of computational data.

Fig. 6 Mass fraction contours of gasifying fuel droplet travelling through quiescent gas.

$$\frac{dab}{dt} = -K'\frac{\bar{\rho}_\infty}{\bar{\rho}^*}u_{mc}a \tag{1}$$

in dimensionless form, where K is a proportional factor and the variables are made dimensionless using the initial droplet radius and speed.

As shown in Fig. 5, the mass center velocity decreases linearly with time while the length of front edge of liquid core increases linearly except at the final stage of gasification. Their variations for different Reynolds and Peclet numbers are also examined and depicted in Fig.5. Based on these computational results, the dependences on time are formulated as

$$u_{mc} = 1 - A\sqrt{LePe}\,t \quad , \quad a = 1 + A\sqrt{LePe}\,t \tag{2}$$

where A is a proportional factor. The typical value of $A\sqrt{LePe}$ for $LePe = 120$ is about 0.1. Substituting Eq.(2) into Eq.(1) and neglecting the second order term, one obtain

$$\frac{dab}{dt} = -K'\frac{\bar{\rho}_\infty}{\bar{\rho}^*} \tag{3}$$

which states that the rate of reduction of the droplet area is a constant and that the dimensionless gasification lifetime becomes independent of Reynolds and Peclet numbers. These results are consistent with the numerical computation.

However, it should be mentioned that, for moderate Reynolds and Peclet numbers, the dimensionless length of front edge, a, decreases at the final stage of gasification, whereas it increases monotonously for larger Reynolds and Peclet numbers. The fuel mass fraction contours for the latter case are shown in Fig. 6. The filamented liquid core is finally segmented into several pieces, to increase the total circumferential length for completion of gasification.

3.3 Combustion

The combustion problem was solved for the same values of parameters as in the pure gasification problem. The figures corresponding to Figs. 1 and 2 are shown in Fig. 7, in which the liquid core region and flame sheet are denoted by heavy lines. The distributions of temperature and fuel mass fraction along the symmetric line of the droplet are also given in Fig.8.

No significant differences from the pure gasification case are observed in Fig. 7 for the gasification process of liquid core. This is because the gasification of a droplet undergoing continuous phase change is governed by the diffusion process alone and almost independent of heat transfer. While additional heating of the mixture introduced by the presence of flame increases the local value of diffusion coefficient, the diffusion coefficient involved in the transport of fuel is more sensitive to changes in fuel mass fraction. In addition, for the case of small fuel-to-oxygen stoichiometric ratio at hand, the flame tends to be located far apart from the droplet and the Shvab-Zeldovich coupling function $\beta_Y = Y_F - \sigma Y_O$ results in, near the droplet, almost the same fuel mass fraction distribution as in the pure gasification case. As a consequence, the transport of heat and mass are decoupled in effect so far as the gasification process

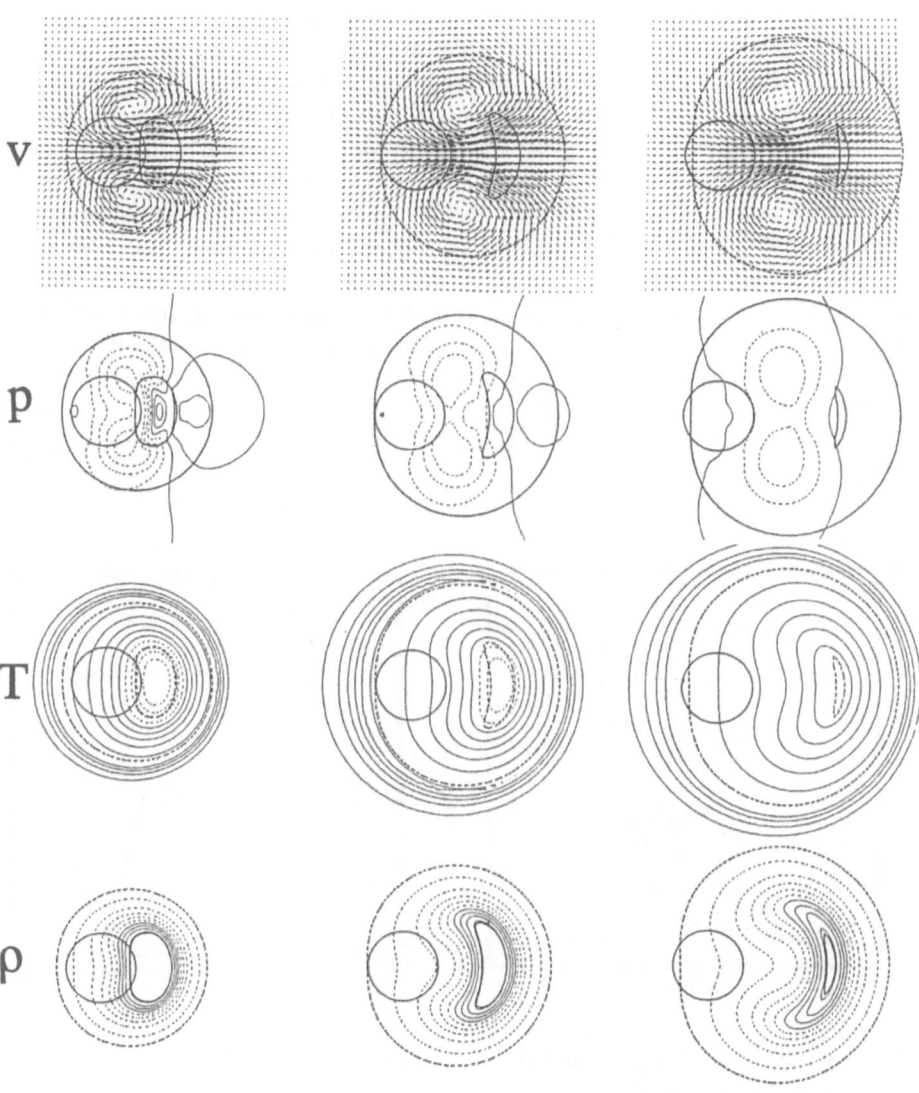

Fig. 7 Combustion of fuel droplet travelling through quiescent gas: $Re = 120$, $Pe = 24$, $Le = 5$, $Y_{o\infty} = 0.1$, $\sigma_T = 7.06$ and $\sigma_Y = 0.33$. N.D. $t = 1, 2, 3$ from left to right. Core region and flame sheet are demonstrated by heavy lines.

is concerned. This point is an important difference from the subcritical case in which gasfication rate, controlled by the phase equilibrium condition, is increased to a great degree by the presence of flame due to a rise in droplet surface temperature.

In Fig. 9, the cross-sectional areas of liquid core and fuel region are plotted against time. Since a step function is used as the initial concentration distribution, the flame sheet initially coincides with the droplet surface. The large fuel concentration gradient pushes the flame away from the droplet for a stoichiometric location and the flame is soon located far ahead of the fuel concentration boundary layer formed in front of the translating droplet. Since the fuel gasified at the front edge of liquid core per unit time is asymptotically proportional to the droplet speed, the gasification rate is not compatible with the combustion rate which is controlled by the diffusion of oxygen from the outside. The excess gasified fuel is, therefore, accumulated inside of the envelop flame. At the same time, part of the fuel vapor is advected behind the droplet and diffuses outward to react with the oxygen in the quiescent gas. Because of the fuel vapor accumulation, the flame continues to expand even after the completion of gasification. It is interesting to note in Fig. 8 that the location of flame behind the droplet is almost fixed in space during the gasification period while the flame ahead of the droplet is departing from the droplet. Obviously, the stationary flame is caused by the suppressed Stefan flow effect due to an adversing flow created by the droplet motion. As the maximum fuel concentration decreases to a low value, the flame begins to shrink.

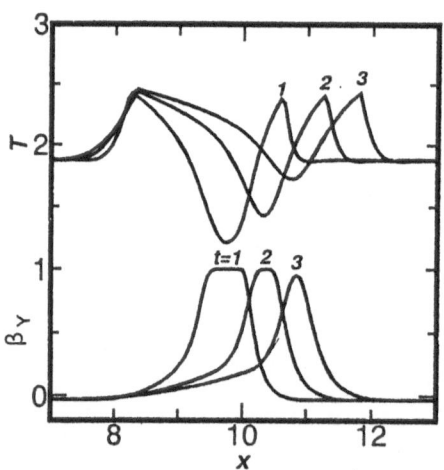

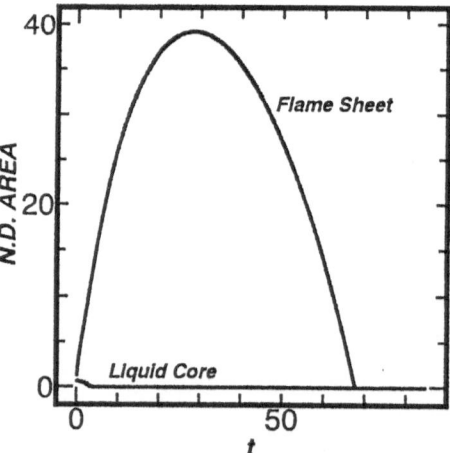

Fig. 8 Temperature and β_Y distributions along symmetric line of burning fuel droplet travelling through quiescent gas. The conditions are the same as in Fig.7.

Fig. 9 Cross-sectional areas of liquid core ($Y_F > 0.9$) and fuel region against time. The conditions are the same as in Fig.7.

The maximum flame radius estimated from Fig. 9 is about eight times as large as the initial droplet radius. This value, based on the 2D calculation, is, however, a little bit too large compared with an experimental result[1] because of different dimensionality. The far flow induced by the translating droplet may be viewed as a potential flow for which the velocity decays inversely proportional to the squared distance from the droplet center. Thus, when it moves away from the droplet to a distance, the flame will be unaffected by the velocity field and behave as if it were in a quiescent gas, resulting in a circular form of flame. In reality, there still survives some effect of convection at the flame location, which makes the flame elliptic with an aspect ratio of about 1.35 at the instant of maximum flame size.

The combustion lifetime is about twenty times longer than the gasification lifetime. Since the Lewis number of the gaseous mixture is about unity, both temperature and concentrations at large distances from the droplet behave similarly. The far-field solution may be obtained analytically by regarding the gasifying droplet as a point source of fuel under the assumptions of constant physical properties and negligible convection effect, yielding the combustion lifetime

$$\bar{t}_b = \frac{\bar{t}_v}{1 - \exp\left(-2\pi\bar{D}\sigma Y_{o\infty}/K\bar{U}\bar{a}_0\right)} \approx \frac{(\bar{\rho}_\infty/\bar{\rho}^*)K\bar{U}\bar{a}_0}{2\pi\bar{D}\sigma Y_{o\infty}}\bar{t}_v \ . \tag{4}$$

Since the gasification lifetime $\bar{t}_v$ is inversely proportional to the droplet speed $\bar{U}$, the combustion lifetime $\bar{t}_b$ becomes independent of $\bar{U}$. The underlying physics are already apparent from the foregoing argument. The combustion would occur as if a diffused puff of fuel vapor were instantaneously placed in the quiescent oxidizing gas. This is what Spalding[13] considered a long time ago, and his analysis based on an instantaneous point-source method provides the combustion lifetime proportional to the cube root of pressure.

4. CONCLUDING REMARKS

Characteristic to the gasification of a translating liquid fuel droplet with continuous phase change in supercritical ambiences is that, although the coupling between heat and mass transfer is relatively weak, the rate of gasification is closely related with the transverse elongation of liquid core which is caused by its own translational movement. The gasification is enhanced by the concentration boundary layer formed in front of the moving droplet, yielding its time-invariant rate proportional to initial droplet speed in the leading order of approximation. The underlying fluid dynamic effect to be noted is that the size of liquid core, normalized by the initial droplet radius, is transversely elongated at the same constant rate (proportional to the square root of the Lewis number times the Peclet number) as the mass center velocity of liquid core, normalized by the initial droplet speed, decelerates.

For the small fuel-to-oxygen stoichiometric ratio case in which the flame tends to be located far away from the droplet, the presence of flame has almost no influence on the gasification process. Most of the rapidly gasified fuel is accumulated inside of the envelop flame, and then burns gradually at the rate controlled by the diffusion of oxygen from the quiescent ambience. Since the flame is, in most of time, outside of the influence range of the flow induced by the droplet motion, the combustion lifetime becomes independent of initial droplet speed. Qualitatively, the combustion lifetime is consistent with the predication from Spaliding's instantaneous point-source theory.

ACKNOWLEDGMENT

The authors wish to thank Mr. Yamaki for his help in numerical calculation.

REFERENCES

[1] Sato, J., Tsue, M., Niwa, M. and Kono, M., Effects of Natural Convection on High-Pressure Droplet Combustion, Combust. Flame, 82, 142-150, (1990).

[2] Umemura, A., Unsteady Transition from Sub-to-Supercritical Evaporation Regime, in Mathematical Modeling in Combustion Science, Lecture Notes in Physics 299, J.D.Buckmaster and T.Takeno eds., Springer-Verlag, Heidelberg, (1988), pp. 52-66.

[3] Shuen, J.S. and Yang, Vigor, Combustion of Liquid Fuel Droplet in Supercritical Conditions, Combust. Flame, 89, 299-319, (1993).

[4] Delplanque, J.P. and Sirignano, W.A., Transient Vaporization and Burning for an Oxygen Droplet at Sub- and Near-Critical Conditions, AIAA-91-0075, 29th Aerospace Science Meeting, Reno Nevada, (1991).

[5] Umemura, A., Fundamental Study for Supercritical Spray Combustion Simulator, Report of Japan Ministry of Education Science and Culture's Grand-in-Aid for Scientific Research (General Research C), (1994).

[6] Umemura, A., Chang, X.Y. and Fujiwara, T., Suprecritical Droplet Evaporation, ICLASS-91, NIST, Gaithersburg, MD, (1991), pp. 89-96.

[7] Li, Y.Q. and Umemura, A., Numerical Simulation of Supercritical Droplet Combustion Trans. Japan Soc. Mech. Engrs., 59, 51-56, (1993), (in Japanese).

[8] Manrique, J.A. and Borman, G.L., Calculation of Steady Droplet Vaporization at High Ambient Pressures, Int. J. Heat Mass Transfer, 12, 1081-1095, (1969).

[9] Umemura, A., Supercritical Liquid Fuel Combustion, Twenty-First Symp. (Intl.) on Combustion, The Combustion Institute, Pittsburgh, PA (1986), pp. 463-471.

[10] Williams, F. A., Combustion Theory (Second Eddition), The Benjamin/Cummings Publishing Company, Menlo Park, CA, (1985), p. 73.

[11] Leonard, B.P., A Stable and Accurate Convective Modeling Procedure Based on Quadratic Upstream Interpolation, Computer Methods in Applied Mech. and Eng., 19, 59-98, (1979).

[12] Lee, H.S., Fernandez-Pello, A.C., Corcos, G.M. and Oppenheim, A.K., A Mixing and Deformation Mechanism for a Supercritical Fuel Droplet, Combust. Flame, 81, 50-58, (1990).

[13] Spalding, D.B., Theory of Particle Combustion at High Pressures, A.R.S. J., 29, 825-835, (1959).

Vapor diffusion flames, their stability, and annular pool fires

I. Fischer [1], J. Buckmaster [2], D. Lozinski [3], & M. Matalon [4]

[1]Exxon Research and Engineering, Florham Park, NJ
[2]Mathematics Department, Hong Kong University of Science and Technology
[3]Department of Mathematics and Statistics, McMaster University
[4]McCormick School of Engineering and Applied Science, Northwestern University

Introduction

We describe here the results of two investigations, one experimental and one numerical. The experimental investigation arose from chance observations by one of us at the Christmas dinner table. The numerical investigation was motivated by the recognition that there appear to be no diffusion flame stability analyses that examine two-dimensional disturbances; and a problem of this nature can be defined that *a piori* considerations suggest might be related to the experimental program. In the sequel the connection is not as strong as we had hoped, but the threads that link the two studies will be described.

The ingredients of both studies are described in detail in the unpublished work of Fischer [1], and only their essential nature will be described here. One set of ingredients derives from observations of annular pools of alcohol dissolved in water, and the flames which they can support. Another set derives from a numerical study of a simple one-dimensional vapor-diffusion flame in which steady solutions are constructed and their stability is examined.

Annular pools differ from simply connected ones in the nature of the observations that they permit. They admit four 'stationary' combustion states:
(i) The quenched condition - no burning - when the state of the system is below the flammability limit;
(ii) Pool burning;
(iii) Front propagation in which the front provides a transition between states (i) and (ii);
(iv) Front propagation in which state (i) prevails on both sides of the front.

In the study of (iv)the annular geometry permits a clear distinction to be made between transient and secular behaviors. Moreover, each time the front circulates it propagates over a slightly modified pool, one that is hotter and for which the alcohol concentration is changed. This can lead to delicate transitions in the burning mode.

The vapor-diffusion flame that we have studied is sketched in Fig. 1. The configuration differs from that studied by Kirkby and Schmitz [2] only in the manner in which the fuel is supplied - as liquid to be vaporized by heat from the flame, rather than as a specified flow of gas through the bottom. We shall consider the steady solution to this problem and the stability of this solution to two-dimensional perturbations.

The configuration shown in Fig. 1 can be thought of as a one-dimensional model of a slice through the annular pool fire. Here the concentration of oxygen at the upper end is fixed by the indicated flow. In the experiment, buoyancy-driven convection is the sustaining mechanism. Perhaps a closer connection between experiment and theory would occur under microgravity conditions.

Observations of annular pool fires

One of us has had ocasion to take brandy (40% alcohol), warm it over the stove, and pour the flaming liquid over a plum pudding contained in a dish. In this way an annular pool of flaming fuel is confined between the pudding and the dish walls. Extinction occurs in this pool when the concentration of alcohol has sufficiently decreased. The fire fragments into strips separated by quenched regions. These strips grow, shrink, divide and meld. The moving edge of a strip corresponds to combustion state (iii), an *ignition front* if moving into the quenched region, a *quenching-front* if moving away from the quenched region. We have been able to duplicate this behavior in the controlled experiments reported here. Also, in the annual experiment, it has happened that the final burning state is (iv). This transition from (ii) to (iv) was not observed in the controlled experiments, but not a great deal of effort was spent seeking it.

The controlled experiments were carried out in annular regions with two geometries: inner diameter 12ins, annular gap 0.45 ins; and inner diameter 15ins, annular gap 1.0ins. Mixtures of ethanol and water, usually at room temperature, were added to the annulus to a depth of 0.4ins. Volume fractions (before mixing) of 30, 40, 50 and 60 per cent ethanol were used.

30% mixture, small apparatus

Ignition led to flame propagation (type (iv)) for at most 1/4 revolution.

40% mixture, small apparatus

Initial attempts at ignition resulted in two flames (type (iv)) that traveled around the annulus in opposite directions from the ignition point. These met on the far side of the annulus and were extinguished. Several additional ignitions were attempted before this event reoccurred. In this trial one of the flames extinguished, and the other continued for almost an entire revolution. The next ignition resulted in a flame of type (iv) that circulated many times (> 10) before self-extinguishing. Traveling waves of this kind were reproduced by further ignition. All ignition/ignition attempts were carried out without replenishment of the fuel, and with minimal delay. These type (iv) flames traveled at an average 1.71revs/sec, for an average 13.8 revolutions.

50% mixture, small apparatus

Generation of type (iv) flames followed the same sequence of events as for the 40% mixture, but the long term behaviour of the flames was different. Transition to total pool burning occurred in the following fashion. During successive revolutions the flame lengthened, exhibited yellowing, and then a *backward* propagating flame split off from the main flame so that there were two counterrotating type (iv) flames. These collided on the far side of the annulus to form two type (iii) flames (i.e. the region between the separating fronts was burning). When these collided at the approximate location of the original flame split, a complete pool fire was established. This type (iv) to type (ii) transition occurred after 10-15 revolutions.

60% mixture, small apparatus

Ignition led to two counter rotating flames with burning in between (type (iii)), culminating in total pool burning. A used mixture, corresponding to concentrations less than 60%, could require flame collision to effect the transition.

Effects of preheating and changes in annular width

Preheating the fuel or increasing the width of the annulus was equivalent to increasing the concentration of alcohol in the mixture. For example, heating the 30% mixture to 52°C led to an ignition response like that of the 60% mixture at room temperature; a 40% mixture at room temperature, but in the large annulus, behaved like the 60% mixture in the small annulus. It is natural to assume that the width of the annulus controls the heat loss to the walls, so that a small gap tends to quench the flame.

One-dimensional vapor-diffusion flame

The configuration is shown in Fig. 1 and its connection to the annular pool fire, although a rough one, has been noted. The following equations were solved.

$$\rho\frac{\partial Y_F}{\partial t} + M\frac{\partial Y_F}{\partial z} = \rho D_F\frac{\partial^2 Y_F}{\partial z^2} - r, \tag{1a}$$

$$\rho\frac{\partial Y_O}{\partial t} + M\frac{\partial Y_O}{\partial z} = \rho D_O\frac{\partial^2 Y_O}{\partial z^2} - br, \tag{1b}$$

$$\rho C_p\frac{\partial T}{\partial t} + MC_p\frac{\partial T}{\partial z} = k\frac{\partial^2 T}{\partial z^2} - \triangle H_r - u(T - T_w), \tag{1c}$$

$$r = A\rho^2 Y_F Y_O \exp(-E/RT), \quad \rho = \text{const.}, \quad M = M(t). \tag{1d, e, f}$$

It need only be noted that the term in u in Eqn. (1c) represents heat losses to the side walls.

Boundary conditions are:

$$\underline{z = L} \quad Y_F = 0, \quad Y_O = Y_{O_L}, \quad T = T_L; \tag{2a, b, c}$$

$$\underline{z = 0} \quad 0 = MY_O - \rho D_O\frac{\partial Y_O}{\partial z}, \tag{2d}$$

$$MY_{F_i} = MY_F - \rho D_F\frac{\partial Y_F}{\partial z}, \tag{2e}$$

$$Mh_{f_g} = -k\frac{\partial T}{\partial z}, \quad T = T_s. \tag{2f, g}$$

Here MY_{F_i} is the mass flux fraction of evaporating mixture that is fuel. Y_{F_i} differs from the fuel concentration in the liquid because the two components do not evaporate as their proportions. A more accurate boundary condition than (2g) might have been worthwhile, but was not used. T_s is the temperature expected in the equilibrium or Burke-Schuman limit and one consequence of this choice is that the lower branch of the familiar S-shape response is elminated when $T_L < T_s$.

Non-dimensional parameters defined by this system are:

$$\tilde{M} = \frac{ML}{\rho D_F}, \quad D_a = \frac{A\rho b Y_{F_i} L^2}{D_F}, \quad Le = \frac{k}{\rho C_p D_F}, \quad D = \frac{D_O}{D_F},$$

$$\sigma = \frac{EC_p}{R(-\triangle H)Y_{F_i}}, \quad U = \frac{uL^2}{\rho C_p D_F}, \quad \phi = \frac{(-\triangle H)Y_{F_i}}{h_{f_g}},$$

$$\gamma_{O_L} = \frac{Y_{O_L}}{bY_{F_i}}, \quad \tilde{T}_j = \frac{T_j C_p}{(-\triangle H)Y_{F_i}} \quad (j = L, s, w).$$

Steady Solutions

Steady solutions were constructed using finite differences, Newton iteration, an adaptive mesh, and arc length continuation, strategies described by Giovangigli & Smooke [3]. Parameter values were chosen to represent a 40% ethanol mixture, viz

T_L	T_s	T_w	ϕ	σ	γ_{0_L}	D	Le
.0165	.0201	.0221	15.9	1.19	0.106	0.62	1.69

Different values of D and Le from these were adopted for a small number of calculations. Solutions were checked for accuracy by evaluating a global energy balance.

Variations of $\tilde{M}$ and $\tilde{T}_{\max}$ (the non-dimensional maximum temperature) with D_a are shown in Figs. 2 & 3 for different values of U. Note the unexpected behavior of $\tilde{M}$ when $U = 10$. $\tilde{M}$ also behaves in an unconventional fashion when $Le = 1, D = 1, U = 0.5$ (Fig. 4). Here the upper branch of the $\tilde{M} - D_a$ response does not rise montonically, but displays a maximum $\tilde{M}$ at finite D_a. But in all cases the $\tilde{T}_{\max}$ response is conventional, as is that of the flame location (not shown). And the minimum Damkohler number at the turning point (static extinction point) increases with U, as expected.

Symmetry breaking

In pursuing a connection between the theoretical study and the experimental one it is natural to associate the upper branch of Fig. 3 with pool burning (state (ii)). The quenched state (i), not discussed in the theoretical treatment, lies below the lower branch. Where could the seeds of front propagation lie in this description? Two possibilities come naturally to mind.

In a contribution to this Proceedings, Weber, Mercer, Gray, and Watt [4] examine a thermal ignition problem with an S-response. They show that for any value of the input parameter (measured on the abscissa) for which there is a middle-branch solution, there is a wave-like solution − a propagating front − joining the lower branch solution to the upper one. If the input parameter lies close to the static extinction point this wave is a *quenching front*, otherwise it is an *ignition front*. Similiar behavior, including the front dichotomy, has been calculated by Dold *et al* [5] for counterflow diffusion flames. We speculate here that the present system has similar behavior, and this has its counterpart in the ignition and quenching fronts observed in the annular pool prior to total extinction.

A second way in which traveling waves can arise is via bifurcation of an unsteady solution branch from the upper branch of the S, near the static extinction point. The location of such a bifurcation point can be identified by means of a stability analysis.

Stability

We have examined the linear stability of the steady solutions by formulating the appropriate matrix eigenvalue problem governing the perturbed system. The perturbations contain the factor $e^{\lambda t + ikx}$ where λ is the growth rate, k is a prescribed wave-number, and x is measured orthogonal to z. Previous stability calculations for diffusion flames do not appear to have examined non-zero values of k.

It is not suprising, in view of the results of Kirkby and Schmitz [2], that for $Le = 1.69$ the neutral stability point lies on the upper branch to the right of the static extinction point. We first identified the point where $\mathrm{Re}(\lambda) = 0, k = 0$ and noted that $\mathrm{Im}(\lambda) \neq 0$ there. We then examined the behaviour of $\mathrm{Re}(\lambda)$ with k at this point, and always found that $\mathrm{Re}(\lambda) < 0$ when $k \neq 0$. In other words the planar mode is the most unstable one. To the left of the neutral stability point $\mathrm{Re}(\lambda)$ is positive for sufficiently small k and monotonically decreases with k, Fig. 5. When $Le = 1, U = 0$ the neutral stability point is at the turning point, as expected, and did not move with modest increases in U ($U \leq 0.5$). The results of Kirkby & Schmitz suggest that larger values of U might give different results.

If an unsteady solution branch bifurcates from the stationary solution it will do so at a neutral stability point such as that identified above. One possibility is a stable, supercritical bifurcation (Fig. 6, curve A) in which a decrease in the input parameter I will cause the system to follow the paths $P - A$. Then for $I < I_u, (I_u - I)$ small, small disturbances would be superimposed on the steady solution and these would have a frequency and wave-number close to that defined at P by the stability analysis. If P was characterised by neutrally stable disturbances of non-vanishing wave-number, the bifurcated solution would exhibit traveling waves. But we have found that $k = 0$ at P and so the bifurcated solution, should it exist, would display planar pulsations. Chan and Tsien [6] carried out experiments on flames in circular tubes, configured like Fig. 1, and failed to see planar instabilities[+].

A second possibility, an unstable subcritical bifurcation, is indicated in Fig. 6, curve B. Passage of I through I_u leads to a jump to the unsteady solution identified with the point P_1. Because P_1 is removed from P the properties of the linear stability eigenfunctions near P can not be used to infer anything about the expected behaviour at P_1.

In summary, the linear stability analysis provide no insight into the matter of front propagation, apart from speculative possibilities.

[+] They observed instabilities under near-limit conditions, but these were manifest as fluctuations in the stand-off distance between the flame edge and the wall

Acknowledgement

This work was partly supported by AFOSR and by the NASA-Lewis Research Center by grants administered by the University of Illinois. And by NSF and the NASA-Lewis Research Center by grants administered by Northwestern University. Hong Kong University of Science & Technology provided travel funds supplementary to those provided by NSF for the workshop which made it possible for JB to travel to the workshop site from Hong Kong. DL's contributions were made during a post-doctoral appointment at the University of Illinois, partly supported by NSERC Canada.

References

[1]   Fischer, I. MS thesis. College of Engineering, University of Illinois, 1994.

[2]   Kirkby, L.L. and Schmitz, R.A. *Combustion and Flame*, **10**,1966, p. 205.

[3]   Giovangigli, V., and Smooke, M.D. *Applied Numerical Mathematics*, **5**, 1989, p. 305.

[4]   Weber, R.O. Mercer, G.N., Gary, P. and Watt, S. D. This Proceedings, 1995.

[5]   Dold, J.W., Hartley, L.J., and Green, D. *In Dynamical Issues in Combustion Theory*, IMA Vol. Math. Appl. ed. P.C. Fife, A. Linan, F. Williams, **35**, 83. New York: Springer-Verlag, 1991.

[6]   Chan, W. Y., and Tsien, J. S. *Combustion Science and Technology*, **18**, 1978, p. 139.

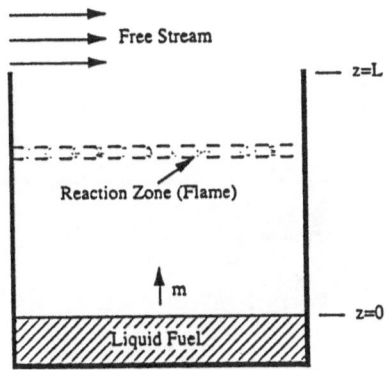

Fig. 1. The combustion system model

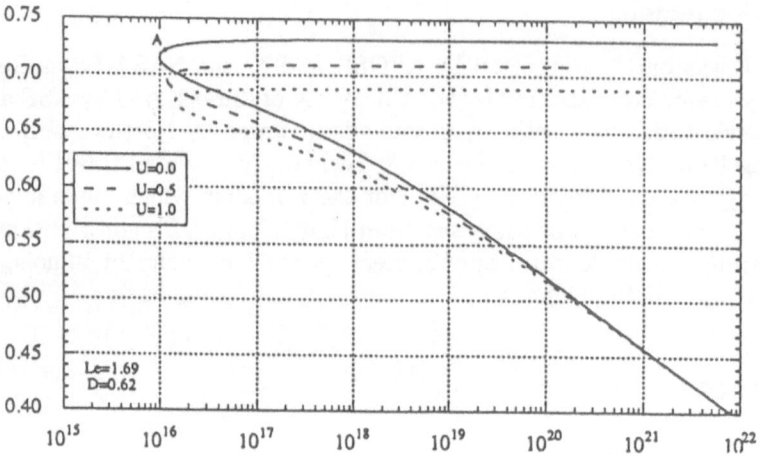

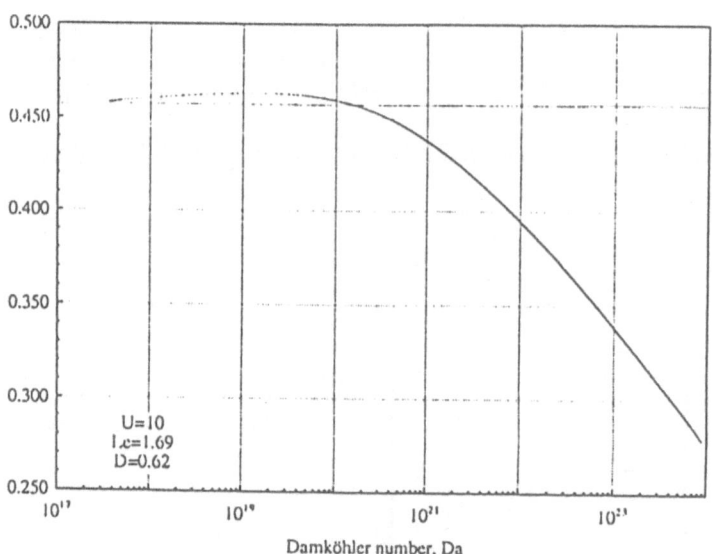

Damköhler number, Da

Fig. 2. Mass flux vs. Damköhler number

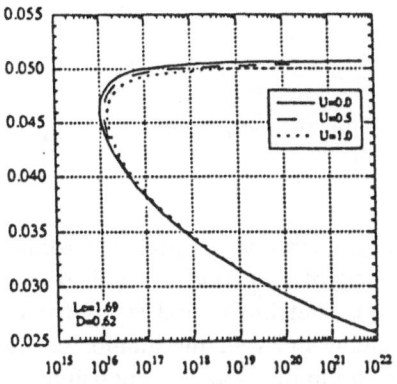

Fig. 3. Maximum temperature vs. Damköhler number

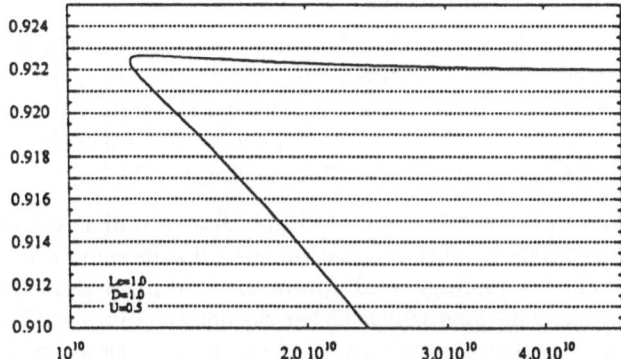

Fig. 4. Mass flux vs. Damköhler number

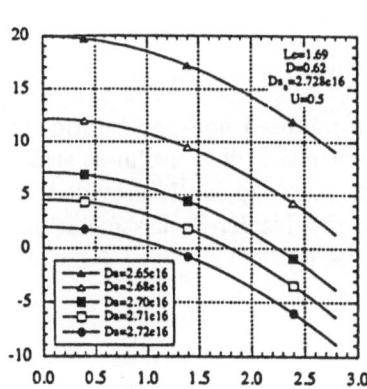

Fig. 5. Growth rate vs. wave number

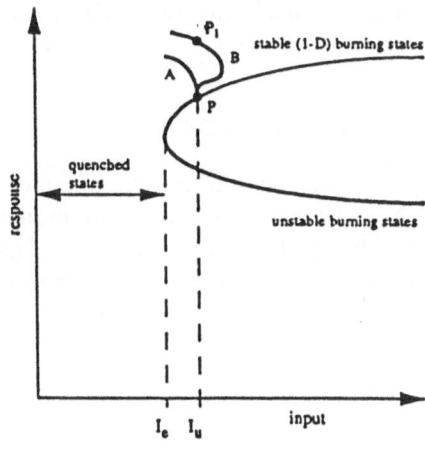

Fig. 6.

Flames in Vortices & Tulip-Flame Inversion

J. W. Dold

School of Mathematics, University of Bristol, Bristol BS8 1TW, Britain

Abstract. This article summarises two areas of research regarding the propagation of flames in flows which involve significant fluid-dynamical motion [1]–[3]. The major difference between the two is that in the first study the fluid motion is present before the arrival of any flame and remains unaffected by the flame [1, 2] while, in the second study it is the flame that is responsible for all of the fluid dynamical effects [3]. It is currently very difficult to study flame-motion in which the medium is both highly disturbed before the arrival of a flame and is further influenced by the passage of the flame.

The first study highlights the role of stretched Navier-Stokes vortices (bearing some resemblance to vortices observed in direct numerical simulations of turbulence) in distorting a flame and influencing its passage through a fluid in non-uniform motion [2]. The vortices used for this purpose are taken from a recently discovered class of exact steady Navier-Stokes solutions [1].

In the second study, the role of vorticity generated at a flame-front is considered in arriving at an approximate evolution equation (which is second-order in time) for a flame's movement. A more general approach is outlined than that used in [3]. Unlike first-order equations which either do not take into account vorticity-production or which neglect its stabilising effect, this equation successfully mimics the process by which an overstretched tulip flame inverts itself. The inversion of such hydrodynamically sculptured flames has been observed to play a major role in some combustion-driven acoustic instabilities [4].

1 Flames in Vortices

In examining the way in which flames respond to imposed non-uniformities in a flow-field (as opposed to flame-induced non-uniformities) it is useful to select a suitably idealised flow for study. One candidate for adoption is provided by a recently discovered class of exact solutions of the Navier-Stokes equations describing periodically distributed stretched vortices [1].

1.1 Periodic Stretched Vortices

In order to understand the nature of these vortices, we can consider a two-dimensional strain-rate field (in, say, the y-z plane, uniform in the x-direction) with additional velocity contributions restricted to the converging direction of the underlying strain and the third spatial dimension (the x-direction). Thus, for a positive strain-rate λ we take the velocity to have the form

$$\boldsymbol{u} = \begin{bmatrix} 0 \\ -\lambda y \\ \lambda z \end{bmatrix} + \begin{bmatrix} U(x,y) \\ V(x,y) \\ 0 \end{bmatrix}. \tag{1}$$

Incompressibility, $\nabla \cdot \boldsymbol{u} = 0$, requires that $U_x + V_y = 0$ which can be satisfied by defining a *disturbance* stream function $\Psi(x,y)$ such that

$$U = \Psi_y, \quad V = -\Psi_x. \tag{2}$$

Conservation of the vorticity, $\omega = V_x - U_y = -(\Psi_{xx} + \Psi_{yy})$, requires that

$$U\omega_x + V\omega_y = \lambda y\,\omega_y + \lambda\omega + \nu(\omega_{xx} + \omega_{yy}) \tag{3}$$

in which, without loss of generality, the viscosity ν can be scaled to take the value unity by using the length-scale for momentum diffusion $(\nu/\lambda)^{1/2}$ and an assumed wavelength $2\pi/K$ of periodicity in the x-direction to construct a dimensionless strain-rate parameter $\lambda' = \lambda/\nu K^2$ (the prime being dropped subsequently, with ν set to 1 and other variables suitably rescaled, as necessary).

Periodic solutions can now be sought in the form

$$\omega = \sum_{k=-\infty}^{\infty} \alpha_k(y)\,e^{ikx}, \quad \Psi = \sum_{k=-\infty}^{\infty} \beta_k(y)\,e^{ikx} \quad \text{with} \quad \alpha_k = \alpha^*_{-k}, \quad \beta_k = \beta^*_{-k} \tag{4}$$

requiring solution of the infinite family of ordinary differential equations, for any k:

$$\left.\begin{aligned} \alpha_k'' + \lambda y\alpha_k' + (\lambda - k^2)\alpha_k &= \sum_{n=-\infty}^{\infty} i\left[(k-n)\beta_n'\alpha_{k-n} - n\beta_n\alpha_{k-n}'\right] \\ \beta_k'' - k^2\beta_k + \alpha_k &= 0. \end{aligned}\right\} \tag{5}$$

Regarding boundary conditions, the requirement that U and V tend to zero as $y \to \pm\infty$ (being localised disturbances) demands that $\alpha_k(\pm\infty) = \beta_k(\pm\infty) = 0$. Since the nonlinear terms play little role in these limits, it is readily seen that these boundary conditions serve to eliminate growing behaviour of the form $e^{|ky|}$ for $\beta_k(y)$ and, provided $k^2 \geq \lambda$, behaviour of the form $|y|^{(k^2-\lambda)/\lambda}$ for $\alpha_k(y)$.

When $k^2 < \lambda$ all forms of asymptotic behaviour of the solution for $\alpha_k(y)$ decay as $y \to \pm\infty$. The slowest decay takes the form $\alpha_k \propto |y|^{-(\lambda-k^2)/\lambda}$ which, in terms of any particle moving with the advective velocity $-\lambda y$ in the y-direction corresponds precisely to an exponential growth in time brought about by an imbalance between the stretching and diffusing terms $\lambda\omega$ and ω_{xx} on the right side of equation (3). Thus, this algebraically decaying behaviour represents an incoming distribution of vorticity which grows with time, because of stretch, as it is convected inwards, from infinitesimally small values at infinity.

By admitting a non-zero coefficient for the algebraically decaying asymptotic form of at least one mode, and by truncating the system (5) at a finite number of modes, standard numerical techniques for ordinary differential equations can be used to obtain solutions. It is found that rapid convergence is achieved as the number of modes is increased, justifying a truncation of the system at an appropriate order. Interestingly [1], provided $\lambda > 1$ so that at least one mode is available having $k^2 < \lambda$, solutions are found for any arbitrarily large or small "amplitude" A, where we define the amplitude to be $A = \frac{1}{2}[\max(\Psi) - \min(\Psi)]$. One such solution is illustrated in figure 1. There are no solutions for $\lambda \leq 1$.

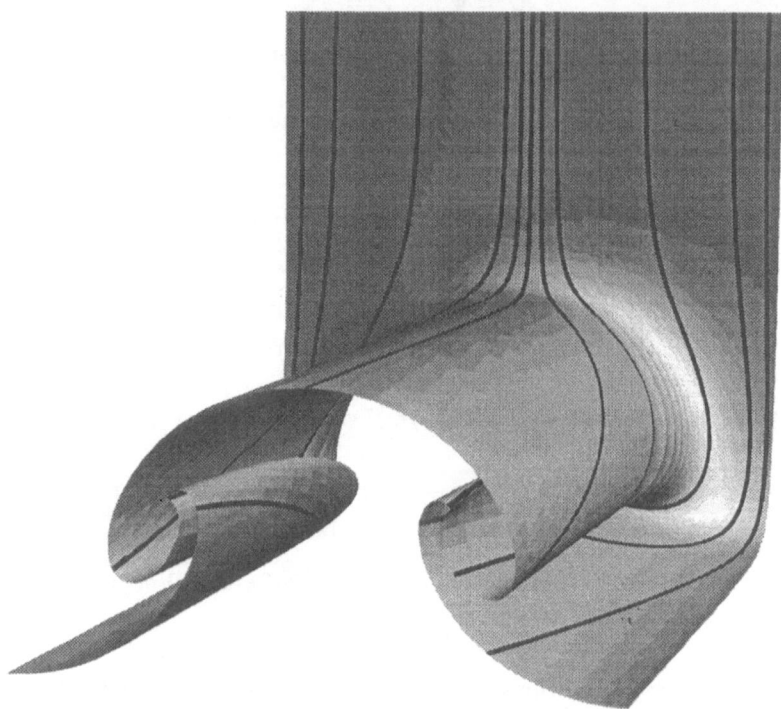

Fig 1. (from [1]). The distortion of a sheet of fluid arriving from $y = +\infty$ as it enters a region with periodic, steady counter-rotating vorticity in a stretched flow-field satisfying the Navier–Stokes equations. The dark lines follow the movement of individual fluid elements, showing that almost all fluid is sucked towards the core of a vortex.

1.2 Flame Propagation through these Vortices

A convenient means of representing the propagation of a thin constant-density flame is obtained by making use of an eiconal representation. If a flame-surface is taken to lie along an isopleth of a function f, i.e.

$$f(\boldsymbol{x}, t) = c \tag{6}$$

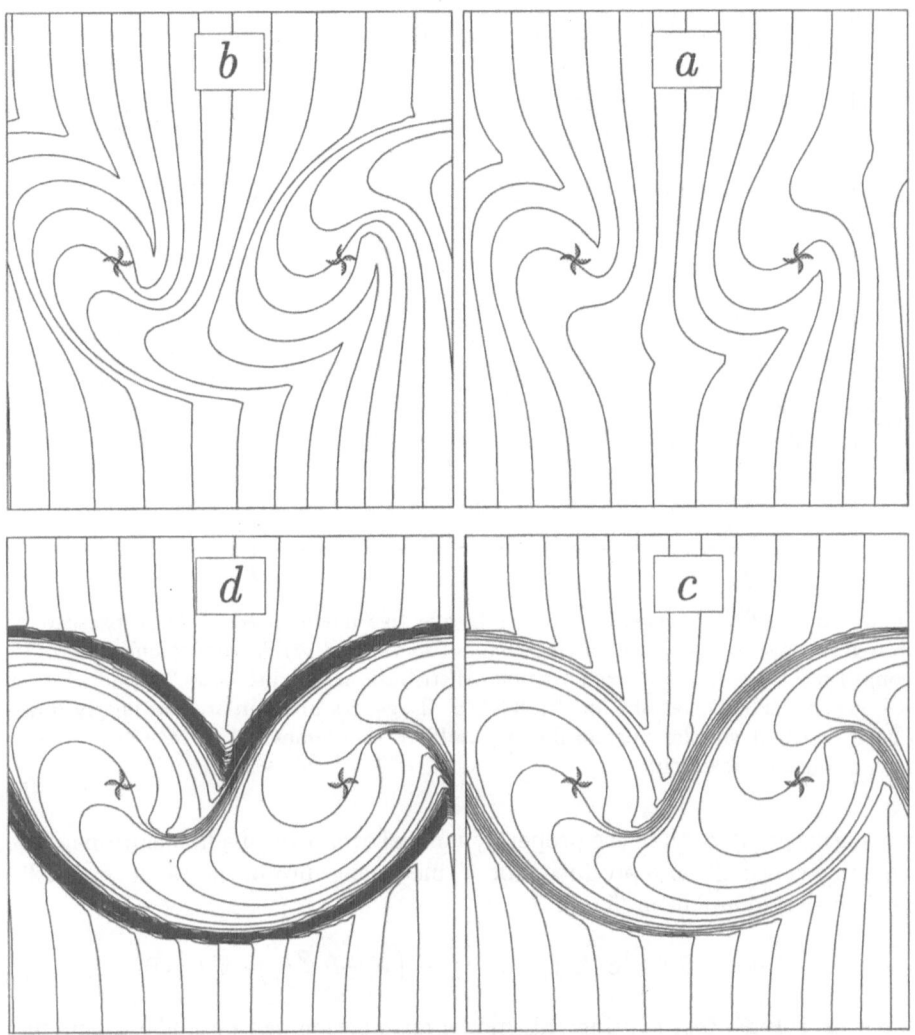

Fig 2. (from [2]). Profiles of flames travelling from right to left through a layer of steady periodic vorticity distribution, having $\lambda = 6$ and the moderate amplitude $A = 20$, at a constant normal flame speed of $S = 8$ (as calculated using the eiconal Eq. (7) which represents flames as isopleths of f). Contours of f at different dimensionless times are shown when: (a), $t = \frac{1}{9}$; (b), $t = \frac{1}{3}$; (c), $t = 1$; and (d), $t = 3$. The centres of regions of counter-rotating vorticity are marked by small fans. The development of a corridor of enhanced flame passage is clearly seen.

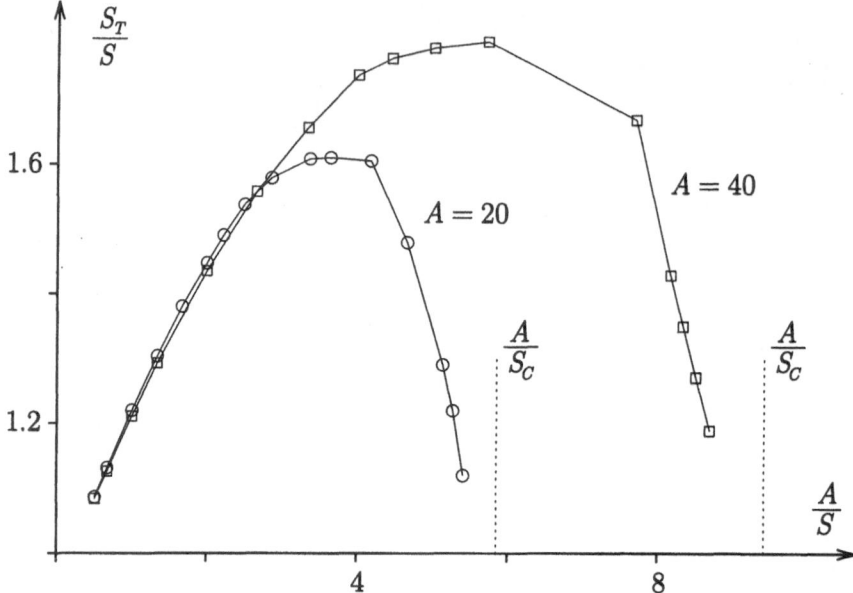

Fig 3. (from [2]). Fractional increases in effective mean-speed of flame passage S_T through the layer of vorticity distribution, as functions of A/S, from calculations of the eiconal Eq. (12) using two fixed vorticity distributions, having $A = 20$ and $A = 40$, with $\lambda = 6$. The variations for small A/S (large S) are similar, and nearly linear, in both cases; at smaller normal flame-speeds S the flames fail to take advantage of the non-uniform flow-field.

for some constant c, then the proper dynamical evolution of a flame, propagating at a flame-speed S, is reproduced for f increasing into burnt gas if f satisfies the eiconal field equation [5, *pg. 8*]

$$f_t + \boldsymbol{u} \cdot \boldsymbol{\nabla} f = S \, |\boldsymbol{\nabla} f| \quad \text{or} \quad f_t + \left(\boldsymbol{u} - S \frac{\boldsymbol{\nabla} f}{|\boldsymbol{\nabla} f|}\right) \cdot \boldsymbol{\nabla} f = 0. \tag{7}$$

Moreover, solving for f using this model formulation can be made to simulate many different possible flames in the same calculation since any isopleth of f, that is $f(\boldsymbol{x}, t) = c$ for *any value* of c, can represent a flame. Solutions of (7) are thus able to provide a great deal of information about the combined advection and propagation of flames around vortices [6]–[9].

It is clear from the second form of equation (7) that the eiconal equation is a form of characteristic equation in which the velocity of a characteristic at a point is given by $\boldsymbol{u} - S \, \boldsymbol{\nabla} f / |\boldsymbol{\nabla} f|$. Using this as the basis for a numerical scheme for solving the eiconal equation (employing a high-order upwinded template to minimise numerical diffusion) families of constant-speed flames propagating through periodic vortices are shown at four successive times in figure 2.

These results show the clear development of a long tongue of flame that advances well ahead of the flames propagating outside of a corridor surrounding the vortices. Within this corridor, a leading tip of flame is able to take advantage

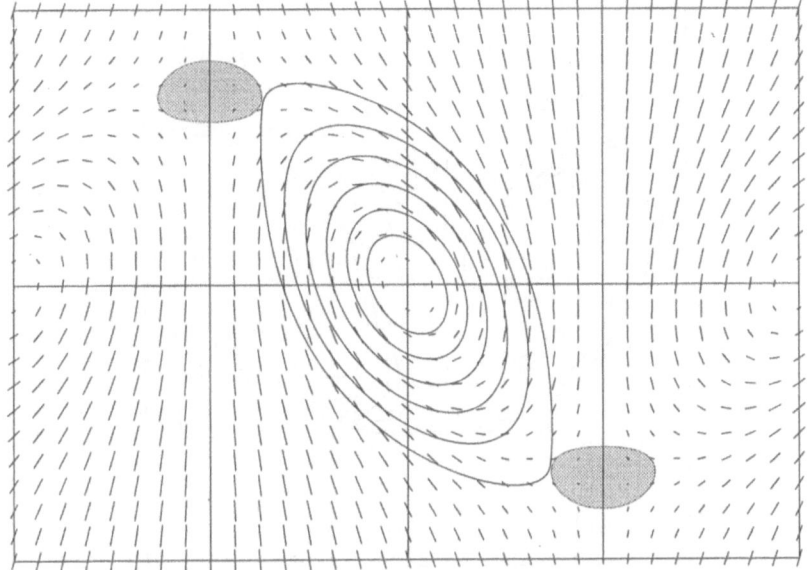

Fig 4. (from [2]). Profiles of stationary flames trapped by the inflow around a region of concentrated vorticity (having $\lambda = 6$ and $A = 10$) with flow directions and magnitudes illustrated by the short lines. Trapped flames are shown for normal propagation speeds of $S = 1$, 1.5, 2, 2.5, 3 and 3.42. Two regions for which $|\boldsymbol{u}| < S$ $(= 3.42)$, within which a flame of speed S could not be held stationary by the flow, are shown as shaded, and a flame of speed $S = 3.42$ is seen to just avoid touching these regions. This demonstrates that trapped flames do not exist (for this flow-field) at speeds of propagation above a critical value close to $S = 3.42$.

of the advection around each vortex, leap-frogging from one vortex to the next. The mean rate of advancement of this tongue of flame, denoted by S_T, depends both on the nature and intensity of the vortices (λ and A) and on the flame-speed S. In figure 3, the relative increase in flame advancement is shown as a function of A/S for two different (but relatively moderate) amplitudes of vorticity. When A/S is small (i.e. S is large) the enhancement grows linearly with A/S, but for sufficiently large values of A/S (i.e. small enough values of S) the trend is reversed. It appears that all advantage disappears at some critical level.

1.3 Trapped Flames

In order to understand this phenomenon, it is worth considering classes of stationary flame solutions. If a flame is stationary, then its propagation speed is balanced exactly by the normal component of the flow velocity coming into the flame. That is, if $\widehat{n}(x, y)$ is a unit normal to a stationary flame that passes

Fig 5. (from [2]). Profiles of flames travelling from right to left, under the same conditions as in figure 2, but for a constant normal flame speed of $S = 3\frac{1}{2}$. Isopleths of f are plotted at the dimensionless time $t = 1$. The bottlenecking inbetween the vortices, of contours at the nose of the flame, demonstrates the difficulty that a flame of this propagation speed experiences in escaping from the sucking effect of each vortex, resulting in a much reduced mean-speed of flame passage and flame profiles that resemble, for a while, the shapes of trapped flames.

through the point (x, y), pointing into unburnt gas, then $\widehat{n}$ must satisfy

$$u \cdot \widehat{n} = -S. \tag{8}$$

In general, this equation has either two solutions for $\widehat{n}$ if $|u| \geq S$, or none if $|u| < S$. By following integral curves determined by tangents corresponding to these normals, it turns out that closed stationary-flame solutions are obtainable for sufficiently small flame-speeds S. Sequences of such solutions are shown in figure 4 for increasing values of S until, at a critical flame speed S_c, the stationary flame meets a region (shown grey) within which solutions are impossible because $|u| < S$.

These critical points are also shown in figure 3 where they appear to be closely related to the inability of a propagating flame to hitch a ride on the advection around successive vortices. By contrast, the suction into any vortex tends to trap the flame and to hinder its advancement. The simulation shown in figure 5 illustrates this phenomenon. For a flame-speed slightly in excess of S_c large lobes of flame are created around any vortex. These lobes are close to the trapped-flame solutions of figure 4 and only manage to advance slowly against the inwards flow into a vortex.

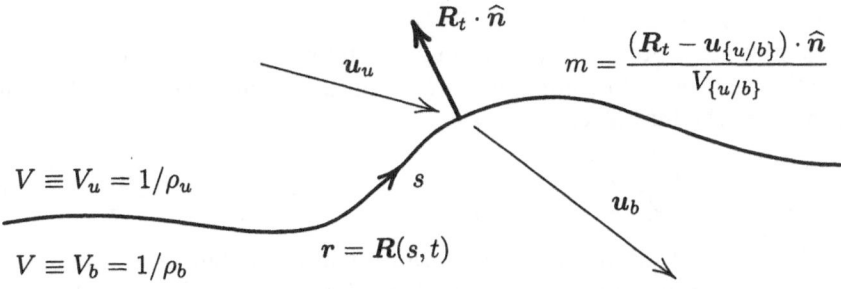

Fig 6. Illustration of the conditions prevailing at an interface $r = R(s,t)$ across which the density falls abruptly from ρ_u to ρ_b, as it propagates with a mass-flux m (per unit area) of material passing through it at any point.

2   Evolution of a Propagating Density Jump

A thin flame which does not conserve volume can be approximated by a propagating interface across which the density decreases abruptly, as sketched in figure 6. At low Mach numbers, when acoustic length scales are much greater than the scale of any geometrical distortion of the interface, a convenient model for the fluid flow induced by the volume generated at the interface consists of the incompressible Euler equations applied separately to both the burnt and unburnt gases.

2.1   Euler Model for a Thin Flame

For simplicity we restrict attention to two-dimensions having velocity components $u(x,y)$ and $v(x,y)$. Incompressibility (applied separately to each fluid) and the inviscid advection of vorticity require that

$$\left.\begin{aligned}
\boldsymbol{\nabla} \cdot \boldsymbol{u} &= & u_x + v_y &= 0 \\
\boldsymbol{\nabla} \times \boldsymbol{u} &= & v_x - u_y &= \omega \\
\dot{\omega} &= & \omega_t + \boldsymbol{u} \cdot \boldsymbol{\nabla}\omega &= 0.
\end{aligned}\right\} \tag{9}$$

With the interface defined to lie at $r = R(s,t)$ (s being an arclength parameter), the mass-flux m of material passing through it is given, as shown in figure 6, by

$$m = \rho_u (\boldsymbol{R}_t - \boldsymbol{u}_u) \cdot \widehat{\boldsymbol{n}} = \rho_u (\boldsymbol{R}_t - \boldsymbol{u}_b) \cdot \widehat{\boldsymbol{n}} \tag{10}$$

where ρ is the density, and subscripts u and b indicate evaluation in the unburnt and burnt regions, immediately ahead and behind the interface, respectively. Since only the normal component of velocity changes across the interface, the jump in velocity is given by the condition

$$\boldsymbol{u}_u - \boldsymbol{u}_b = m\,(V_b - V_u)\,\widehat{\boldsymbol{n}} \tag{11}$$

in which the specific volume is defined as $V = 1/\rho$.

For a given mass-flux $m(s,t)$ through the interface, only one further piece of information is needed to complete the formulation, namely the rate of vorticity production at the interface, which gives

$$\omega_b = \frac{V_u}{V_b}\,\omega_u + \frac{(V_b - V_u)^2}{V_b}\,m_s + \frac{V_b - V_u}{V_b}\frac{P_{us}}{m} \quad \text{where} \quad \frac{P_{us}}{m} = \frac{\widehat{s}\cdot(g - \dot{u}_u)}{S} \tag{12}$$

where P_{us} is the tangential pressure gradient in the unburnt gas immediately ahead of the interface. Inviscid conservation of momentum provides the expression shown for P_{us} in terms of the unit tangent $\widehat{s} = R_s$, the gravitational acceleration vector g, the acceleration of the fluid immediately ahead of the interface $\dot{u}_u$ and the propagation speed of the interface relative to unburnt gas $S = mV_u$.

A convenient model for the mass-flux m is provided by a form of Markstein's relation [5, *pg. 22 et seq.*]

$$m = m_0(1 + L_\kappa\,\kappa) + L_\sigma\frac{\widehat{n}\cdot u_{nu}}{V_u} \quad \text{where} \quad \kappa = -\widehat{s}\cdot\widehat{n}_s \tag{13}$$

in which L_κ is a Markstein length relating changes in mass-flux to the curvature κ of the interface, and L_σ is a Markstein length relating changes in mass-flux to the normal component of the strain-rate in the flow immediately ahead of the interface, $\widehat{n}\cdot u_{nu}$. We assume that these two lengths are positive (stabilising) and of comparable magnitude.

2.2 Stability of a Flat Flame

If we now consider the linear stability of a flat interface propagating downwards (taking $g = -g\,\nabla y$), in the form $y = -S_0\,t + \phi(x,t)$ with ϕ small enough to justify linearisation and given by $\phi = \varepsilon\,e^{\lambda t + ikx} + O(\varepsilon^2)$, the following dispersion relation is obtained (c.f. [10, *pp. 498–503*])

$$\beta(1+|k|L_1)\frac{\lambda^2}{S^2} + (1+|k|L_2)\,|k|\frac{\lambda}{S} - \alpha(1-|k|L)\,k^2 + \frac{1-\mu}{2S^2}\,g|k| = 0 \tag{14}$$

in which, for simplicity, S_0 is abbreviated to S. The various constants appearing in this relation are

$$\left.\begin{array}{l} \mu = V_u/V_b \quad \text{giving} \quad 0 < \mu < 1, \quad \beta = \dfrac{1+\mu}{2}, \quad \alpha = \dfrac{1-\mu}{2\mu}, \\[2mm] L_1 = L_\sigma\dfrac{1-\mu}{1+\mu} \quad L_2 = L_\kappa + L_\sigma\dfrac{1-\mu}{\mu} \quad \text{and} \quad L = L_\sigma + L_\kappa\dfrac{2}{1-\mu}. \end{array}\right\} \tag{15}$$

When the density ratio μ is close to one, making α small, the latter three lengths are very disparate in size, having $L_1 \ll L_2 \ll L$, with

$$L_1 = O(\alpha^2 L), \quad L_2 \sim \alpha\mu L \quad \text{and} \quad L \sim \frac{L_\kappa}{\alpha} \quad \text{as} \quad \alpha \to 0. \tag{16}$$

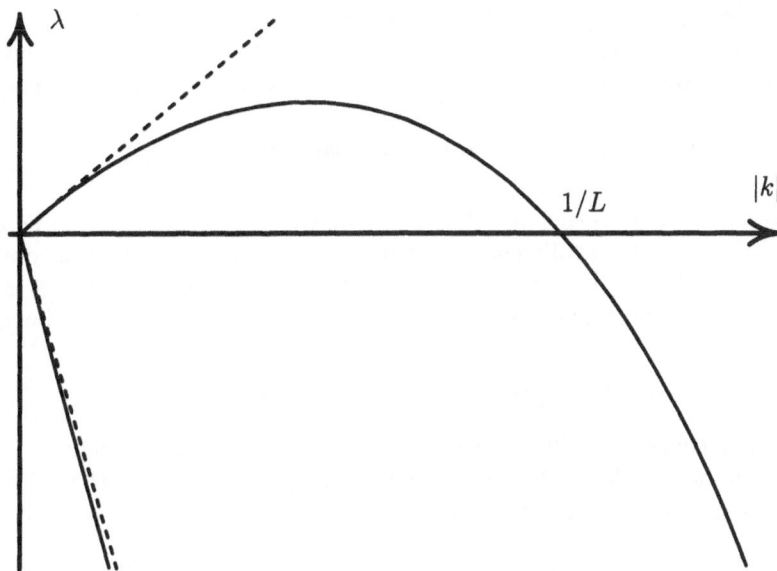

Fig 7. Roots of the Darrieus–Landau dispersion relation (17), shown dashed, and of the more general dispersion relation (14) for $\alpha = \frac{1}{2}$ with $L_1 = L_2 = g = 0$.

For a constant mass-flux m and zero gravity, the dispersion relation obtained by Darrieus and Landau [11, 12] is reproduced

$$\beta \frac{\lambda^2}{S^2} + |k| \frac{\lambda}{S} - \alpha\, k^2 = 0. \tag{17}$$

The roots of this equation and of equation (14) are sketched in figure 7.

2.3 Evolution Equation

In the absence of gravity ($g = 0$), the highest root of equation (14) changes sign when $|k| = 1/L$, being positive (unstable) for smaller wavenumbers. When α is small, we anticipate that the positive unstable root, which has the magnitude $\lambda = O(\alpha S/L)$, is most significant in any evolution. However, we also note that the quadratic term containing λ^2/S^2 is then of order α^2/L^2.

Retaining only terms of order α^2 or larger, we can thus infer that $\phi(x,t)$ should satisfy a linear evolution equation of the form

$$\beta\phi_{tt} + I\left(\phi_t + \alpha\mu L\, I(\phi_t) - \alpha\, I(\phi) - \alpha L\, \phi_{xx} + \alpha\mu g\phi \right) = 0 \tag{18}$$

in which $I(\cdot)$ is a linear operator defined such that $I(e^{ikx}) = |k|e^{ikx}$, and in which a simple rescaling of time and of g has been used to set S to unity, without loss of generality. Another rescaling (of both time and lengths, *including* ϕ) could be used to set L to any convenient value.

Anticipating that typical nonlinear evolutions would involve weak gradients of ϕ for small density changes, $\phi_x = O(\alpha)$, invariance to weak rotation of the interface can now be invoked to produce the weakly nonlinear evolution equation

$$\beta\phi_{tt} + I\left(\phi_t + \tfrac{1}{2}\phi_x{}^2 + \alpha\mu L\, I(\phi_t) - \alpha\, I(\phi) - \alpha L\, \phi_{xx} + \alpha\mu g\phi\right) = 0 \qquad (19)$$

in which, again, only terms of order α^2 or larger are retained. It can be argued that, because the L_1 coefficient is much smaller than L when α is small, the dynamics should not be affected significantly if the term $\alpha\mu L\, I(\phi_t)$ is omitted. Also neglecting gravity, this leads to the slightly more simplified evolution equation

$$\beta\phi_{tt} + I\left(\phi_t + \tfrac{1}{2}\phi_x{}^2 - \alpha\, I(\phi) - \alpha L\, \phi_{xx}\right) = 0 \qquad (20)$$

the general structure of which was deduced using heuristic arguments in [3] based on the stability result of Darrieus and Landau [11, 12] and on the Michelson–Sivashinsky evolution equation [13]

$$\phi_t + \tfrac{1}{2}\phi_x{}^2 - \alpha\, I(\phi) - \alpha L\, \phi_{xx} = 0. \qquad (21)$$

It is worth noting that the left hand side of this equation appears as an argument of the operator $I(\,\cdot\,)$ in equation (20), the result of which determines the acceleration of the interface ϕ_{tt}.

2.4 Roles of Vorticity and Negative Roots

The major difference between the Michelson–Sivashinsky equation (21) and equation (20), or its bigger sibling (19), is the appearance of the second time derivative ϕ_{tt}. This time-derivative is a direct result of the role played by vorticity in the dynamics of the interface. In particular, vorticity introduces an inertial or memory effect since, at any one time, the vorticity that was created earlier at the interface and that has been advected behind the interface continues to contribute to the flow-field at the interface.

In this way, an instantaneously flat interface $\phi = $ constant, which is an exact equilibrium of the Michelson–Sivashinsky equation, is not necessarily an equilibrium of equations (19) or (20). These equations admit the possibility that $\phi_t \neq 0$ at any equilibrium of the Michelson–Sivashinsky equation, reflecting the fact that the vorticity distribution behind the interface need not be uniform. Indeed, close to an equilibrium of the Michelson–Sivashinsky equation we have

$$\beta\,\phi_{tt} + I(\phi_t) \sim 0 \quad \Rightarrow \quad \overline{\phi}_t \sim A(k)\exp\left(-\frac{|k|}{\beta}t\right) \qquad (22)$$

where $\overline{\phi}(k,t)$ is the Fourier transform of $\phi(x,t)$, which shows that any existing modes of ϕ_t relax towards zero, with lower wavenumbers decaying more slowly.

The inertial term ϕ_{tt} in these equations is directly related to the λ^2 term of the dispersion relation (14), and hence is related to the fact that equation (14)

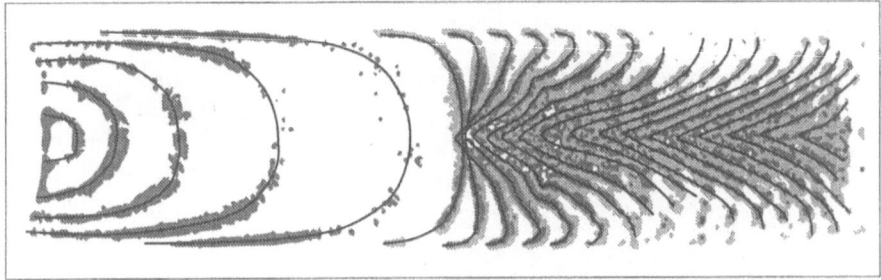

Fig 8. Tulip-flame inversion, redrawn following Ellis (1928) [14]: a stroboscopic history of flame-shapes is shown following an ignition at one end of a closed tube of diameter 5 cm and length 19.5 cm, containing CO and O_2 (ratio 10:1) saturated with water-vapour. Approximate flame-shapes are superimposed as solid lines. The early tulip shape inverts itself such that the leading part of the tulip (along the centerline) becomes the trailing part of two separate half-tulips.

has two roots. One of these is negative (and relatively large when $\alpha \ll 1$), making the corresponding linear normal-mode of dubious physical relevance when seen in isolation. However, this does not justify neglecting the λ^2 term which still plays an important role in determining the unstable root. Indeed, when α is small this term is of similar magnitude to other terms that appear in equations (19)–(21).

2.5   Tulip-Flame Inversion

One phenomenon that is captured qualitatively well by equation (20) is the inversion of tulip flames, a process that has been known for a long time but that is still not fully understood. An experimental observation [14] of the inversion of a tulip flame is shown in figure 8. It can be seen that a much elongated tulip-shaped flame collapses rapidly, passing through an almost flat shape, to assume another tulip structure in which noses and tails are reversed.

A numerical solution of equation (20) is shown in figure 9, calculated for $\alpha = \frac{1}{2}$ and $L = \frac{1}{2}$, with an initial tulip-shape of $\phi = 4\cos(x)$ having $\phi_t = 0$. As can be seen, the normal propagation term $\frac{1}{2}\phi_x{}^2$ leads to a sharpening of the tail of the flame until a very high curvature is created (resembling the shapes seen in figure 8 when the flame reaches the walls of the tube). At this point, the $I(\alpha L \, \phi_{xx})$ term dominates and leads to a strong forwards acceleration of the tail of the flame which subsequently moves rapidly ahead, passing right through a flat profile, to emerge as the leading part of an inverted tulip-shaped flame.

This form of evolution is observed for any large enough initial distortion of the interface. Small enough distortions do not lead to an inversion [3], and instead, the progress of the interface is then observed to follow reasonably closely an evolution modelled by the Michelson–Sivashinsky equation. In such cases,

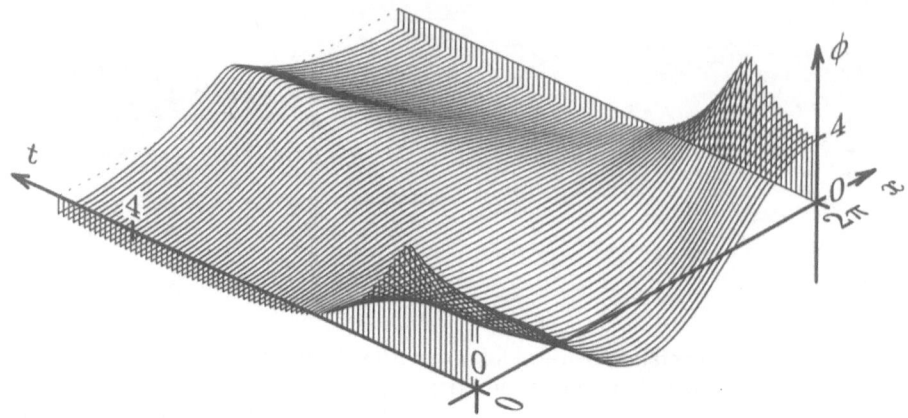

Fig 9. A numerical (spectral collocation) solution of equation (20) taking $\alpha = L = \frac{1}{2}$ with, initially $\phi = 4\cos x$ and $\phi_t = 0$. Sequences of profiles of the flame-shape ϕ (normalised to have zero mean) are plotted, up to a time of about $t = 5$, for a flame that is propagating downwards. An inversion of the initial shape of the flame is seen to occur after a considerable sharpening in the shape of the interface in the tail of the flame.

the larger negative root for λ leads to the Michelson–Sivashinsky equation as a centre manifold reduction of equation (20)—it is worth remarking on the fact that equations (20) and (21) share the same set of equilibrium solutions, having $\phi_t = \phi_{tt} = 0$.

Where sufficiently large or rapid disturbances excite behaviour on the faster time-scale of the stable root, as provided by large enough initial values or by (say) acoustic oscillations [4] (which are equivalent to having an oscillating "gravitational" acceleration g), the second-order models (19) or (20) are more physically meaningful.

3 Conclusions

Another method of introducing disturbances is through nonuniformities in the flow-field ahead of the flame, as were examined in section 1. On the one hand, without density change, stretched vortices or any given velocity field can be seen to provide a number of means by which flame and flow interactions can enhance or hinder the passage of a flame through the flow-field. On the other hand, density changes in a flame lead to both hydrodynamical instabilities and an inertial behaviour which is capable of responding actively to disturbances imposed on the evolution of a flame-interface.

It is currently an open challenge to find a way of combining these two extremes in the modelling of flames in non-uniform flows. Which effect is likely to dominate in physically realistic circumstances remains an open question.

Acknowledgements. The author is grateful to the E.P.S.R.C. for its support during the conduct of the research summarised in this article through the provision of an Advanced Research Fellowship. Many valuable discussions on various aspects of the ideas related here were conducted with Guy Joulin and Oliver Kerr.

References

[1] Kerr, O.S. and Dold, J.W. (1994) J. Fluid Mech. **276** pp. 307–325.

[2] Dold, J. W., Kerr, O. and Nikolova, I. P. (1994) *Flame Propagation through Periodic Vortices.* Combustion and Flame, to appear.

[3] Dold, J. W. and Joulin, G. (1994) *An Evolution Equation Modelling Inversion of Tulip Flames.* Combustion and Flame, to appear.

[4] Searby, G. and Rochwerger, D. (1991) J. Fluid Mech. **231** pp. 529–543.

[5] Markstein, G.H. (1964) *Nonsteady Flame Propagation.* Pergamon Press.

[6] Ashurst, W.T. and Sivashinsky, G.I. (1991) Combust. Sci. Tech. **80** pp. 159–164.

[7] Kerstein, A.R., Ashurst, W.T. and Williams, F.A. (1988) Phys. Rev. A **37** p. 2728.

[8] Ashurst, W.T., Sivashinsky, G.I. and Yakhot, V. (1988) Combust. Sci. Tech. **62** p. 273–284.

[9] Ashurst, W.T. and Williams, F.A. (1990) 23rd International Symposium on Combustion, The Combustion Institute. pp. 543–550.

[10] Zeldovich, Ya. B., Barenblatt, G. I., Librovich, V. B. and Makhviladze, G. M. (1985) *The mathematical theory of combusion and explosions.* Plenum Press.

[11] Darrieus, G. Unpublished work communicated at *La Technique Moderne,* Paris (1938) and *6eme Conférence Internationale de Mécanique Appliquée,* Paris (1946).

[12] Landau, L. D. (1944) Acta Physicochimica USSR **19** pp. 77–85.

[13] Sivashinsky, G. I. (1977) Acta Astronautica **4** pp. 1177-1206.

[14] Ellis, O. C. de C. (1928) J. of Fuel Sci. **7** pp. 502–508.

Flame Ignition of Premixed Methane-Air Flow

T. Sano
Department of Mechanical Engineering,Tokai University
Hiratsuka, Kanagawa, 259-12,Japan

Keywords. laminar flow,numerical model,full kinetics,ignition

1 Introduction

The flame ignition of combustible gas mixtures by a hot plate and a pilot flame
has been of importance for practical use in commercial combustors, but the details of ignition kinetics have not been well understood. Unsteady phenomena
in the process of flame ignition of combustible gas mixtures are controlled by
the chemical reactions as well as the transport processes of heat and mass in the
gas flow. While the natural gas, of which main component is methane, has been
widely used in combustors including the automobile engine, it is well known for
the gas to be hardly ignited. In order to understand the kinetics of ignition
by hot plate or pilot flame, the flame ignition process of premixed methane-air
mixture by high-temperature plate or hot-burned gas is numerically studied in
a two-dimensional laminar flow by considering detailed chemical reactions and
transport properties.

2 Numerical model

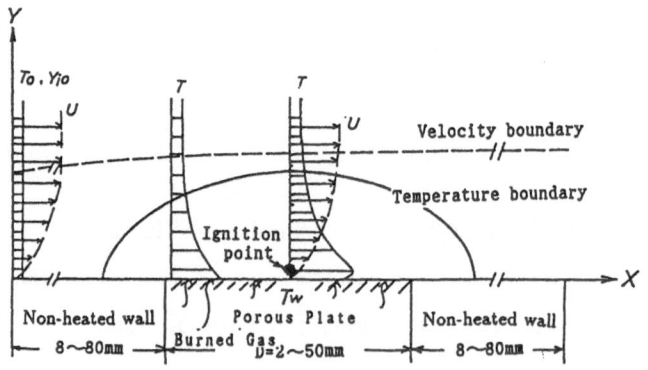

Fig.1. Numerical model.

The reactive flow system of ignition studied is illustrated in Fig.1 and modeled
as follows [1] : (1) A combustible gas mixture of methane and air flows over

a flat and permeable porous plalte, (2) constructing a two-dimensional laminar boundary-layer flow.(3) The plate is heated at a constant temperature during a distance between two non-heated parts of the flat plate. (4) The hot burned gas is permeated through the heated part and is injected with a constant mass flow rate into the combustible gas mixture of methane-air. (5) Two types pf plate surface for the active species are studied. One is the surface at which the active species are completely deactivated to have a vanishing concentration, and the other is an inert surface which has no influence on the active species. (6) The pressure of flow is constant. In this study, the positive catalytic effect of the plate surface on ignition is not discussed.

As mentioned before, the pressure is assumed to be constant in the flow field, so that the boundry conditions are as follows.

$$x = 0 \quad : \quad u = u_0, T = T_0, Y_i = Y_{i0}$$

$$x = \infty \quad : \quad \frac{\partial u}{\partial x} = 0, \frac{\partial T}{\partial x} = 0, \frac{\partial Y_i}{\partial x} = 0$$

$$y = 0 \quad : \quad u = 0, \ v = (\rho v)_b / \rho_g$$

$$T = T_w \quad \text{at heated walls}$$

$$\frac{\partial T}{\partial y} = 0 \quad \text{at non-heated wall}$$

$$Y_i = 0 \quad \text{for active species at deactivated walls}$$

$$\frac{\partial Y_i}{\partial y} = 0 \quad \text{for inactive speces at deactivated walls}$$

$$Y_{ig}(\rho v)_g = Y_{ib}(\rho v)_b + \rho D_i \frac{\partial Y_i}{\partial y} \quad \text{at inert or permeable walls}$$

$$y = \infty \quad : \quad \frac{\partial u}{\partial y} = 0, \ \frac{\partial T}{\partial y} = 0, \ \frac{\partial Y_i}{\partial y} = 0$$

The transport properties of μ, λ and D_i are calculated theoretically [2] as a function of temperature, species composition and pressure. As the composition of combustion gases of methane, 18 species are considered: CH_4, CH_3, CH_2, CH, CH_3O, $HCHO$, CHO, CO, CO_2, H_2, H_2O, H, HO_2, H_2O_2, OH, O, O_2 and N_2. As the oxidation reaction of methane with air, 61 elementary reactions are taken into account [1]. As far as the initial conditions of computations, the velocity u_0 is assumed of a Blasius profile, and the temperature and the mass fraction of combustible gas mixture are distributed. Under the above initial and boundary conditions, the governing equations of the reactive laminar flow are solved using the numerical algorithm of Sano and Kotake [3] in order to deal with the stiffness problem associated with chemical reactions.

3. Results and Discussion

3.1 Ignition by a hot plate

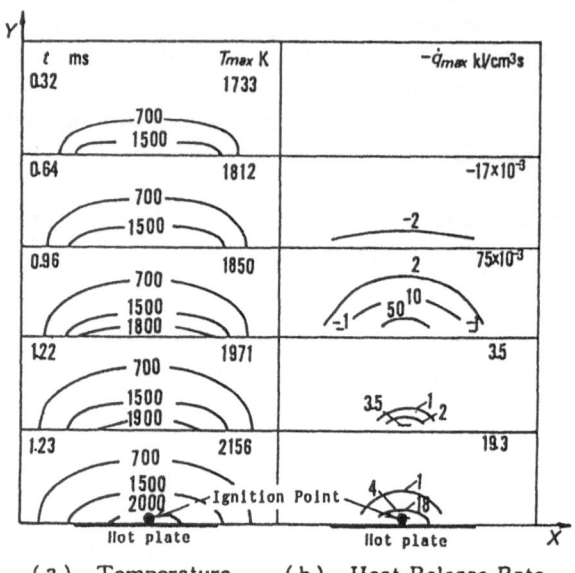

(a) Temperature   (b) Heat Release Rate

Fig.2   Contors of temperature and heat release rate near plate wall.
T_w=2000K, u_∞=5cm/s, ϕ=1, p=0.1MPa, D=5mm,
Inert surface, $(\rho v)^*$=0

Figures 2(a) and 2(b) show the time history of contor of the temperature and
heat release rate, respectively near the wall. The flow conditions are as follows;
the equivalence ratio of 1.0, the surface temperature of 2000 K without injection
of combustion gas through the porous plate, the pressure of 0.1 MPa, the main
flow velocity of 5 cm/s and the hot surface length of 5mm. The plate surface is
assumed to be chemically inert for all species. The contors of temperature and
heat release rate are nearly symmetrical with respect to the center of the plate
in the range of the main flow velocity from 1cm/s and 100cm/s. The temperature of the gas adjacent to the center of the hot plate, where ignition occurs,
increases, having the maximum temperature in the combustible gas mixture and
rises above the surface temperature. In Fig.2, the maximum gas temperature
and heat release rate in the contor are noted by the figures T_{max} and $-\dot{q}_{max}$,
respectively.

Figure 3 shows the time histories of methane consumption rate $-\dot{m}$, methane
concentration CH_4, heat release rate $-\dot{q}$, hydrogen atom concentration H, and
temperature T at the ignition point. The flow conditions are the same as those

in Fig.2. The values of $-\dot{m}$, $-\dot{q}$, H and T increase gradually with time, and after attaining the maximum values, they decrease rapidly. First, the methane consumption rate attains a maximum value, then heat release rate reaches its maximum and finally, the H concentration and the temperature take their maxima. In the present study, the ignition time is defined as the time when the methane consumption rate $-\dot{m}$ takes the maximum value. For such an unsteady phenomena, this definition of the ignition time would be the most proper.

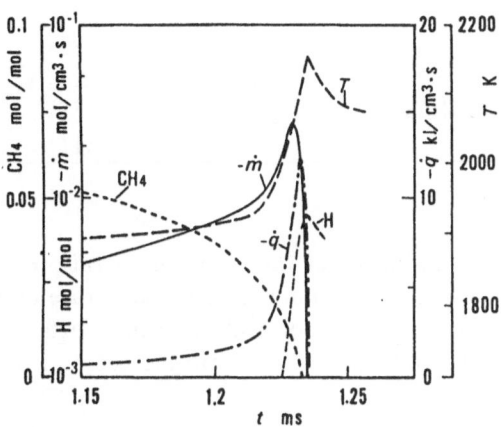

Fig.3   Time histories of methane consumption rate $-\dot{m}$, methane concentration (CH_4), heat release rate($-\dot{q}$), H concentration (H) and temperature (T) at ignition point.
T_w=2000K, u_∞=5cm/s, ϕ=1, p=0.1MPa, D=5mm,
Inert surface, $(\rho v)^*$=0

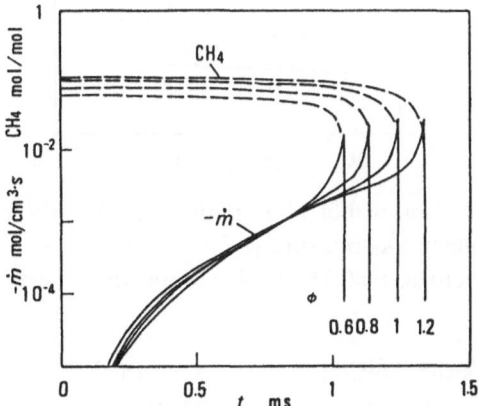

Fig.4   Effect of equivalence ratio on ignition delay.
T_w=2000K, u_∞=5cm/s, p=0.1MPa, D=5mm,
Inert surface, $(\rho v)^*$=0

Table 1: Effect of chemical conditions of plate surface. $T_w=2000K$, $u_\infty=5cm/s$, $\phi=1$, $p=0.1MPa$, $D=5mm$, Inert surface, $(\rho v)^*=0$

Chemical conditions of plate surface	Ignition delay ms
Inert surface for all species	1.23
Deactivated surface for species i	
H	1.29
O	1.24
OH	1.25
HO_2	1.24
H,O,OH,HO_2	1.31
CH_3	4.11

Figure 4 shows the influence of the equivalence ratio on the ignition delay in which the time histories of methane consumption rate and methane concentration at the ignition point are shown at the same surface temperature. The ignition is delayed with the equivalence ratio increasing from 0.6 to 1.2 at the same plate temperature.

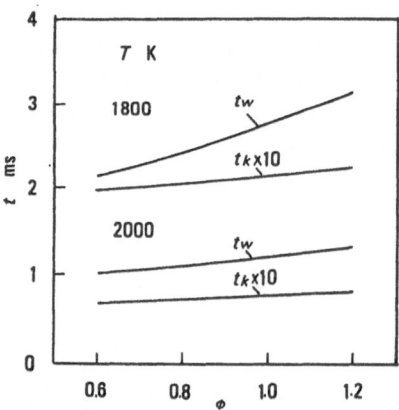

Fig.5   Comparison of hot plate ignition (t_w) with ignition in a stationary gas mixture (t_k).
$u_\infty=5cm/s$, $p=0.1MPa$, $D=5mm$, Inert surface, $(\rho v)^*=0$

Figure 5 compares the ignition delay between the hot plate ignition (t_w) with flow and the ignition in a stationary gas mixture (t_k) (that is, considering only chemical reactions) at the same temperature and pressure. Both ignition times t_w and t_k increase with the equivalence ratio. The ignition time by the hot plate

Table 2: Compositions of injection gases.

ϕ	(a)		(b)	
	1	1.4	1	1.6
Tc K	2197	1975	2310	2077
N_2	7.062×10^{-1}	6.486×10^{-1}	6.420×10^{-1}	5.339×10^{-1}
H_2	5.999×10^{-3}	5.671×10^{-2}	2.181×10^{-1}	1.633×10^{-1}
O_2	5.456×10^{-3}	1.534×10^{-5}	8.513×10^{-3}	2.580×10^{-4}
H_2O	1.808×10^{-1}	1.719×10^{-1}	3.122×10^{-1}	2.790×10^{-1}
H	8.885×10^{-4}	1.704×10^{-3}	4.096×10^{-3}	1.940×10^{-2}
O	3.969×10^{-4}	1.093×10^{-5}	1.290×10^{-3}	3.296×10^{-4}
OH	3.809×10^{-3}	4.968×10^{-4}	9.911×10^{-3}	3.839×10^{-3}
HO_2	4.529×10^{-7}	2.404×10^{-9}	1.214×10^{-6}	5.067×10^{-7}
CO	1.378×10^{-2}	7.043×10^{-2}	0	0
CO_2	8.231×10^{-2}	5.010×10^{-2}	0	0

ignition (t_w) is ten times larger than that in a stationary gas mixture (t_k) at in a wide range of the equivalence ratios and temperature. The difference between t_w and t_k is increased with the equivalence ratio. This is more evidently observed at lower temperatures. It means that the hot plate ignition is largely affected by the diffusion processes of heat and mass, although it is fundamentally controlled by chemical reactions.

Table 1 shows the effect of the chemical conditions of plate surface-chemically deactivated or inert for active species on the ignition time. If the CH_3 radical is deactivated at the plate surface, the ignition time is about 3.5 times longer than that of the inert surface. This is due to cutting the main path of combustion reactons of methane by deactivating CH_3. Among active species such as H, O, OH and HO_2, II has the largest effect of deactivation on ignition delay.

3.2 Ignition by hot burned gas

As the burned gas injected from the hot porous plate into the combustible gas mixture of methane, two kinds of burned gas are used as shown in Table 2: the one is the burned gas of CH_4-air at the equivalence ratio of 1 and 1.4 (Table 2(a)) and the other is the burned gas of H_2-air at the equivalence ratio of 1 and 1.6 (Table 2(b)).

Figure 6 shows the time history of methane consumption rate $-\dot{m}$ and the influence of the injection gases on the ignition delay. In the figure, H_2-1 $(T_c=2310K)$ means that the burned gas of H_2-air at the equivalence ratio of 1 and the flame temperature of 2310K is injected from the porous plate and in the caption, $T_w = T_c$ means that the plate temperature is the same as the flame temperature T_c of injection gases. The mass flow rate $(\rho v)^*$ of injection gas is nondimensionalized by the main flow rate. At higher plate temperatures, the ignition delay time is more reduced regardless of the composition of injection gases.

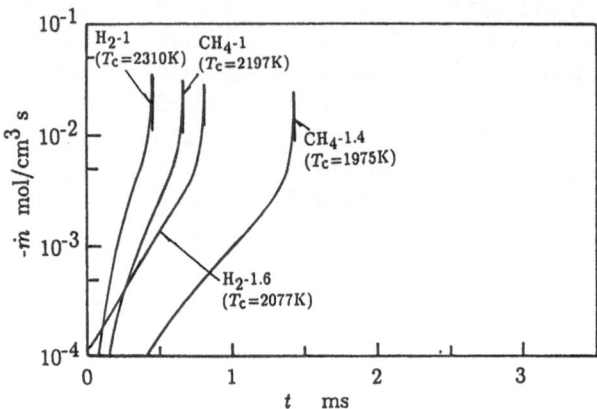

Fig.6   Effect of injection gas on ignition delay.
$T_w = T_c, u_\infty = 5$cm/s, $\phi = 1, p = 0.1$MPa, $D = 5$mm, $(\rho v)^* = 0.1$

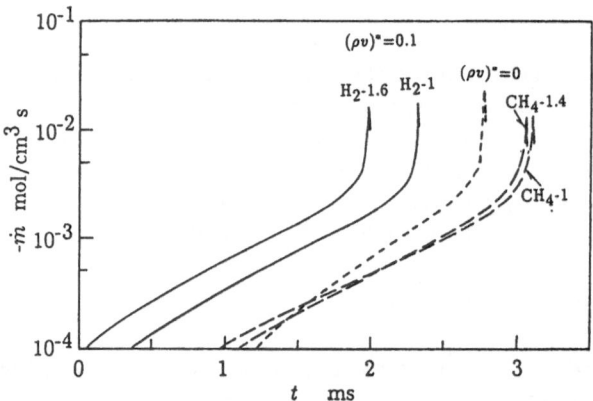

Fig.7   Effect of injection gas on ignition dealy.
$T_w = 1800$K, $u_\infty = 5$cm/s, $\phi = 1, p = 0.1$MPa, $D = 5$mm, $(\rho v)^* = 0.1$

Figure 7 also shows the time history of methane consumption rate. The composition of the injection gases is the same as that shown in Fig.6, but the plate temperature is kept at 1800K for all cases of the injection gases which take different flame temperatures. By injection of burned gases of H_2-air, the ignition time is more reduced than that without injection of burned gas $((\rho v)^* = 0)$

Table 3: Effect of active species on ignition delay time. $T_w=1800\text{K}, u_\infty=5\text{cm/s}, \phi=1, p=0.1\text{MPa}, D=5\text{mm}$

injection gas	$(\rho v)^*$	$t^*=t_{ig}/t_{ig0}$
no injection	0.0	1
1% H in N_2	0.1	0.89
1% O in N_2	0.1	0.87
1% OH in N_2	0.1	0.92
1% CH_3	0.1	0.92

at the same plate temperature, and more effectively for the injection of richer burned gases. However, the injection of CH_4-burned gases makes the ignition more delayed than that of no-injection of burned gas, and less effectively for the injection of leaner burned gases.

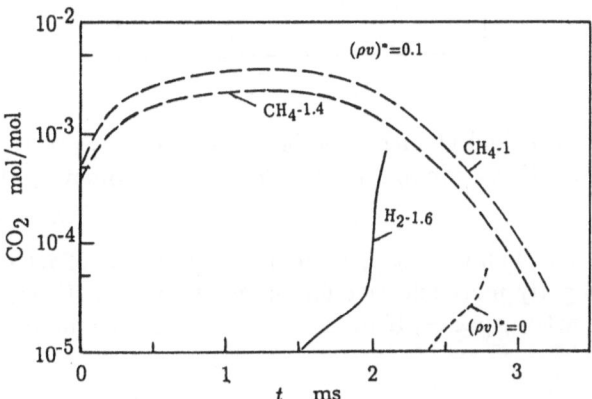

Fig.8   Time histories of CO_2 concentration at ignition point. $T_w=1800\text{K}, u_\infty=5\text{cm/s}, \phi=1, p=0.1\text{MPa}, D=5\text{mm}, (\rho v)^*=0.1$

Figure 8 shows the time history of CO_2 concentration at the ignition point. The plate temperature is 1800K. With increase in the gas temperature of combustible gas flow at the ignition point, CO_2 in the burned gas injected from the porous plate is dissociated to CO through the exothermic reaction of CO_2 by H, and reduces its concentration. Whereas, without injection of burned gas, CO_2 increases its concentration with increase in the gas temperature. The dissociation of CO_2 by injection of methane combustion gas causes the ignition time more delayed than that of H_2 combustion gas.

3.3 Ignition by active species

Table 3 shows the effect of the active species injected from a porous plate on the ignition delay time. At small concentrations of active species in N_2 gas, O species is the most effective and OH the lowest on the ignition. Figure 9 compares the ignition delay of O and H species with respect of the concentration of the species in N_2.

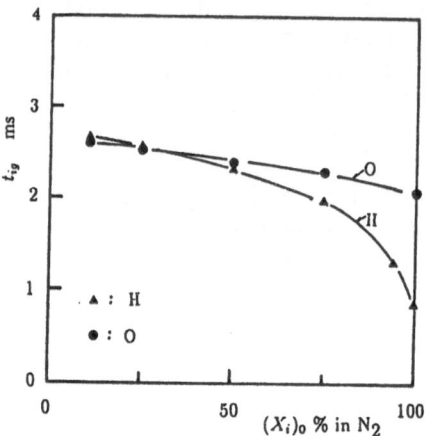

Fig.9   Effect of H and O species on ignition time.
$T_w=1800$K,$u_\infty=5$cm/s, $\phi=1$,$p=0.1$MPa, $D=5$mm, $(\rho v)^*=0.001$

As shown in Table 3 and in Fig.9, at small concentration of active species less than 25%, O is slightly more effective for the ignition than H, but at increased concentrations of active species, H reduced the ignitin delay more actively than O does.

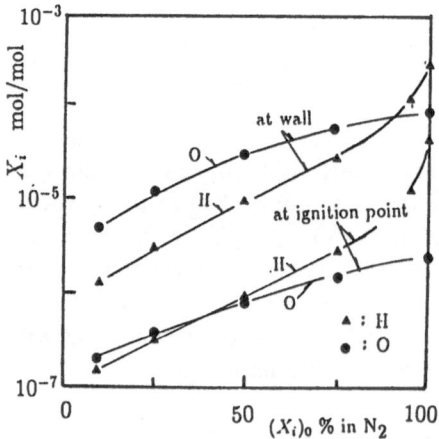

Fig.10   Concentration of active species at wall and ignition point.
$T_w=1800$K,$u_\infty=5$cm/s, $\phi=1$,$p=0.1$MPa, $D=5$mm, $(\rho v)^*=0.001$

Figure 10 shows the concentration of O or H at the wall and the ignition point at the initial stage of 0.4ms versus the concentration in the injection gas. At the initial stage of 0.4ms, the active species of O and H at the ignition point are mainly transfered by diffusion from the wall. At the wall, the concentration of H species is lower than that of O species for the injection gases including active species less than 80%, whereas at the ignition point, the concentration of H becomes higher than that of O for cases of the injection gas including active species higher than 30%. This causes for H to reduce more effectively the ignition time than O at relatively high concentrations of H species in the injection gas.

4 Conclusions

The flame igniiton process of premixed methane-air mixtures by a high temperature plate and a hot-burned gas injection is numerically studied in a two-dimensional laminar flow by considering 18 species with 61 elementary reactions.

- At the ignition point, the methane consumption rate initially attains a maximum value, then the heat release rate reaches its maximum and finally, the temperture takes a maximum.

- The ignition delay time is influenced by the chemical conditions of the surface, that is , whether the surface is inert or deactivating for active species. Among active species such as H, O, OH and HO_2, H has the greatest effect on the ignition.

- By injection of burned gases of H_2-air from the porous plate, the ignition time is more reduced than that without the injection at the same plate temperature, more effectively with the richer burned gases. Whereas, by injection of burned gases of CH_4, the ignition time is more delayed than that without the injection at the same plate temperature, less effectively with the leaner burned gases.

- At small concentrations of active species, O is the most effective for reduction of the ignition time, whereas at increased concentrations of active species, H tends to be the most effective.

References

[1] Sano,T. and Yamashita,A.,JSME Int.J.(B),vol.37-1,(1994), 180.
[2] Sano,T. Combust. Sci. Technol., vol.29, (1982), 261.
[3] Sano,T. and Kotake, S.,Numerical Methods in Thermal Problems, vol.5, (1987), 869. Pineridge.

6. Combustion Waves

Combustion Waves: Non Adiabatic

R.O. Weber[1] G.N. Mercer[1], B.F. Gray[2] and S.D. Watt[1]

[1] Department of Mathematics, University College UNSW
Australian Defence Force Academy, Canberra 2600, Australia

[2] Department of Applied Mathematics, University of Sydney
Sydney, NSW 2006, Australia

Abstract. A non-adiabatic thermal model of combustion is presented. The model admits ignition and extinction waves as exact solutions. The shape of these waves and their speed are determined numerically. Phase plane equations are used to understand the uniqueness of the wavespeeds and to determine watershed criteria for initial conditions.

Keywords. Combustion, wave speed, critial initial condition

1   Introduction

Combustion waves have been extensively studied for many years (Frank-Kamenetskii [1], Buckmaster [2]) This includes contributions to purely thermal theories (e.g. Gill *et al* [3]) and more recent asymptotic techniques (Bush and Fendell [4], Williams [5]). Surprisingly, most of the research has focussed on the adiabatic case where the infamous "cold boundary problem" always recurs. In the present paper we wish to present a two dimensional model of combustion in a solid layer of material and its reduction to a one dimensional model. The boundary conditions imposed by us on the 'upper' and 'lower' surfaces of the layer result in a volumetric heat loss term in the one dimensional model. There results multiple thermal equilibria due to the balance between exothermic heat release and volumetric heat loss. It then becomes possible for initial conditions to evolve into 'combustion waves' which effect transitions between different states of thermal equilibrium. For our one dimensional model this is possible in a mathematically precise sense, with the determination of wavespeeds and threshold initial conditions. The extension to include fuel consumption is discussed elsewhere (Mercer *et al* [6]).

2 One Dimensional Model

For a two dimensional layer of combustible material a simple model without reactant consumption can be derived from the conservation of energy, assuming heat conduction according to Fourier's Law and a reaction rate according to Arrhenius's Law. In non-dimensional coordinates (see e.g. Frank-Kamenetskii [1])

$$\frac{\partial \theta}{\partial t} = \frac{\partial^2 \theta}{\partial x^2} + \frac{\partial^2 \theta}{\partial y^2} + \delta e^{\theta/1+\epsilon\theta} \tag{1}$$

Here θ is the non-dimensional temperature, δ is the non-dimensional energy release rate, and ϵ is a non-dimensional parameter related to the activation energy and the ambient temperature. Spatial coordinates x and y are non-dimensionalised such that $x \in (-\infty, \infty)$ and $y \in [0, 1]$. The time coordinate t is also non-dimensionalised and $t \in [0, \infty)$. To solve the partial differential equation (1), initial and boundary conditions must also be specified. Here we consider linear heat loss from the upper boundary,

$$\frac{\partial \theta}{\partial y}(x, 1, t) = -h\theta(x, 1, t); \tag{2}$$

an insulated lower boundary,

$$\frac{\partial \theta}{\partial y}(x, 0, t) = 0; \tag{3}$$

temperatures to be prescribed as $x \to \pm\infty$ and an arbitrary initial condition,

$$\theta(x, y, 0) = f(x, y). \tag{4}$$

As shown by Watt *et al* [7], the two dimensional equation (1) with the boundary conditions (2) and (3) can be systematically reduced to a one dimensional equation for the average layer temperature

$$\bar{\theta}(x, t) = \int_0^1 \theta(x, y, t)\, dy \tag{5}$$

Dropping the bar over $\bar{\theta}(x, t)$, the one dimensional equation is (to first order in h)

$$\frac{\partial \theta}{\partial t} = \frac{\partial^2 \theta}{\partial x^2} + \delta e^{\theta/1+\epsilon\theta} - h\theta. \tag{6}$$

Note that the coefficient of the heat loss term in equation (6) is $-h$. To first order this is the first eigenvalue of the associated linear boundary value problem

$$\varphi'' + \lambda\varphi = 0, \quad \varphi'(0) = 0, \quad \varphi'(1) = -h\varphi(1). \tag{7}$$

An instructive, independent way to derive equation (6) is from first principles for a one dimensional region subject to heat transfer by conduction, volumetric heat generation due to combustion and volumetric heat loss. The initial condition to be satisfied in conjunction with equation (6) is found by substituting (4) into (5):

$$\theta(x,0) = \int_0^1 f(x,y)\,dy \tag{8}$$

The results we shall obtain for equation (6) in the following sections are to be compared with those obtained by Yang [8], Gray and Kordylewski [9], [10], Adler [11] and Buounincontri and Hagstrom [12]. It is then evident that analysing the one dimensional model, obtained by dimensional reduction, is much simpler than working with the two dimensional model and provides considerable insight into nonadiabatic combustion waves.

3   The Reaction Function $\delta e^{\theta/1+\varepsilon\theta} - \theta$

A simple rescaling of x and t in equation (6) yields the compact form of the governing equation

$$\frac{\partial\theta}{\partial t} = \frac{\partial^2\theta}{\partial x^2} + \delta e^{\theta/1+\varepsilon\theta} - \theta \tag{9}$$

Note that δ in this equation has a factor of $\frac{1}{h}$ absorbed into it.

The reaction function in this one-dimensional model is

$$F(\theta) = \delta e^{\theta/1+\varepsilon\theta} - \theta, \tag{10}$$

For $\varepsilon \in (0, 0.25)$ there exists a minimum and a maximum value of δ (δ_{min} and δ_{max}) so that for each $\delta \in (\delta_{min}, \delta_{max})$ the reaction function $F(\theta)$ has exactly three zeros. For ε and δ outside these ranges, $F(\theta)$ has two, one or no zeros. This is demonstrated in Figure 1 where the solutions of $F(\theta) = 0$ are shown as a function of δ with ε held fixed at 0.15. The 'S-shaped' curve, very familiar from studies of bifurcation in combustion, implies that there exists an intermediate/impotent value of δ, δ_{imp}, for which $\int_{\theta_1}^{\theta_3} F(\theta)\,d\theta = 0$. Here θ_1 refers to the lower θ value found when solving $F(\theta) = 0$ at fixed ε and δ, while θ_3 refers to the upper θ value. In the following table, the values of $\delta_{min}, \delta_{imp}$ and δ_{max} are given for representative ε values. (Corresponding θ values can be found using $F(\theta) = 0$.)

Because of the 'lop-sided' nature of the cubic-like $F(\theta)$, the δ_{imp} value is always quite close to δ_{min}. The values for $\varepsilon = 0$ and $\varepsilon = 0.25$ are included for reference.

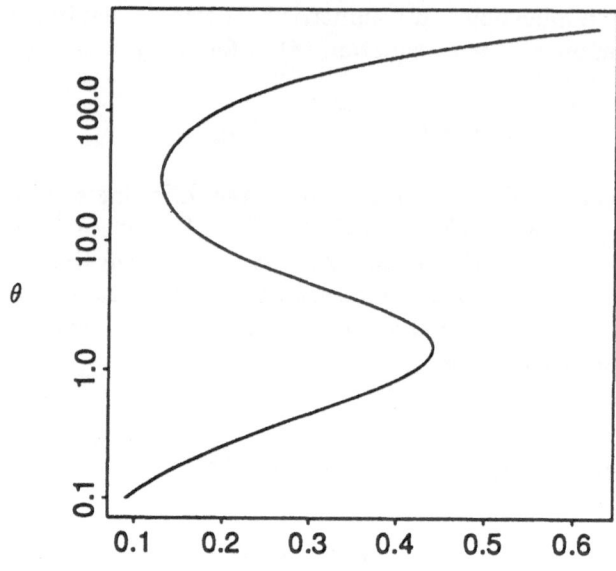

Figure 1. *The 'S-shaped' bifurcation diagram for the reaction function $\delta e^{\theta/1+\epsilon\theta} - \theta$ for $\epsilon = 0.15$.*

ϵ	δ_{min}	δ_{imp}	δ_{max}
0	0	-	0.36788
0.05	2.1260×10^{-6}	2.3316×10^{-6}	0.38780
0.10	0.011027	0.011927	0.41153
0.15	0.12830	0.13619	0.44086
0.20	0.35128	0.36359	0.47953
0.25	0.54134	0.54134	0.54134

Table 1: *Significant values of δ for the reaction function $\delta e^{\theta/1+\epsilon\theta} - \theta$, for representative values of ϵ.*

In summary, it is clear that for $\epsilon \in (0, 0.25)$ and $\delta \in (\delta_{min}, \delta_{max})$, there are three solutions of $F(\theta) = 0$ and hence three uniform equilibria of the partial differential equation (9). A simple linearised stability analysis reveals that of these three equilibria, the intermediate one, θ_2, is unstable; while the lower one, θ_1, and upper one, θ_3, are both stable.

4   Combustion Waves

Transitions between equilibria can occur. Physically only stable → stable transitions are expected, although it would be possible to begin with an unstable equilibrium as the initial state in a numerical experiment. Here we are particularly interested in wavelike structures which result in transitions between equilibria. Therefore, begin by defining $z = x - ct$ and assuming

$$\theta(x, t) = \theta(z) = \theta(x - ct), \qquad (11)$$

where c is the (assumed) constant speed of a wave.

Substituting this into equation (9) yields the ordinary differential equation

$$\theta'' + c\theta' + F(\theta) = 0 \qquad (12)$$

Multiplying this equation by θ' and integrating over z from $-\infty$ to ∞, one obtains an equation for the speed:

$$c = \frac{\int_{\theta_1}^{\theta_3} F(\theta)\, d\theta}{\int_{-\infty}^{\infty} (\theta')^2\, dz} \qquad (13)$$

In writing down equation(13) it was assumed that $\theta(z) \to \theta_3$ as $z \to -\infty$ and $\theta(z) \to \theta_1$ as $z \to +\infty$. For different equilibria as $z \to \pm\infty$ one simply substitutes other values in the integration limits in the numerator.

Notice that the denominator in equation (13) is positive definite. Hence the direction of the wave is determined by the area under the reaction function between the upper and lower equilibria. Refering to the earlier table it is apparent that for any fixed $\varepsilon \in (0, 0.25)$, an "ignition" wave will occur for $\delta \in (\delta_{imp}, \delta_{max})$ and an "extinction" wave will occur for $\delta \in (\delta_{min}, \delta_{imp})$. For $\delta = \delta_{imp}$ there is a stationary wave structure. As δ_{imp} is close to δ_{min}, for nearly all of the relevant parameter range "ignition" waves are the only stable to stable transition that occurs. That is, a wave propagates into the lower/colder equilibrium, θ_1, leaving behind the upper/hotter equilibrium, θ_3.

5   Phase Plane Analysis

Further analysis of the behaviour of solutions to equation (12) can be performed in the phase plane. That is, let $p = \theta'$. Then equation (12) is equivalent to the autonomous system of first order ordinary differential equations

$$\theta' = p \tag{14}$$

$$p' = -cp - F(\theta) \tag{15}$$

The critical points of this system are $(\theta, p) = (\theta_1, 0), (\theta_2, 0), (\theta_3, 0)$. For any given $\varepsilon \in (0, 0.25)$ and $\delta \in (\delta_{min}, \delta_{max})$ one can prove that there exists a unique trajectory from the saddle at $(\theta_3, 0)$ to the saddle at $(\theta_1, 0)$ with a uniquely determined speed c. For $\delta \in (\delta_{min}, \delta_{imp})$ the uniquely determined speed is negative $(c < 0)$, implying an "extinction" wave. For $\delta \in (\delta_{imp}, \delta_{max})$ the uniquely determined speed is positive $(c > 0)$, implying an "ignition" wave. The proof of uniqueness of c for general cubic-like reaction functions can be found in the literature (see e.g. Grindrod [13] pp.34-49).

Of greater interest than merely verifying our observations in section 4, is the ability to use the phase plane equations (14 and 15) to enumerate all possible waves and/or spatial structures. The usual (linearised) classification of the critical point $(\theta_2, 0)$ reveals it to be either a spiral, a centre (when $c = 0$), or a node. Furthermore, for the ε and δ range where ignition/extinction waves occur there always exists associated waves from θ_2 to θ_1 or to θ_3.

As these associated waves are represented in the phase plane as a trajectory from a spiral/centre/node to a saddle there is a range of speeds for which they can occur, rather than a uniquely determined speed. This is similar to the standard analysis for the Fisher-Kolmogorov equation (see e.g. [13] p.39). It varies from the analysis for the Fisher-Kolmogorov equation in that oscillatory behaviour around $(\theta_2, 0)$ can occur without violating any physically sensible condition. Therefore it is (numerically) possible to construct waves from θ_2 to θ_1 or θ_3 which have temperature oscillations, although it is unlikely that these could ever be experimentally observed.

6 Wave Speed

Ultimately, the wave speed and its dependence upon δ and ε must be determined by numerical methods. Both explicit and implicit schemes have been used with satisfactory results, see e.g. Tang *et al* [14]. A Crank-Nicholson scheme including a variable time step, was used by Mercer and Weber [15] to accurately determine the shape of the wavefront. An example of an ignition wavefront evolving from an initial condition is given for $\varepsilon = 0.15$ and $\delta = 0.2$ in figure 2. The final speed is $c = 0.9516$. The calculations can be carried out for the whole range of δ applicable to any particular choice of ε. The results for $\varepsilon = 0.05, \varepsilon = 0.10$ and $\varepsilon = 0.15$ were found by Mercer and Weber [15] and are reproduced in Figure 3.

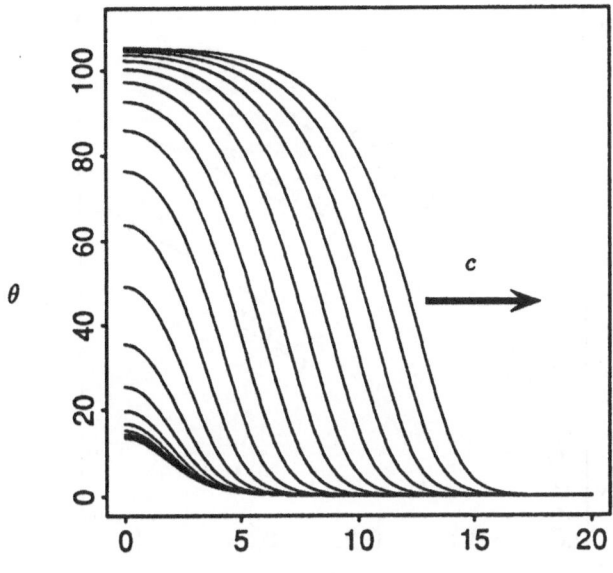

Figure 2. *The evolution of an initial condition into an ignition wavefront for $\varepsilon = 0.15$ and $\delta = 0.2$. The final speed of the wavefront is 0.9516. The initial condition used is actually the watershed of section 7 plus a tiny Gaussian.*

In the small δ range where an extinction wave exists, $\delta \in (\delta_{min}, \delta_{imp})$, Mercer and Weber [15] have computed the wavefront and speed. As an example, for $\varepsilon = 0.15$ and $\delta = 0.13$ the speed is found to be $c = -0.2698$. The negative value indicates that the wave is moving from right to left.

7   Watershed Initial Condition

There exists a stationary, non-uniform equilibrium with boundary conditions $\theta(x) \to \theta_1$ as $x \to \pm\infty$. This is evident from the phase plane equations (14 and 15) with $c = 0$:

$$\theta' = p \tag{16}$$
$$p' = \delta e^{\theta/1+\varepsilon\theta} - \theta \tag{17}$$

The trajectories are now symmetric under $p \to -p$, $x \to -x$, and there exists a unique homoclinic orbit on the critical point $(\theta_1, 0)$ for each $\varepsilon \in (0, 0.25)$ and $\delta \in (\delta_{imp}, \delta_{max})$. It is conjectured that this stationary, non-uniform equilibrium is a "watershed" for initial conditions used in the solution of the time dependent partial differential equation (9). See also Fife [16] Theorem 4.9 p.94.

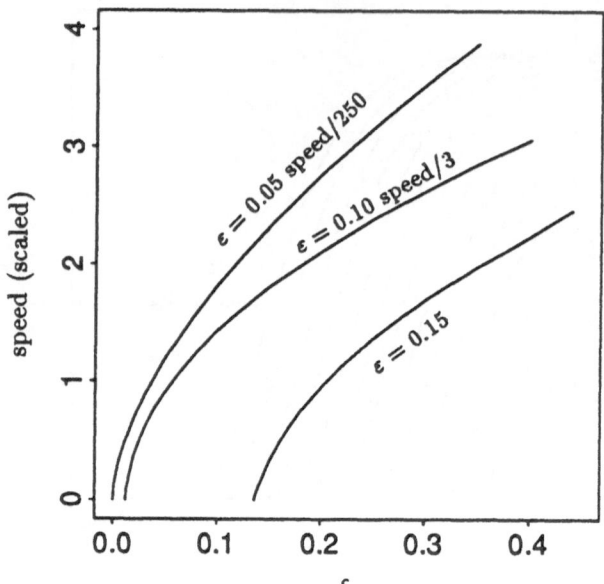

Figure 3. *Ignition wave speed dependence upon δ for ε = 0.05, 0.10 and 0.15. Note that the curve for ε = 0.05 is divided by 250 and the curve for ε = 0.10 is divided by 3 in order to fit all the curves onto one diagram.*

This watershed conjecture has been tested numerically over the ε, δ range and found to be correct. Examples showing the evolution from this non-uniform equilibrium for ε = 0.15 and δ = 0.2 are shown in Figures 2 and 4. The addition of any small positive perturbation results in the development of an ignition wave. The subtraction of any small positive perturbution results in the decay to the unifrom equilibrium θ = θ₁.

Further evidence in support of this wateshed conjecture can be obtained by a linear stability analysis. One begins by denoting the stationary, non-uniform equilibrium by $\theta_1(x)$ and writing

$$\theta(x,t) = \theta_1(x) + e^{-\lambda t}u(x). \tag{18}$$

As $x \to \pm\infty$, $\theta(x,t) \to \theta_1$ and $\theta_1(x) \to \theta_1$. Hence $u(x) \to 0$ as $x \to \pm\infty$. Substituting this ansatz, (18), into the partial differential equation (9) and linearising in $u(x)$ gives

$$u'' + \left[\lambda + \frac{\delta e^{\theta_1(x)}}{(1 + \varepsilon\theta_1(x))^2} - 1\right] u = 0 \tag{19}$$

subject to

$$u(\pm\infty) = 0. \tag{20}$$

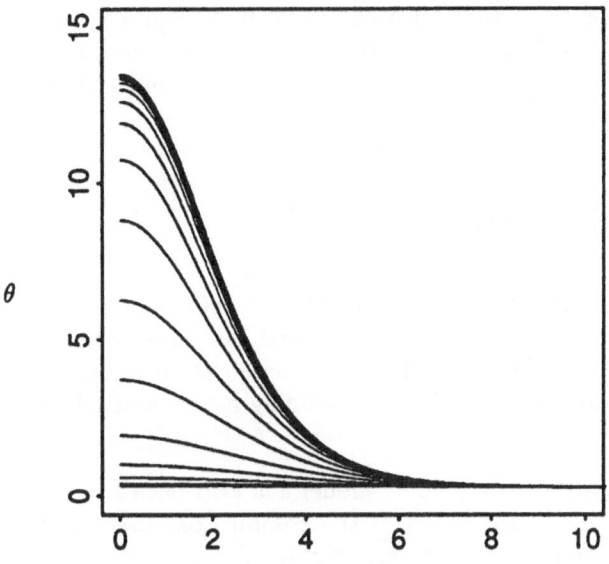

Figure 4. *Evolution of the watershed initial condition minus a tiny Gaussian into the uniform steady state $\theta(x,t) = \theta_1$. Compare with Figure 2.*

This linear boundary value problem for λ and $u(x)$ determines the local, linearised stability of $\theta_1(x)$. Indeed, one is only required to find the sign of the eigenvalues λ to determine the stability. As table 2 shows, the lowest (non-zero) eigenvalue is negative for all parameter values tested. Hence the stationary, non-uniform equilibrium $\theta_1(x)$ does indeed appear to be a watershed for initial conditions.

ε	δ	λ_{min}
0.10	0.2	-1.6393
0.10	0.25	-1.3759
0.10	0.3	-1.0823
0.15	0.2	-0.6102
0.15	0.25	-0.7254
0.15	0.3	-0.7199

Table 2. *The lowest eigenvalue of equations (19 and 20), determining the instability of the watershed $\theta_1(x)$.*

It should also be noted that for each $\varepsilon \in (0, 0.25)$ and $\delta \in (\delta_{min}, \delta_{imp})$, there exists a unique homoclinic orbit on $(\theta_3, 0)$. This stationary, non-uniform equilibrium is likely to be a "watershed" for initial conditions that may evolve into an extinction wave.

8 Concluding Remarks

We have described a non-adiabatic, thermal model of combustion which possesses exact travelling waves as solutions. The model can be obtained by dimensional reduction from a two dimensional energy conservation equation, or it can be obtained directly in one dimension by including a volumetric heat loss.

The adiabatic limit of the present model has also been considered. It seems that (formally at least) the speed of the combustion waves will then become infinite as the model is not well formulated for this extreme case.

The extension of the present model to include reactant consumption is being considered by Mercer *et al* [6]. It seems that a natural definition of flame thickness and extinction may emerge. Further extension to multiple species and porous media would also be of merit.

9 Acknowledgements

The Australian Research Council provided partial funding for this work. The organisers and participants of the workshop provided helpful feedback and encouragement.

10 References

1. D.A. Frank-Kamenetskii (1969) *Diffusion and Heat Transfer in Chemical Kinetics*, 2nd ed. Plenum Press, New York.
2. J. Buckmaster (1986) The Contribution of Asymptotics to Combustion. *Physica* **20D** pp. 91-108.
3. W. Gill, A.R. Shouman and A.B. Donaldson (1984) The Frank-Kamenetskii Problem Revisited, Part III: The Moving Flame Front. *Combust. Flame* **55** pp. 105-115.
4. W.B. Bush and F.E. Fendell (1970) Asymptotic Analysis of Laminar Flame Propagation for General Lewis Numbers. *Combust. Sci. Tech.* **1** pp. 421-428.

5. F.A. Williams (1986) Lectures on Applied Mathematics in Combustion. *Physica* **20D** pp. 21-34.

6. G.N. Mercer, R.O. Weber, B.F. Gray and S.D. Watt (1994) Combustion pseudo-waves in a system with reactant consumption and heat loss. University College UNSW preprint, to appear.

7. S.D. Watt, A.J. Roberts and R.O. Weber (1994) Dimensional Reduction of a Bushfire Model. *Math. Comput. Modelling*, to appear.

8. C.H. Yang (1961) Burning Velocity and the Structure of Flames near Extinction Limits. *Combust. Flame* **5** pp. 163-174.

9. P. Gray and W. Kordylewski (1985) Standing Waves in Exothermic Systems. *Proc. Royal Soc. Lond.* **A398** pp. 281-288.

10. P. Gray and W. Kordylewski (1988) Travelling Waves in Exothermic Systems. *Proc. Royal Soc. Lond.* **A416** pp. 103-113.

11. J. Adler (1988) The Minimum Local Temperature Disturbance for Ignition in a Slab. *Proc. Royal Soc. Lond.* **A417** pp. 245-254.

12. S. Buonincontri and T. Hagstrom (1989) Multidimensional Travelling Wave Solutiions to Reaction-Diffusion Equations. *IMA J. Applied Math.* **43** pp. 261-271.

13. P. Grindrod (1991) *Patterns and Waves*. Clarendon Press, Oxford.

14. S. Tang, S. Qin and R.O. Weber (1993) Numerical Studies on 2-Dimensional Reaction-Diffusion Equations. *J. Austral. Math. Soc.* **B35** pp. 223-243.

15. G.N. Mercer, R.O. Weber, B.F. Gray and S.D. Watt (1994) Combustion pseudo-waves in a system with reactant consumption and heat loss. University College UNSW preprint, to appear.

16. P.C. Fife (1979) Mathematical Aspects of Reacting and Diffusing Systems. Lecture Notes in Biomathematics **28** Springer-Verlag.

DETERMINATION OF LAMINAR FLAME SPEEDS FROM NOZZLE-GENERATED COUNTERFLOW FLAMES

B. H. Chao[1] and F. N. Egolfopoulos[2]

[1] Department of Mechanical Engineering, University of Hawaii, Honolulu, HI 96822, USA
[2] Department of Mechanical Engineering, University of Southern California, Los Angeles, CA 90089, USA

Abstract – The accuracy of the counterflow, twin-flame technique for the determination of laminar flame speeds was examined analytically and numerically. The analysis was conducted by using multiple-expansion, high activation energy asymptotics while the numerical simulation incorporated detailed chemistry and transport. In both approaches the solutions were obtained in a finite domain and with plug flow boundary conditions in order to better simulate the actual experiments. Results show that linear extrapolation of the minimum velocity to zero stretch over-estimates the true laminar flame speed. This over-estimate, however, can be reduced by using smaller ratios of the flame thickness to the nozzle separation distance. Numerical results indicate that for typical paraffin/air mixtures, nozzle separation distances of the order of 14 to 22 mm yield laminar flame speeds accurate to within the uncertainty range of the experiment. The results obtained herein thus provide further support for the viability of the counterflow technique, when the influence of the nozzle separation distance is properly accounted for. An alternate technique for the determination of laminar flame speeds, based on the variation of flow velocity at a constant temperature near the upstream boundary of the flame with stretch, suggest that the over-estimation by linear extrapolating to zero stretch is smaller compared to the minimum velocity approach.

1. Introduction

In the study of premixed combustion phenomena, the laminar flame speed, which is the steady propagation speed of a one-dimensional, planar, adiabatic, flame in a given mixture of doubly-infinite extent, is of particular fundamental and practical interest. This global property, indicating the diffusivity, reactivity, and exothermicity of the mixture, is frequently adopted as the referencing parameter to quantify many premixed flame characteristics. The knowledge of the laminar flame speed is also essential in the modeling of more complex combustion processes such as turbulent flames, and the design of practical devices. Recognizing the fundamental significance of the laminar flame speed, its determination has been the subject of numerous investigations. A major difficulty in the determination arises from the fact that the conditions required to define the laminar flame speed cannot be readily achieved in the laboratory. Furthermore, since the flame has a certain thickness, there is considerable uncertainty in defining the flamefront at which the flame speed is to be evaluated.

Over the years, the major experimental techniques used to determine the laminar flame speeds have been those of the Bunsen burner, flat-flame burner, soap-bubble, and spherical-bomb. The data from these methods, however, appear to have a large scatter [1] which is beyond the experimental uncertainty of each technique. It is consequently recognized that in most of these techniques the flames do not satisfy the one-dimensional requirement, and hence are subjected to aerodynamic stretch which affect their propagation [2,3]. For example, while in a Bunsen flame the flame suffers negative stretch due to the compressive nature of its curved surface, an outwardly-propagating flame in the soap-bubble and spherical bomb methods suffers positive stretch. There are also additional limitations associated with each technique. For example, the Bunsen flame is not adiabatic close to the nozzle rim and is also not one-dimensional especially in the tip region. For the soap-bubble and spherical-bomb methods, the flame shape can be significantly distorted from spherical symmetry due to buoyancy and severely wrinkled due to the intrinsic flame-front instability. The flames obtained by using the flat-flame burner are one-dimensional but are inherently nonadiabatic because they are stabilized through heat loss to the burner surface. Although data on laminar flame speeds using the flat-flame burner have been reported by extrapolating to zero heat loss [4], there are still concerns about the possibility of radical loss to the burner surface which is not accountable at present.

Recently, Law and co-workers [5–9] proposed a new method to determine the laminar flame speed by measuring the minimum upstream speed of the symmetric counterflow stretched flame. As shown in Fig. 1, in this method, two identical axisymmetric nozzles are placed facing each other with a separation distance $2L$. Combustible premixtures having the same concentration, discharge velocity u_L and temperature T_L emerge from both nozzles to yield a symmetric counterflow so that two identical stationary flames exist. Because of symmetry, the downstream, burned region between the two flames is adiabatic. Moreover, since L is much larger than the characteristic flame thickness, there is no heat loss to the burners. Thus the only nonadiabatic source is the intrinsic radiative loss which, however, affects minimally the flame response with the exception of near-limit conditions. Due to thermal expansion, the axial flow velocity is increased after passing

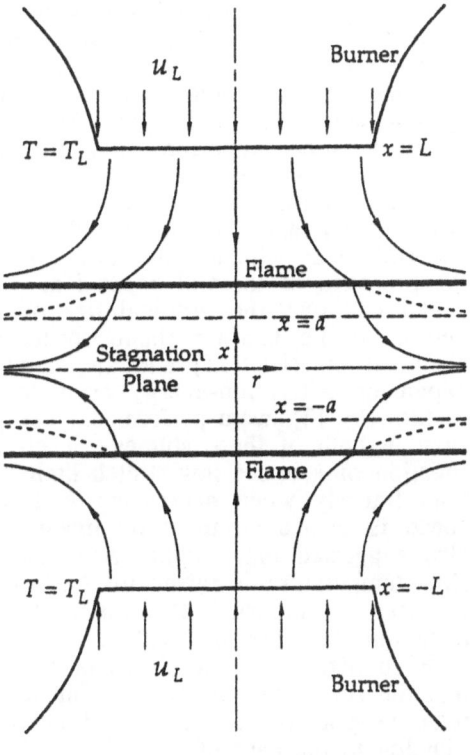

Figure 1 Schematic representation of the system.

through the flames so that there is a discontinuity in the gradient of the streamlines. The dashed curves denote the streamlines if there is no thermal expansion. The broken lines at $x = \pm a$ are then the apparent stagnation planes when observed from the upstream of the flames.

On each side of the stagnation plane, the flow velocity decreases from the burner toward the flame, attaining a local minimum near the leading edge of the flame. It subsequently increases in the flame region because of thermal expansion. A local maximum is then reached near the downstream boundary of the flame, beyond which the velocity decreases again to approach zero at the stagnation plane. The local minimum velocity in front of the flame along the centerline, U_{min} is measured using LDV as a function of the local strain rate k, which is the axial velocity gradient immediately upstream of the minimum velocity point. The quantity U_{min} is adopted as a reference (upstream) flame speed which decreases with decreasing k except for very large Le flames. Theoretically, by continuously decreasing k, the flame behaves more like a laminar flame; and in the non-strained limit, the upstream velocity gradient approaches zero so that U_{min} becomes the upstream laminar flame speed. The measured value of U_{min} is then plotted versus k to yield the laminar flame speed. However, there is a minimum value of k below which flashback occurs. Thus the laminar flame speed can not be actually measured.

Experimental data obtained from the low stretch rate limit indicates that U_{min} varies quite linearly with k and, therefore, by linearly extrapolating to the zero stretch rate the laminar flame speed is determined. This is further supported by the constant density theory [10] which suggests that the upstream flame speed is a linear function of the stretch rate when the stretch rate is low. The advantages of this method are that the measured data are accurate and easily obtainable and that the effect of stretch is systematically subtracted out. However, it is not certain whether U_{min} is really a linear function of k for small k. Moreover, although only quantitative difference is observed between the constant density theory and the more realistic, variable density theory, for moderate stretch rates (Buckmaster and Mikolaitis, 1982; Libby and Williams, 1984), there is uncertainty as whether the former is applicable in the low stretch limit in the presence of thermal expansion.

To examine the applicability of the linear extrapolation approach to determine the laminar flame speed, a numerical study by using detailed chemical kinetics [11] and an analytical study [12] by using a multiple-expansion activation-energy asymptotics, first introduced by Matalon and Matkowsky [2], were performed. The potential flow was adopted as the base flow in both of these studies. Their results show that U_{min} is a nonlinear function of k in the low stretch limit. Because U_{min} decreases more rapidly than linearly when decreasing k, it was concluded that the laminar flame speed is over-determined by linear extrapolation. Tien and Matalon [12] also suggested that another reference point, namely the location where the flow temperature is raised by 1% of the overall temperature increase, can be used to measure the flame speed. The same linear extrapolation technique can then be used to determine the laminar flame speed.

Recognizing the usefulness of the results of Tien and Matalon [12], and that the flow field in the experiment resembles more the plug flow (Fig. 1) than the potential flow, it is of interest to extend the analysis of Tien and Matalon to the case of plug flows. There are additional motivations for this study. First, because the potential flow is created by a point source at infinitely faraway from the stagnation plane, there is no characteristic hydrodynamic length scale and hence the effect of burner separation distance cannot be obtained. Furthermore, Dixon-Lewis [13] studied the problem in a finite domain by using the plug flow conditions and detailed chemistry and transport. It was found that linear extrapolation of the minimum velocity to zero stretch indeed yields a higher flame speed. However, the results also show that the reference flame speed varies linearly with the stretch rate, and the nozzle separation distance has no effect on the stretch rate dependence of the reference flame speed. If the latter were correct, then there is a fundamental limitation of the counterflow technique which can not be improved.

In view of the above considerations, we have conducted an analytical and numerical study on nozzle-generated counterflow flame, aiming to gain a thorough understanding of the symmetric counterflow premixed flame propagation in the low stretch limit, and the applicability of the counterflow method in the determination of laminar flame speed. In the analytical investigation, the multiple-expansion activation-energy asymptotics similar to that used by Tien and Matalon [12] is adopted. The numerical investigation is performed by using detailed chemistry and transport similar to that of Dixon-Lewis [13]. For both the analytical and numerical investigations, the problem was solved in a finite domain with plug-flow boundary conditions to better describe the experimental conditions. The results are then compared and applicability of the method is assessed.

2. Formulation

The system to be analyzed is the twin premixed flames created by impinging two identical premixture streams into each other (Fig. 1). The burners with separation distance $2L$ discharge steady, uniform flows of speed u_L along the axial direction so that there is no initial transverse velocity at the burner exits. The flames are situated at where the propagation speed dynamically balances the flow speed. Due to symmetry at the stagnation plane, only half of the domain needs to be analyzed. Because the flow speed is small compared to the sonic speed, constant pressure and negligible energy dissipation are assumed in the energy equation. A single-reactant, first-order, one-step Arrhenius reaction is assumed.

The nondimensional equations governing the overall mass conservation, momentum conservation in both the axial direction x and radial direction r, energy conservation, and species conservation are given in the cylindrical coordinate system, with the origin at the stagnation point, by

$$\frac{\partial}{\partial \tilde{x}}(\tilde{\rho}\tilde{u}) + \frac{1}{\tilde{r}}\frac{\partial}{\partial \tilde{r}}(\tilde{r}\tilde{\rho}\tilde{v}) = 0 , \tag{1}$$

$$\tilde{\rho}\tilde{u}\frac{\partial \tilde{u}}{\partial \tilde{x}} + \tilde{\rho}\tilde{v}\frac{\partial \tilde{u}}{\partial \tilde{r}} - \delta Pr\left\{\frac{4}{3}\frac{\partial^2 \tilde{u}}{\partial \tilde{x}^2} + \frac{1}{\tilde{r}}\frac{\partial}{\partial \tilde{r}}\left[\tilde{r}\left(\frac{\partial \tilde{u}}{\partial \tilde{r}} + \frac{1}{3}\frac{\partial \tilde{v}}{\partial \tilde{x}}\right)\right]\right\} = -\frac{\partial \tilde{p}}{\partial \tilde{x}} , \tag{2}$$

$$\tilde{\rho}\tilde{u}\frac{\partial \tilde{v}}{\partial \tilde{x}} + \tilde{\rho}\tilde{v}\frac{\partial \tilde{v}}{\partial \tilde{r}} - \delta Pr\left\{\frac{\partial^2 \tilde{v}}{\partial \tilde{x}^2} + \frac{4}{3}\frac{\partial}{\partial \tilde{r}}\left[\frac{1}{\tilde{r}}\frac{\partial}{\partial \tilde{r}}(\tilde{r}\tilde{v})\right] + \frac{1}{3}\frac{\partial^2 \tilde{u}}{\partial \tilde{x}\partial \tilde{r}}\right\} = -\frac{\partial \tilde{p}}{\partial \tilde{r}} , \tag{3}$$

$$\tilde{\rho}\tilde{u}\frac{\partial \tilde{T}}{\partial \tilde{x}} + \tilde{\rho}\tilde{v}\frac{\partial \tilde{T}}{\partial \tilde{r}} - \delta\left[\frac{\partial^2 \tilde{T}}{\partial \tilde{x}^2} + \frac{1}{\tilde{r}}\frac{\partial}{\partial \tilde{r}}\left(\tilde{r}\frac{\partial \tilde{T}}{\partial \tilde{r}}\right)\right] = \frac{Da}{\delta}\tilde{\rho}\tilde{Y}\exp\left(-\frac{\tilde{E}}{\tilde{T}}\right) , \tag{4}$$

$$\tilde{\rho}\tilde{u}\frac{\partial \tilde{Y}}{\partial \tilde{x}} + \tilde{\rho}\tilde{v}\frac{\partial \tilde{Y}}{\partial \tilde{r}} - \frac{\delta}{Le}\left[\frac{\partial^2 \tilde{Y}}{\partial \tilde{x}^2} + \frac{1}{\tilde{r}}\frac{\partial}{\partial \tilde{r}}\left(\tilde{r}\frac{\partial \tilde{Y}}{\partial \tilde{r}}\right)\right] = -\frac{Da}{\delta}\tilde{\rho}\tilde{Y}\exp\left(-\frac{\tilde{E}}{\tilde{T}}\right) , \tag{5}$$

with the equation of state being $\tilde{\rho}\tilde{T} = \tilde{T}_L$ and the boundary conditions

$$\tilde{x} = 0 \quad : \quad \tilde{u} = 0 \;, \quad \partial \tilde{T}/\partial \tilde{x} = 0 \;, \quad \partial \tilde{Y}/\partial \tilde{x} = 0 \;; \tag{6}$$

$$\tilde{x} = -1 \quad : \quad \tilde{u} = \tilde{u}_L \;, \quad \tilde{T} = \tilde{T}_L \;, \quad \tilde{Y} = 1 \;, \quad \tilde{v} = 0 \;, \quad \tilde{p}(\tilde{r} = 0) = 0 \;. \tag{7}$$

The various nondimensional quantities are given by

$$\tilde{Y} = \frac{Y}{Y_L} \;, \quad \tilde{x} = \frac{x}{L} \;, \quad \tilde{r} = \frac{r}{L} \;, \quad \tilde{u} = \frac{u}{s^{\circ}} \;, \quad \tilde{v} = \frac{v}{s^{\circ}} \;, \quad \tilde{\rho} = \frac{\rho}{\rho_L} \;, \quad \tilde{p} = \frac{p}{\rho_L R T_L} \;,$$

$$\tilde{T} = T \Big/ \left(\frac{q Y_L}{c_p} \right) \;, \quad \tilde{E} = E \Big/ \left(\frac{q Y_L}{c_p} \right) \;, \quad \delta = \left(\frac{\lambda}{\rho_L c_p s^{\circ}} \right) \Big/ L \;, \quad \varepsilon = \frac{\tilde{T}_{ad}^2}{\tilde{E}} \;,$$

$$Le = \frac{\lambda}{\rho D c_p} \;, \quad Pr = \frac{\mu c_p}{\lambda} \;, \quad Da = \frac{B \lambda}{\rho_L c_p (s^{\circ})^2} \;.$$

where T is the temperature, Y the mass fraction of the controlling species, q the heat of combustion per unit mass of this species, u the axial flow velocity, v the transverse flow velocity, s° the laminar flame speed, ρ the density, c_p the constant pressure specific heat, p the pressure perturbed from its constant leading order value ($\rho_L R T_L$), λ the thermal conductivity, D the mass diffusivity, μ the viscosity, B the pre-exponential factor, R the gas constant, E the activation temperature, $T_{ad} = 1 + T_L$ the adiabatic flame temperature, Le the Lewis number considered close to unity, Pr the Prandtl number, Da the Damköhler number, and λ, μ, c_p and ρD are considered constants. The characteristic flame thickness $\lambda/(\rho_L c_p s^{\circ})$ is considered, realistically, to be much smaller than L so that $\delta \ll 1$. Moreover, the characteristic reaction region is much thinner than the characteristic flame thickness because of the high activation energy and hence $\varepsilon \ll 1$.

Solving the system [14] by a multiple-expansion analysis that was first developed by Matalon and Matkowsky [2] yields the solution of the flame temperature Θ^- and axial flow velocity U^- in the preheat region, given by

$$\tilde{T}_f = \tilde{T}_{ad} - \delta \, \varepsilon [\beta \ell \tilde{T}_L \, I(-\infty) + O(\varepsilon)] + O(\delta^2) \;, \tag{8}$$

$$\Theta^- = \left[\Theta_{0,0}^- + O(\varepsilon^2) \right] - \delta \left[\Theta_{1,0}^- + \varepsilon \Theta_{1,1}^- + O(\varepsilon^2) \right] + O(\delta^2) \;, \tag{9}$$

$$U^- = U_{0,0}^- - \delta \left[U_{1,0}^- + \varepsilon U_{1,1}^- + O(\varepsilon^2) \right] + O(\delta^2) \;, \tag{10}$$

where $\Theta_{0,0}^- = \tilde{T}_L U_{0,0}^- = \tilde{T}_L + \exp(\xi)$, $\Theta_{1,1}^- = \tilde{T}_L U_{1,1}^- = \ell \beta \tilde{T}_L \, I(-\infty) \exp(\xi)$, and

$$\Theta_{1,0}^- = \beta \left\{ \left[\left(\alpha - 1 + \frac{\xi}{2} \right) \xi + I(\xi) - \tilde{T}_{ad} \ell n \left(\frac{\tilde{T}_{ad}}{\tilde{T}_L} \right) \right] \exp(\xi) + \Theta_{0,0}^- \ell n \left(\frac{\Theta_{0,0}^-}{\tilde{T}_L} \right) \right\} \;, \tag{11}$$

$$U_{1,0}^- = \beta \left\{ \left[\alpha(\xi + 1) + \frac{\xi^2}{2} + I(\xi) - \tilde{T}_{ad} \ell n \left(\frac{\tilde{T}_{ad}}{\tilde{T}_L} \right) \right] \frac{\exp(\xi)}{\tilde{T}_L} + \alpha + \xi \right\} \;. \tag{12}$$

In the above, $\ell = (Le - 1)/\varepsilon$ is the perturbation of Lewis number from unity,

$$I(\xi) = \int_{\xi}^{0} \ell n \left[1 + (e^{\xi}/\tilde{T}_L) \right] d\xi \;, \quad \alpha = \tilde{T}_{ad} \ell n (\tilde{T}_{ad}/\tilde{T}_L) + [\ell \tilde{T}_L I(-\infty)/2] \;,$$

$$\beta = 2 \left[1 + \tilde{T}_L \tilde{u}_L + \sqrt{\tilde{T}_L \tilde{T}_{ad} \tilde{u}_L (\tilde{u}_L - 1)} \right] \sqrt{(\tilde{u}_L - 1)} \Big/ \left\{ \sqrt{\tilde{T}_L} \left[\sqrt{\tilde{T}_{ad} \tilde{u}_L} + \sqrt{\tilde{T}_L (\tilde{u}_L - 1)} \right] \right\} \;.$$

The stretch rate experienced by the flame, defined as the local gradient of the axial flow velocity at the leading edge of the flame by Law and co-workers, can not be clearly defined because the upstream boundary of the flame can not be defined. However, it can be defined to leading order as

$$\kappa = -(d\bar{u}^- / d\tilde{x})_{\tilde{x}=-\tilde{x}_f} = -\delta^{-1}(dU^- / d\xi)_{\xi \to -\infty} = \beta + O(\delta) , \qquad (13)$$

where $\kappa = k(s^\circ/L)$ is the nondimensional stretch parameter. When $\bar{u}_L = 1$, we have $\tilde{x}_{f,0} = 1$ and $\kappa = 0$, which is the stretchless limit. In this limit, the upstream of the flame is attached to the burner exit and the flow is uniform.

3. Numerical Methodology

Numerical simulation of the counterflow configuration was conducted by solving the conservation equations of mass, momentum, species concentrations, and energy along the stagnation streamline. The stagnation-point flow has been numerically investigated quite extensively during the past ten years (see, e.g., [11,15,16]). The original stagnation code [16] uses the radial pressure gradient and the velocity gradient at the exit of the nozzle as the two parameters characterizing the flow. The code has since been modified for more efficient convergence, more accurate calculation of the mass diffusion velocities, and the ability to obtain solutions through continuation for increasing and decreasing nozzle separation distances.

The one-dimensional flames were simulated by using the code of Kee *et al.* [17] to compare with the extrapolated results from the stretched flames. Both flame codes were integrated to the Chemkin-II [18] and Transport [19] codes which provide the detailed chemistry and transport information. The kinetic scheme used was a hierarchically-developed C_2-mechanism [20] which satisfactorily predicts a wide range of oxidation properties of hydrogen, carbon monoxide, methane, ethane, ethylene, acetylene, and methanol.

Results were obtained for atmospheric methane/air flames, for various stoichiometries and nozzle separation distances, for zero velocity gradient at the nozzle exit, and with thermal diffusion included in the mass diffusion velocity calculations. For each stoichiometry and nozzle separation distance, a solution was obtained at the lowest possible nozzle exit velocity. The exit velocity was subsequently increased to the state of extinction by using the arc-length continuation technique. This exit velocity imposes the stretch rate which was determined by using the linear velocity gradient ahead of the location of the minimum velocity. Because the code does not place many grid points near the minimum velocity region where the gradients are small, special care was taken so that a smooth variation of the minimum velocity with the stretch rate can be achieved. Smooth variation is important when small differences in velocities need to be resolved. The nozzle separation distances used correspond to those of the experiments, namely 7 and 22 mm for all stoichiometries.

4 Results and Discussions

Numerical calculations are performed by adopting $\tilde{T}_L = 0.2$, $\tilde{E} = 12.5$ and $Pr = 1$. Figure 2 shows the relation between the burner discharge velocity $\bar{u}_L$ and the stretch parameter k, which is given by Eq. (13). In the experiments, $\bar{u}_L$ is the controllable quantity while k is a flame response. Thus the results will be presented using $\bar{u}_L$ as the independent variable although the results by Law and co-workers were presented in terms of k. Figure 2 can then be used to convert the dependence of other flame responses from $\bar{u}_L$ to k.

Figure 3 shows the dependence of the reaction sheet location $\tilde{x}_f$ on $\bar{u}_L$

for unity Lewis number ($\ell = 0$) and specified δ, which is the ratio of characteristic flame thickness to the distance between the burner and the stagnation surface. The curve in the limit of $\delta = 0$, which implies that the flame be infinitely thin as compared to L, is simply $\tilde{x}_f = \tilde{x}_{f,0}$. For any given δ, $\tilde{x}_f$ decreases with increasing $\tilde{u}_L$. This means that the flame moves toward the stagnation plane with increasing burner discharge velocity, as expected. In the limit $\tilde{u}_L = 1$, we have $\tilde{x}_f = 1$ so that the flame is attached to the burner exit and is unstretched. This limit cannot be reached unless $\delta \to 0$ because $\tilde{x}_f$ is the location of the reaction sheet and the preheat region will be in the burner for all $\delta > 0$. Figure 3 also shows that the flame is closer to the stagnation surface (smaller $\tilde{x}_f$) for larger δ, which implies a correspondingly smaller burning intensity. As to the effect of Lewis number, it is well known (see, e.g. [3]) that for a stretched flame, the flame temperature is lower (higher) for Lewis numbers which are greater (smaller) than unity (Eq. 8). Thus for flames with positive ℓ, the burning intensity is lower and the flame is closer to the stagnation surface.

The minimum flow velocity, U_{min}, which is the quantity measured in the experiments, is plotted versus $\tilde{u}_L$ for $\ell = 0$ and selected values of δ in Fig. 4 by using Eq. (10), with the higher order terms neglected. The variation of U_{min} with k is plotted in Fig. 5 for comparison. It can be seen that for any specified δ, U_{min} increases with increasing $\tilde{u}_L$ or k, as observed experimentally. Moreover, there exist a broad region away from small values of $\tilde{u}_L$, over which U_{min} is relatively linear. This weak nonlinearity has not been observed in the experiments because the variation is smaller than

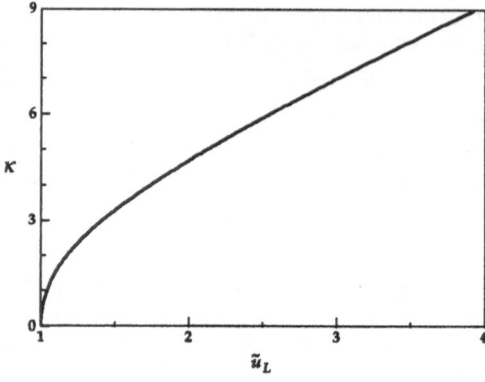

Figure 2 Stretch parameter as a function of the burner discharge velocity.

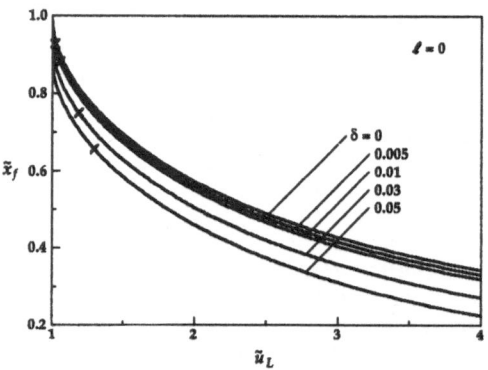

Figure 3 Reaction sheet location versus the burner discharge velocity for $\ell = 0$.

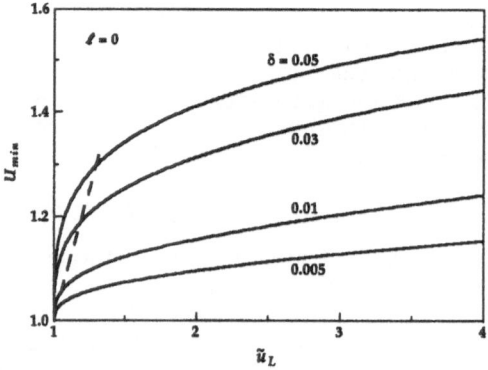

Figure 4 Upstream minimum flow velocity versus the burner discharge velocity for $\ell = 0$.

the range of experimental error, typically ±5%. However, when $\tilde{u}_L \to 1$ or $k \to 0$, U_{min} starts to curve downward and the nonlinearity becomes more prominent. Therefore the laminar flame speed obtained from linear extrapolation renders a value higher than its real value, which is consistent with previous studies by adopting potential flow [11,12] and plug flow [13].

The effects of the burner separation distance on U_{min} are also shown in Figs. 4 and 5. As mentioned earlier, the flame behaves as a freely-propagating flame in the limit of $\delta \to 0$ so that $U_{min} = 1$ for all $\tilde{u}_L$. For all $\delta > 0$, flow divergence occurs in the flame region. However, since δ is small in weakly-stretched flows, the axial mass flux is still uniform to the leading order. The correction appears in the $O(\delta)$ terms. Because of flow divergence, U_{min} now increases with increasing $\tilde{u}_L$. For a larger δ, the flame is thicker as compared to L, so that the effect of flow divergence is stronger, and the deviation from the unstretched limit is larger. Therefore, although the laminar flame speed s° is always higher than the true value by linear extrapolation, the discrepancy can be reduced by increasing the burner separation distance. If the separation distance is made such that the discrepancy is less than 5%, then the s° obtained from linear extrapolation can be considered to be the real s° because it is within the range of experimental error. This is perhaps the most significant result in this study. This conclusion, however, is contrary to that of Dixon-Lewis [13], in which it was concluded that the separation distance has no effect on the determination of U_{min}. Moreover, this result cannot be obtained from the studies adopting the potential flow [11,12] because the characteristic hydrodynamic length scale, namely L, does not exist. Of course, further improvements in the determination of s° can be obtained through a nonlinear extrapolation using Eq. (10) in the manner of Tien and Matalon [12].

Figures 4 and 5 also show, in broken curves, that there exists a state at which $U_{min} = \tilde{u}_L$. By decreasing $\tilde{u}_L$ from this limit, $U_{min} > \tilde{u}_L$ and flame flashback will occur. Flame flashback is observed in the experiments by continuously decreasing $\tilde{u}_L$ from a larger value. This may impose an additional limitation in experimentally obtaining very small values of U_{min} and consequently extrapolation is necessary. It must nevertheless be cautioned that since the analysis originally assumed only the upstream edge of the flame can reach the burner exit in the limit of $\tilde{u}_L = 1$, while U_{min} is within the flame region, the solution may not be applicable when $U_{min} = \tilde{u}_L$ is approached. The effect of Lewis number on U_{min} is shown in Fig. 6, for $\delta = 0.01$ and three specified values of $\ell = 3, 0, -3$, each representing the cases of Lewis number larger than, equal to and smaller that unity. As discussed earlier, the burning intensity is lower for a larger ℓ, so that the flame is closer to the stagnation surface. Therefore for a fixed $\tilde{u}_L$, U_{min} is lower for the flame with a larger ℓ. The broken line again represents the limit of $U_{min} = \tilde{u}_L$.

An alternate approach towards determination of the laminar flame speed by using the counterflow

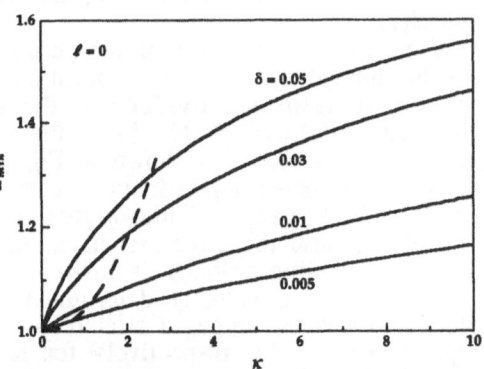

Figure 5 Upstream minimum flow velocity versus the stretch rate for $\ell = 0$.

technique, as suggested by Tien and Matalon [12], is to measure the flow velocity at the position of a fixed temperature, say where the temperature is raised by 1% of the total temperature increase. The similar extrapolation technique can then be conducted and the result of $s°$ is converted to that at the reference temperature through the equation of state. The axial velocity where $\Theta^- = 0.21$, a 1% temperature increase, occurs as a function of $\tilde{u}_L$ is shown in Fig. 7 for the same δ and l as used in Fig. 4. Due to the effect of thermal expansion, all the curves converge to $U^- = 1.05$ at $\tilde{u}_L = 1$. Comparing Figs. 4 and 7, it is seen that the flame response at $\Theta^- = 0.21$ is similar to that at U_{min}, although the amount of over-determination is reduced. The effect of the Lewis number is qualitatively similar to Fig. 6. It may be concluded that the result will be more accurate if the flow velocity is measured at where $\Theta^- = 0.21$ occurs. However, there are experimental difficulties in being able to accurately measure the temperature and velocity

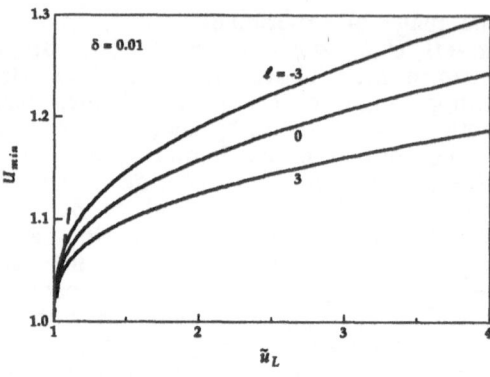

Figure 6   Upstream minimum flow velocity versus the burner discharge velocity for $\delta = 0.01$.

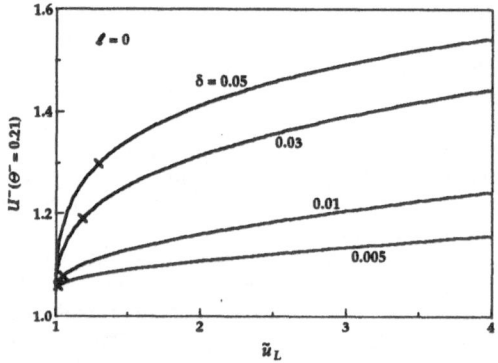

Figure 7 Flow velocity at 1% temperature increase versus the burner discharge velocity for $l = 0$.

simultaneously. Moreover, in the region near $\Theta^- = 0.21$, the temperature variation with the change of spatial location is very gradual and hence a small inaccuracy in the temperature measurement may result in a large error in determining its position and hence the corresponding velocity.

Conceptually, the idea of measuring the velocity at constant temperature can be applied not only at the location of 1% temperature increase but at all other temperatures, provided that the thermal expansion effect is properly adjusted. Following this idea, the axial velocity at a 5% temperature increase ($\Theta^- = 0.25$) is shown in Fig. 8, which shows a better improvement because the percentage of over-estimation is further decreased compared to $\Theta^- = 0.21$ case. This higher temperature is experimentally more feasible because the error associated in the determination of its location and the corresponding velocity is lower for a higher temperature gradient.

The minimum velocity determined by detailed numerical calculations is plotted versus the imposed stretch rate, k, and the Karlovitz number, Ka, in Figs. 9 10 and 11, respectively for lean, stoichiometric, and rich mixtures with $L=7$ and 22 mm. Because the Karlovitz number is defined as $Ka = k l_d / s°$, we need to specify the laminar flame speed $s°$ and the characteristic flame thickness l_d. In the computation, $s°$ is given by the one-dimensional code while ℓ_d is calculated from the temperature profile of the one-dimensional flame using the tangent definition. The values of ℓ_d are 0.786, 0.436 and 0.928 mm for equivalence ratios of 0.7, 1.0 and 1.45 respectively. Similar to the analytical results, these results show that U_{min} varies with the separation distance L and curves downward at low stretch rates. Thus linear extrapolations of U_{min} for Ka larger than about 0.5 can lead to substantial over-estimates of s°. However, linear extrapolations with $Ka < 0.5$ lead to values of s° for the 22 mm separation distance within 2.5-5% of the that obtained from the 1-D code. This difference is within the noise level of the experimentation, especially with regard to the accuracy of the LDV measurements. These results can thus be considered to be sufficiently accurate. The extrapolated values of s° for $L = 35$ mm are slightly lower than those for 22 mm, while for $L = 14$ mm they are between those for 7 mm and 22 mm.

Similar to the asymptotic analysis, the numerical results reported herein are contrary the numerical results reported herein are contrary to those of Dixon-Lewis [13] in terms of the dependence of U_{min} on the nozzle separation distance. It may also be noted that the present results are in complete qualitative agreement with the experimental measurements. It is significant from the numerical results that while the minimum velocities do depend on L, the extrapolated values of s° also appear to converge to those from the 1-D code by increasing L. It is therefore reasonable to state that the concept of minimum velocity extrapolation is viable, and that the extent of over-estimation can be made arbitrarily small by increasing L. Experimentally, because the flow tends to

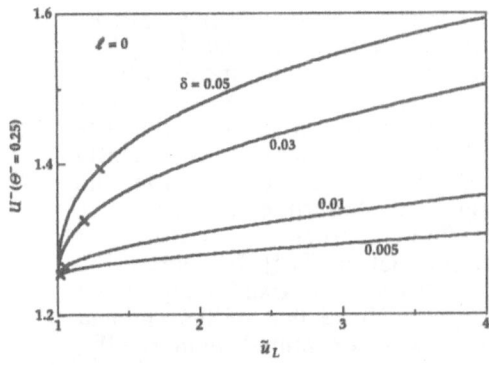

Figure 8 Flow velocity at 5% temperature increase versus the burner discharge velocity for $\ell = 0$.

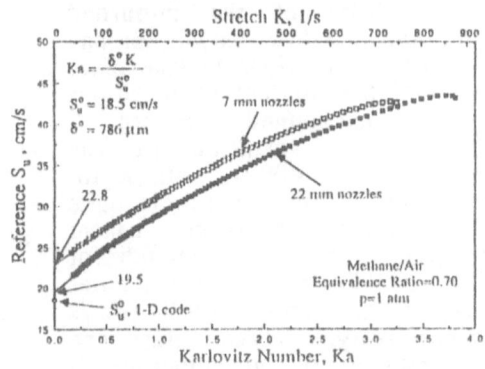

Figure 9 Numerically determined U_{min} versus the stretch for the methane/air flame with $\phi = 0.7$.

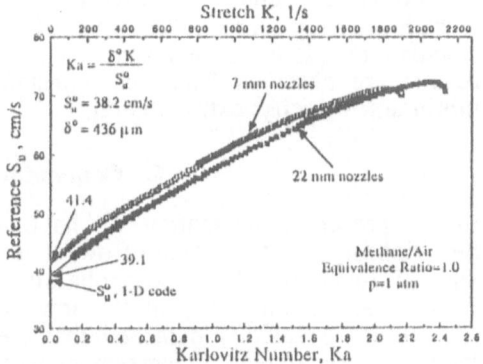

Figure 10 Numerically determined U_{min} versus the stretch for the methane/air flame with $\phi = 1.0$.

become turbulent for larger L, there is a limitation on how close the laminar flame speed can be approached. The present study shows that a separation distance of 14 to 22 mm is sufficiently adequate when considering the inherent experimental errors.

The alternate mode of extrapolation [12] by measuring velocities at a fixed temperature slightly above that of the ambient was also examined numerically. We select 400 K as the temperature because it is high enough for the flow properties to be experimentally distinguishable from those of the unburned mixture, and low enough for any significant reactions to take place so that the mixture compositions remain unchanged. Figure 12 shows the variation of the velocities at 400 K, with k for equivalence ratio of 1.0, is quite linear for a wide range. Linear extrapolation to zero stretch for L=7 and 22 mm yields velocities of 51.5 and 50.2 cm/s respectively. These values can be related to the velocities at 300 K through the mass conservation ρU = constant. This relation yields values for s° equal to 38.6 and 37.7 cm/s respectively,

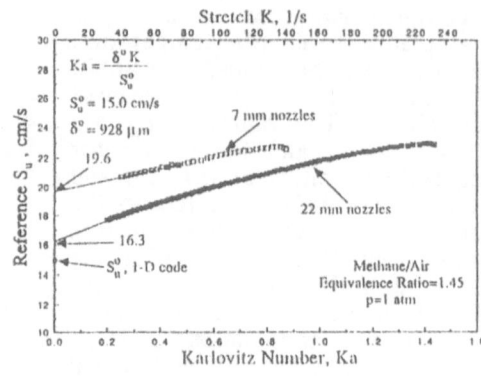

Figure 11 Numerically determined U_{min} versus the stretch for the methane/air flame with $\phi = 1.45$.

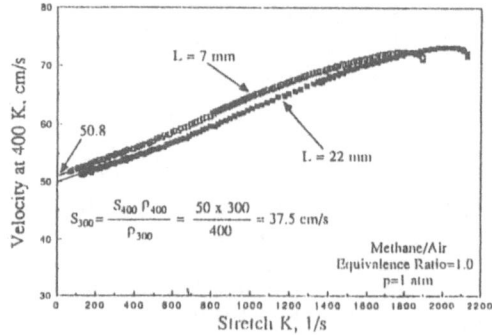

Figure 12 Numerically determined $U(400K)$ versus the stretch for the methane/air flame with $\phi = 1.0$.

which are very close to the value of 38.2 cm/s given by the one-dimensional calculations. It is of interest to note that even for $L = 7$ mm, more accurate results are obtained from this constant temperature extrapolation than the minimum velocity extrapolation.

6. Concluding Remarks

In the present investigation we have analytically and numerically assessed the viability of the counterflow, twin-flame technique for measuring the laminar flame speed by extrapolating the minimum velocity to zero stretch. Results demonstrate that in the low stretch regime there is a curvature in the minimum velocity data and that the extrapolation can lover-estimate the laminar flame speeds, in agreement with some previous studies. It is found, however, that the extent of over-estimation depends on the ratio of the flame thickness to the nozzle separation distance and that it can be arbitrarily reduced by gradually decreasing this ratio. The qualitative agreement between the two independently conducted studies indicates that the counterflow technique is viable in the determination of laminar flame speed provided that the largest possible nozzle separation distances are used. For typical paraffin/air mixtures, nozzle separation distances of 22 mm provide sufficient accuracy.

It was further demonstrated that the constant temperature extrapolation technique appears to be more accurate. Its experimental implementation, however, could be more involved because it requires the additional accurate measurement of the flame temperature in the preheat region, especially if a non-intrusive technique is to be used.

Acknowledgments – This work has been supported by the Air force Office of Scientific Research and the Office of Basic Energy Sciences of the Department of Energy.

References

1. Andrews, G. E. and Bradley, D., *Combust. Flame* 18:133 (1972).
2. Matalon, M. and Matkowsky, B. J., *J. Fluid Mech.* 124:239 (1982).
3. Law, C. K., *Twenty-Second Symposium (International) on Combustion*, The Combustion Institute, Pittsburgh, 1989, p. 1381.
4. Botha, J. P. and Spalding, D. B., *Proc. Roy. Soc. London* A225:71 (1954).
5. Wu, C. K. and Law, C. K., *Twentieth Symposium (International) on Combustion*, The Combustion Institute, Pittsburgh, 1985, p. 1941.
6. Yu, G., Law, C. K. and Wu, C. K., *Combust. Flame* 63:339 (1986).
7. Egolfopoulos, F. N., Cho, P. and Law, C. K., *Combust. Flame* 76:375 (1989).
8. Zhu, D. L., Egolfopoulos, F. N. and Law, C. K., *Twenty-Second Symposium (International) on Combustion*, The Combustion Institute, Pittsburgh, 1989, p. 1941.
9. Egolfopoulos, F. N., Zhu, D. L. and Law, C. K., *Twenty-Third Symposium (International) on Combustion*, The Combustion Institute, Pittsburgh, 1991, p. 471.
10. Buckmaster, J. D. and Mikolaitis, D., *Combust. Flame* 47:191 (1982).
11. Stahl, G., Warnatz, J. and Rogg, B., in *Dynamics of Reactive Systems, Part I : Flames* (A. L. Kuhl, J. R. Bowen, J.-C. Leyer and A. Borisov Eds.), AIAA, Washington, D.C., 1988, p. 195.
12. Tien, J. H. and Matalon, M., *Combust. Flame* 84:238 (1991).
13. Dixon-Lewis, G., *Twenty-Third Symposium (International) on Combustion*, The Combustion Institute, Pittsburgh, 1991, p. 305.
14. Chao, B. H., Egolfopoulos, F. N. and Law, C. K. Structure and Propagation of Counterflow Premixed Flames in a Finite Domain. Manuscript in preparation, (1994).
15. Miller, J. A., Kee, R. J., Smooke, M. D. and Grcar J. F., The Computation of the Structure and Extinction Limit of a Methane-Air Stagnation Point Diffusion Flame. 1984 Spring Meeting of the Western States Section of the Combustion Institute, Boulder, Colorado, (1984).
16. Kee, R. J., Miller, J. A., Evans, G. H. and Dixon-Lewis, G., *Twenty-Second Symposium (International) on Combustion*, The Combustion Institute, Pittsburgh, 1989, p. 1479.
17. Kee, R. J., Grcar, J. F., Smooke, M. D. & Miller, J. A., Sandia Report SAND 85-8240 (1985).
18. Kee, R. J., Rupley, F. M. and Miller J. A., Sandia Report SAND 89-8009 (1989).
19. Kee, R. J., Warnatz, J. and Miller J. A., Sandia Report SAND 83-8209 (1983).
20. Egolfopoulos, F. N., Du, D. X. and Law, C. K., *Comb. Sci. Tech.* 83:33 (1992).
21. Libby, P. A. and Williams, F. A. *Comb. Sci. Tech.* 37:221 (1984).

Some topics in reverse smoulder

John Buckmaster [1], David Lozinski [2]

[1]Mathematics Department, Hong Kong University of Science &
Technology, Clear Water Bay, Kowloon, Hong Kong
[2]Department of Mathematics and Statistics, McMaster University,
Hamilton, Ontario, Canada

Abstract: The role of thermal non-equilibrium and endothermic pyrolysis is discussed in the context of a simple model of reverse smoulder combustion. It is shown that non-equilibrium has little qualitative effect on the nature of the solutions, but endothermic pyrolysis can lead to quenching at sufficiently large blowing rates.

Introduction and the basic model

Smouldering combustion is the heterogeneous exothermic reaction of a porous solid fuel. Reverse or opposed flow smoulder takes place in a nominally one-dimensional configuration in which air is forced through the fuel in the direction opposite to that of propogation of the smouldering front. Recent experiments by Torero et al [1] have examined this situation, and their observations are an important reference point for our theoretical discussion.

The core physics of reverse smoulder can be captured by very simple model equations, and there are several such discussion in the literature, e.g. Dosanjh et al [2]. Lozinski & Buckmaster [3] examined the following equations:

$$C_p M \frac{dT}{dx} = \lambda \frac{d^2 T}{dx^2} + Q\Omega, \tag{1a}$$

$$M_g \frac{dY}{dx} = \rho_g D \frac{d^2 Y}{dx^2} - \Omega, \tag{1b}$$

$$M = M_s + M_g, \quad \Omega = BY e^{-E/RT}, \tag{1c, d}$$

where T is the temperature (of both the solid and gas phases), Y is the mass fraction of oxygen in the airflow, M_g is the applied mass flux of gas, and M_s is the mass flux of solid relative to the reaction front, a quantity to be determined.

A standard asymptotic treatment, valid in the limit $E \to \infty$, yields the formulas

$$\frac{QY_iM_g}{\lambda} = \left(\frac{2B}{\rho_g D}\right)^{1/2} \frac{RT_b^2}{E} e^{-E/2RT_b}, \tag{2a}$$

$$M_g(T_b - T_i) = QY_iM_g, \tag{2b}$$

where T_i and Y_i are supply (cold, fresh) values, and T_b is the temperature (uniform) behind the reaction front. The formula (2a) determines T_b, and then M_s is fixed by the formula (2b). M_s varies with M_g in the manner shown in Fig. 1. This response is consistent with the experimental record of Torrero et al [1] except that, for sufficiently large M_g (beyond the value for which M_s is a maximum, but short of the point where $M_s = 0$) they observed extinction. Thus although equations (1) undoubtedly contain essential ingredients of the problem, they must be deficient in some respect.

Non-equilibrium between the phases

Equations (1) do not distinguish between the temperature of the solid (T_s) and that of the gas (T_g); nor between the value of Y in the main body of the flow (Y_g) and the value at the surface (Y_s). Whether or not equilibrium prevails depends partly on the length scales that characterize the flame structure, and those that characterize the pore structure. Since the reaction zone is thin, the first effects of nonequilibrium will be felt there. A rough accounting for this can be discussed in the following fashion.

When $T_s \neq T_g$ and $Y_s \neq Y_g$ it is necessary to consider the fluxes between the two phases, as well as the fluxes between neightboring portions of the same phase. Suppose f is the heat transmitted to the gas from the solid, per unit volume, by conduction. We assume that f can be written as

$$f = A\lambda_g(T_s - T_g) \tag{3}$$

for some constant A. By analogy, the mass flux of oxygen to the solid surface per unit volume, which is equal to the reaction rate, is

$$A\rho_g D(Y_g - Y_s) = \Omega = BY_s e^{-E/RT_s}. \tag{4a, b}$$

Thus the two temperature equations are

$$C_{p_g}M_g\frac{dT_g}{dx} = \lambda_g\frac{d^2T_g}{dx^2} + f, \tag{5a}$$

$$C_{p_g}M_s\frac{dT_s}{dx} = \lambda_s\frac{d^2T_s}{dx^2} - f + Q\Omega, \tag{5b}$$

replacing (1a); equation (1b) is correct after the replacement $Y \to Y_g$; and (1d) is replaced by (4b). Formally we recover equations (1) in the limit $A \to \infty$. As noted above, this limit would first break down, for finite A, in those parts of the flame where terms other than f attain their largest values, i.e. the reaction zone. We assume that equilibrium prevails outside of this zone.

Because the reaction zone is thin, convection can be neglected there. It is convenient to introduce scaled variables as follows:

$$T_g = T_b(1 + \epsilon \tau_g), \quad T_s = T_b(1 + \epsilon \tau_s), \tag{6a, b}$$

$$Y_g = \frac{\epsilon T_b \lambda_g}{Q \rho_g D} y_g, \quad Y_s = \frac{\epsilon T_b \lambda_g}{Q \rho_g D} y_s, \quad x = \frac{s}{\sqrt{A}}, \quad \epsilon^{-1} = \frac{E}{R T_b}. \tag{6c, d, e, f}$$

Then the reaction zone structure is described by the equations

$$0 = \tau_g + \frac{\lambda_s}{\lambda_g} \tau_s + y_g, \tag{7a}$$

$$y_g - y_s = \frac{B e^{-1/\epsilon}}{A \rho_g D} y_s e^{\tau_s}, \tag{7b}$$

$$0 = \frac{d^2 \tau_g}{ds^2} - \tau_g + \tau_s, \quad 0 = \frac{d^2 y_g}{ds^2} - y_g + y_s, \tag{7c, d}$$

where $\tau_g \to 0, \tau_s \to 0, y_g \to 0, y_s \to 0$ as $s \to \infty$. (7e)

Equations (7) are equivalent to the single equation

$$\frac{1}{2} \int_{-\infty}^{\infty} d\bar{s} \, y_s(\bar{s}) e^{-|s - \bar{s}|} - y_s = \frac{B e^{-1/\epsilon}}{A \rho_g D} y_s \exp \left\{ \frac{-\lambda_g / \lambda_s}{2\sqrt{1 + \lambda_g / \lambda_s}} \int_{-\infty}^{\infty} d\bar{s} \, y_s(\bar{s}) \right.$$
$$\left. \cdot e^{-\sqrt{1 + \frac{\lambda_g}{\lambda_s}} |s - \bar{s}|} \right\}. \tag{8}$$

Since the origin of the coordinate system is arbitrary, we may impose the condition

$$y_s(0) = 1. \tag{9}$$

The classical flame-sheet structure is recovered in the limit $P \to 0$, where P is the parameter $B e^{-1/\epsilon} / A \rho_g D$. To show this we introduce the scalings

$$s = t/\delta, \quad y_s(s) = z(t), \quad \delta \to 0, \tag{10}$$

so that the only important contributions to the integrals in (8) are from the neighborhood of $\bar{s} = s$. In this way we recover the familiar equation

$$\frac{d^2 z}{dt^2} = \frac{P}{\delta^2} z \exp \left[\frac{-z}{1 + \lambda_s / \lambda_g} \right], \tag{11}$$

establishing the connection $\delta \sim \sqrt{P}$. In this limit, the slope as $s \to -\infty$ is given by the formula

$$\frac{dy_s}{ds} = -\left(1 + \lambda_s/\lambda_g\right)\sqrt{2P}, \tag{12}$$

and is to be matched with the gradient of Y at the end of the preheat zone, i.e.

$$\lim_{s \to -\infty} \frac{dy_s}{ds} = \frac{-M_g Y_i Q}{\sqrt{A}\lambda_d T_b \epsilon}. \tag{13}$$

When (12) and (13) are equated, A (correctly) makes no contribution and we recover the result (2a) which defines the flame temperature T_b in terms of the blowing rate M_g.

For non-vanishing P, equation (8) must be solved numerically, and this leads to the curve shown in Fig. 2 which, with (13), determines T_b. Since, for large activation energy, the role of T_b is felt most strongly via the factor $e^{-1/\epsilon}$ in P, an accounting for non-equilibrium introduces an $O(\epsilon T_b)$ increase in flame temperature over the equilibrium value. Qualitatively, the behavior of solutions to (8) is similar to (12) so that this discussion of non-equilibrium fails to identify any significant effects.

Endothermic pyrolysis

As noted earlier, the system (1) fails to predict quenching, and the discussion above suggests that nonequilibrium is not the missing ingredient which could rectify this failure (although more elaborate discussions are certainly possible). Recently Lozinski & Buckmaster [3] have proposed that endothermic pyrolysis is the key to this phenomenon, and we shall briefly review these results.

Endothermic pyrolysis occurs when nitrogen comes into contact with the solid surface at sufficiently high temperatures. In the framework of a flame structure labelled with respect to the exothermic oxidation (1a), pyrolysis will occur in the preheat zone, in the reaction zone, and behind the reaction zone; however, only the latter has been accounted for. Pyrolysis behind the reaction zone leads to negative temperature gradients there, a withdrawal of heat, by conduction, from the back of the reaction zone. Thus the analysis examines whether or not such withdrawal can lead to solutions that mimic the observed behavior. The merit of this analysis is rooted in the reasonable speculation that, should quenching be obtained, it is most unlikely that the additional removal of heat from the preheat zone and the reaction zone could have a countervailing effect. Moreover, since the pyrolysis rate is an increasing function of temperature, it should play a more important role in the post-heat zone than in the preheat zone (the lengths of the zones are comparable). And its effects in the thin exothermic reaction zone are surely to just reduce the effective magnitude of the exothermicity. Thus, although a fuller accounting of the pyrolysis might be useful, the approximate treatment is of fundamental interest, and minimizes the complications inherent in the analysis.

The model described by (1) is now modified by introducing the equation

$$C_p M \frac{dT}{dx} = \lambda \frac{d^2T}{dx^2} - Q_p B_p e^{-E_p/RT} \quad \text{in} \quad x > 0, \tag{14}$$

where the subscript p refers to pyrolysis parameters. Reaction is effectively terminated when the temperature drops significantly below T_*, the flame-temperature (i.e. the temperature at $x = 0$).

The effective Damkohler number depends on the local value of the temperature. For large x, where T is significantly smaller than T_*, the Damkohler number will be small so that there is a convective-reactive balance with

$$\frac{dT}{dx} \sim \frac{Q_p B_p}{MC_p} e^{-E/RT}, \tag{15}$$

a small quantity. On the other hand, when $T \sim T_*$ it is assumed that the Damkohler number is large so that there is a diffusive-reactive balance, whence

$$0 \sim \frac{1}{2} p^2(T) - \frac{1}{2} p^2(T_*) - \frac{Q_p \tilde{B}_p}{\lambda} \int_{T_*}^{T} dT e^{-E_p/RT + E_p/RT_*}, \tag{16a}$$

$$p \equiv \frac{dT}{dx}, \quad \tilde{B}_p \equiv B_p e^{-E_p/RT_*}. \tag{16b, c}$$

When the integral is evaluated using Frank-Kamenetskii's strategy, it follows that

$$\frac{dT}{dx}(0+) = -PT_* e^{-E_p/2RT_*}. \tag{17}$$

Quite apart from the present derivation, this result represents a plausible *ab initio* modeling ingredient: heat is removed from the back of the reaction zone at a rate that depends on the rate of pyrolysis.

It is a straightforward matter to incorporate the condition (17) into the model described by equations (1). The gradient is not assumed small, so that the flame-sheet structure is no longer given by (11), but is, rather, Liñán's premixed flame structure with reactant leakage, [4]. This structure equation must be integrated numerically, but can be written in terms of just a single parameter

$$\gamma \equiv \frac{-dT/dx(0+)}{dT/dx(0-) - dT/dx(0+)}, \tag{18}$$

the conductive heat flux to the post-heat zone, divided by the total heat flux from the reaction zone. γ is non-negative and, for a solution to exist, cannot exceed $1/2$.

Results calculated in this way show a strong sensitivity to the values of the several parameters, and this prompts an observation relevant to one of the themes of this workshop, namely the interaction between analysis and computation. When a problem has a large number of parameters whose experimental values are poorly known – and when, in addition, the qualitative nature of the solution is sensitive to the values of these parameters - then a conventional numerical strategy can easily miss important qualitative behavior. This is less likely to occur with an analytical/numerical strategy of the kind described here, which permits extensive searches over the parameter space with modest effort and can therefore be a valuable forerunner to a more extensive numerical treatment.

A representative solution (M_s vs M_g) reported in Lozinski & Buckmaster (1994) which mimics the experimental behavior is shown in Fig. 3. The upper branch is believed to be the physically relevant one. Thus we have an increase in M_s to a maximum value, followed by a decrease to the quenching point Q. γ is approximately $1/3$ at this point. The issue of stability is discussed in the aforementioned work but is, essentially, an open question, and we do not discuss it here.

Finally, as an example of the parameter sensitivity, solutions for two different values of Q are shown in fig. 4. Only the solutions for $M_s > O$ are physically sensible so that quenching at finite M_s is lost for the smaller value of Q.

Acknowledgement

This work was partly supported by AFOSR and by the NASA-Lewis Research Center by grants administered by the University of Illinois. Hong Kong University of Science & Technology provided travel funds supplementary to those provided by NSF for the workshop which made it possible for JB to travel to the workshop site from Hong Kong. DL's contributions were made during a postdoctoral appointment at the University of Illinois, partly supported by NSERC Canada.

References

[1]   Torero, J. L., Fernandez-Pello, A.C., & Kitano, M. *Combustion Science & Technology*, **91**, 1993, p. 95.

[2]   Dosanjh, S.S., Pagni, P.J., & Fernandez-Pello, A.C. *Combustion and Flame*, **68**, 1987, p. 131.

[3]   Lozinski, D., & Buckmaster, J. To appear, *Combustion and Flame*, 1994.

[4]   Liñán, A. *Acta Astronautica*, **1**, 1974, p. 1007.

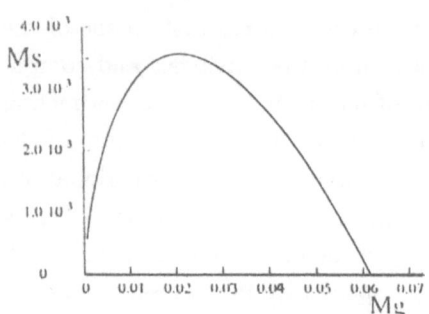

Fig. 1. M_s vs M_g for the basic model

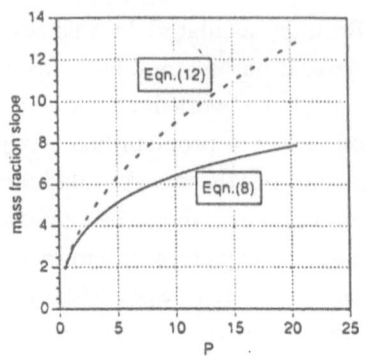

Fig. 2. Slope on the cold side of the flame-sheet vs $P(\lambda_s = \lambda_g)$

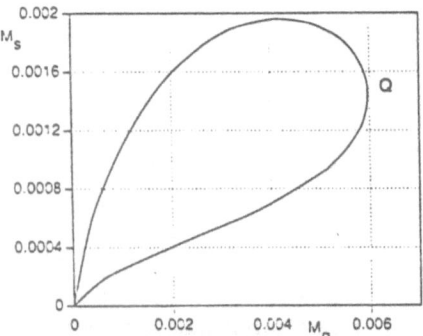

Fig. 3. M_s vs M_g when endothermic pyrolysis is accounted for, from Lozinski & Buckmaster (1994)

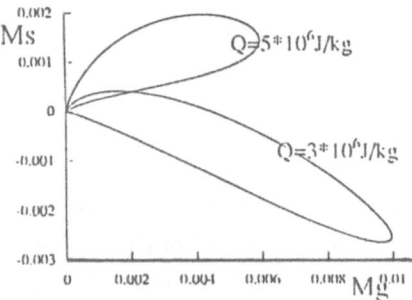

Fig. 4. M_s vs M_g for different Q, from Lozinski & Buckmaster (1994)

Porous Medium Combustion

A.P. Aldushin[1], B.J. Matkowsky[2], D.A. Schult[2],
G.V. Shkadinskaya[1], K.G. Shkadinsky[1], and V.A. Volpert[2]

[1] Institute of Structural Macrokinetics, Chernogolovka, Russia 142432
[2] Department of Engineering Sciences & Applied Mathematics,
Northwestern University, Evanston, IL 60208, USA

1.   Introduction: Filtration Combustion

Smolder waves and SHS (self-propagating high-temperature synthesis) waves are
both examples of combustion waves propagating in porous media. When delivery
of reactants through the pores to the reaction site is an important aspect of the
process, it is referred to as filtration combustion [5]. The two types of filtration
combustion have a similar mathematical formulation, describing the ignition,
propagation and extinction of combustion waves in porous media. The goal
in each case, however, is different. In smoldering the desired goal is to prevent
propagation, whereas in SHS the goal is to insure propagation of the combustion
wave, leading to the synthesis of desired products [11], [12]. In addition, the
scales in the two areas of application may well differ. For example, smoldering
generally occurs at a relatively low temperature and with a smaller propagation
velocity than SHS filtration combustion waves. Nevertheless, the two areas of
application have much in common, so that mechanisms learned about in one
application can be used to advantage in the other. In this paper we discuss
recent results in the areas of filtration combustion.

2.   SHS Waves

We consider filtration combustion of a porous solid layer, surrounded by an
external gaseous environment. The sample, consisting of a finely ground powder
mixture of reactants, is ignited at one end and a high temperature thermal wave,
having a frontal structure, propagates through the sample converting reactants
to products. Both the gas initially in the pores of the solid, and the gas that
filters into the pores from the environment take part in the reaction to form solid
products. We consider various types of gas exchange with the external bath. In
some case the side surface of the sample is assumed to be both heat insulated
and non permeable to gas flow, while in others it is assumed to be open so that
both heat and gas mass flow through the side surface (crossflow filtration). Gas
exchange with the bath can also occur through the ends of the sample. We refer
to the case when the gas filters in the same (opposite) direction as the wave
propagates as coflow (counterflow) filtration (Fig. 1).

   In our model we also account for deformation of the sample. The pressure in
the pores can be either greater or lower than the external pressure, depending on
the processes occurring in the sample. Thus pressure gradients may deform the
hot product. Deformation of the product can also be caused by external forces,
such as gravity or centrifugal forces, acting during the combustion process. Both
dense materials and materials with high porosity are needed in applications. It
is reasonable to attempt to achieve the desired porosity of the product during
the combustion process itself, rather than by post processing the material.

   We consider thin porous layers so that we employ a one dimensional model
[13]-[17]. For simplicity we assume that particles forming the porous solid do

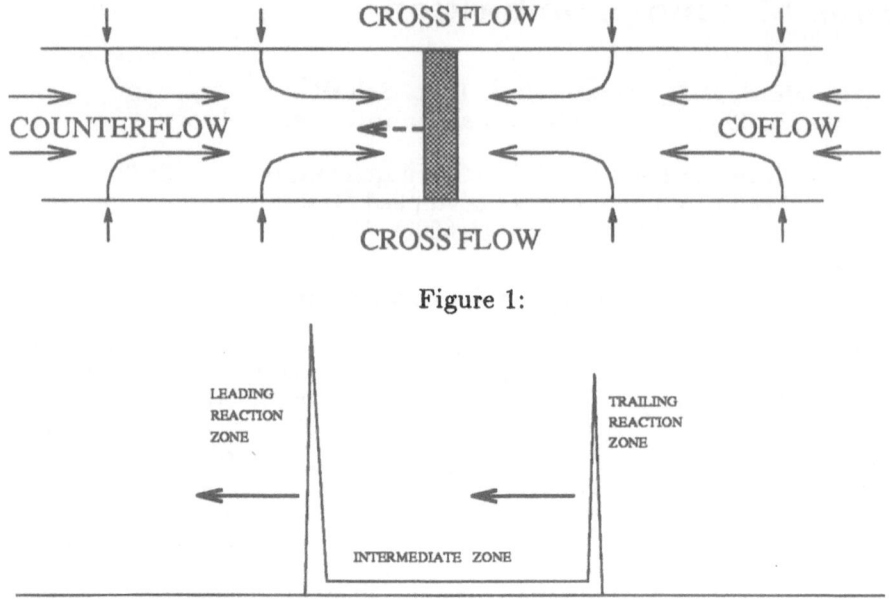

Figure 1:

Figure 2:

not change their size during the course of experiment, so that deformation of the sample is due solely to changes in porosity. The force of deformation is resisted by a viscous force. The effective coefficient of viscosity of the porous solid is lower in the hot product region than in the cold fresh mixture, so that the product is more readily deformable. We assume the simplest chemical kinetics, namely first order reaction with respect to the solid and Arrhenius temperature dependence. If gas participates in the reaction, we assume that the reaction rate is independent of pressure, provided it is not too small. When the pressure is very small the reaction rate goes to zero, so that the reaction ceases when the gas is depleted. We model the crossflow gas mass transfer between the sample and the external environment as being proportional to the difference in the gas pressure between the pores and the environment.

We first consider nondeformable samples, for which we demonstrated the existence of uniformly propagating filtration combustion waves having various structures, including a two front structure [16], and determined their characteristics. In the leading reaction zone (Fig. 2), which determines the propagation velocity, both the gas initially located in the pores, and the gas which enters the pores by counterflow filtration, react, while in the trailing reaction zone, gas which enters by coflow filtration, reacts. In the intermediate zone, chemical conversion occurs very slowly under low pressure, and is limited by the rate of crossflow filtration.

The parameter regime which leads to this wave corresponds to (i) large filtration coefficients, (ii) small initial gas pressure in the pores and in the bath, and (iii) small gas mass transfer coefficients between the sample and the bath. Thus the regime describes combustion synthesis under low pressure conditions. Since the conditions of combustion determine the nature of the resulting product, the new regime described above may offer interesting technological possibilities.

We also considered the effect of the crossflow delivery of the gas in a number of limiting cases in terms of a parameter B that describes the ratio of the gas exchange rate to the reaction rate [14]. We derived explicit analytical expressions for various quantities of interest associated with uniformly propagating solutions, including the combustion temperature, the propagation velocity and the final depth of conversion of the porous solid. We also numerically computed profiles for the reaction rate, temperature, pressure, density and depth of conversion. In addition, we identified a new pulsating instability, associated with the gas mass transfer between the pores and the bath, which occurs for $B_1 < B < B_2$. In a small range of B, we found bistability, in which both the uniformly propagating and pulsating propagating solutions are simultaneously stable, each having its own domain of attraction. Finally, the mechanism for the development of the pulsating solutions was described.

We now describe results for deformable samples in the absence of crossflow filtration and external forces [13]. Deformation is induced solely by the pressure difference of the gas outside and inside the sample which changes the porosity of the sample. The sample is assumed to be sufficiently long that a traveling wave solution can develop. To describe the evolution of porosity we derive an equation which allows us to define a characteristic time of deformation t_d. If t_d is sufficiently smaller than the characteristic time of combustion t_r, the deformation process is sufficiently fast to compensate for pressure gradients, so that pressure is equalized almost instantaneously, and filtration is suppressed. If $t_d \gg t_r$, deformation occurs solely in the product, and does not affect the propagation velocity. We determined various characteristics of a uniformly propagating combustion wave, and the materials produced by it, such as the propagation velocity, combustion temperature, final depth of conversion and final porosity of the product, as a function of the thermophysical parameters of the system. In addition we identified a regime of pulsating propagation, in which the final porosity of the products is periodic in space. We also found that deformation affects stability. In particular, viscosity was found to be stabilizing.

For relatively short samples [17], i.e. samples for which the filtration length is comparable to the length of the sample, we found that deformation prevents the propagation of secondary combustion fronts, often seen in nondeformable samples if the depth of conversion in the primary front is incomplete, so that solid fuel remains which can be reignited. Deformation can result in shrinking the pores so that gaseous oxidizer is impeded from reaching the unburned fuel. Thus, if complete conversion is desired, it must be achieved in the primary reaction front by arranging for a sufficiently high (stoichiometric) initial pressure.

The complete model, which accounts for all types of gas delivery discussed above as well as for external forces, exhibits a rich variety of behavior. For example, in the case when deformation corresponds to an increase in porosity, the sample can become so brittle that it breaks. Two types of breaks were found, which occur at distinctly different rates, and which correspond to different sizes of the pieces broken off [15].

3. Enhancement of SHS Waves

We now discuss combustion models in which features of gasless and filtration combustion processes are combined. This is of interest from the point of view of the practical realization of SHS processes involving weakly exothermic solid-solid reactions in which the heat release is insufficient for the propagation of SHS waves. An auxiliary exothermic solid-gas reaction may significantly increase the reaction temperature in the solid-solid reaction zone thus helping the synthesis reaction to propagate through the sample. For example, arranging for coflow

filtration to occur in an otherwise gasless SHS process can enhance the exothermicity of the reaction due to the superadiabatic effect [1], in which the burning
temperature exceeds that found for the same degree of conversion of an identical
sample which reacts uniformly under adiabatic conditions, throughout.

We studied combustion waves which propagate uniformly through a condensed porous sample formed by cold pressing a powder mixture of reactants
A and S, which react to form the condensed product AS. In addition, the
sample contains a condensed inert component I which does not participate in
the reaction. Finally, gas is blown through the sample. The gas consists of
both an inert component I_g and an active component O which reacts with the
solid A to form the product AO consisting of both solid and gaseous parts.
Thus the initial solid reactant A can be consumed in either of the two competing reactions, one of which is gasless, and the other of filtration combustion
type. Specific examples of such systems include three component systems consisting of metal-carbon-nitrogen or metal-carbon-hydrogen which are used to
synthesize carbonitride and carbohydride metals. Other examples are ceramic-
oxide superconductors and silicon and tungsten carbides (SiC and WC) that
are widely used in industry, though their production by conventional methods
involves large energy consumption. The low heat release and insufficiently high
thermodynamic temperature is the principal difficulty in using the SHS method
for these systems. The possibility of generating high temperatures may, in fact,
be important for the SHS process in general. Currently, SHS employs rather
pure, and therefore relatively expensive reactants, which must first be produced
with an attendant energy cost. It would be desirable to employ less expensive
natural components in variants of the SHS process. This may be possible if a
filtering gas is introduced. For example, in the industrial production of SiC,
pure silicon is not used, but rather the abundant natural mineral SiO_2 (sand).
The direct interaction of SiO_2 with carbon in the combustion synthesis (SHS)
of SiC is not possible without the external supply of large amounts of energy.
However, if active gas is allowed to filter through the system the process may
be carried out in a self-sustained manner without the need for external energy.
The additional heat necessary for synthesis is generated by the reaction between
the excess carbon in the solid mixture and the oxidizer in the flowing gas.

The most effective arrangement for increasing the temperature is coflow filtration, however counter-flow filtration is advantageous in some cases, as discussed
below. The goal of this study was to determine the most effective arrangement
of such processes and to enhance the understanding of the fundamental mechanisms of interaction between gasless and filtration combustion. We studied
combustion waves with two reactions propagating through samples with a prescribed constant incoming gas mass flux. In [3] we considered coflow filtration,
while in [4] we considered counterflow filtration.

For coflow filtration, under the assumption of large activation energies of
both reactions we found a uniformly propagating combustion wave in which
both reactions are localized at a single site at the maximum temperature. The
interaction of the two reactions in the reaction zone arises due to the competition of both reactions to consume the reactant A. This competition, which
is influenced by the gas flux through the product region, is rather complicated
and can result in either nonexistence or multiplicity of solutions. To achieve the
maximal superadiabatic effect, the composition of both the condensed mixture
and the gas should be optimized. For example, in the case of equal exothermicity
of the two reactions, the introduction of a chemically active component to the
gas is not as effective in increasing the burning temperature as is introducing an
inert gas. The gas flux can significantly affect the gasless reaction in the case of

higher caloricity of the solid-gas reaction. The oxidizer concentration should not be high in this case since it would decrease the combustion temperature, as well as the proportion of A which reacts in the solid-solid reaction. We note that the presence of the gasless reaction prevents an inversion of the temperature profile, which often occurs in filtration combustion waves [1], [5].

In coflow filtration the synthesized product may possibly decompose in the high temperature oxide environment. In the case of oxide superconductors this problem does not exist. In systems in which product decomposition is accompanied by the formation of a protective layer, e.g., SiO_2 layer in the production of SiC, it exists but is not too significant. However, in some systems, e.g., the WC system, the problem is of major significance due to the rapid oxidation of the product. Depending on the specific system under consideration, different arrangements can be employed to deal with this problem, e.g., control of oxidizer concentration, flow velocity, or particle size of solid reactants, or the addition of special additives to form protective layers, etc. In some cases counterflow, rather than coflow filtration, which, though less efficient in heat generation, may be more advantageous overall, when the problem of product decomposition exists.

We considered the possibility of enhancing the propagation of weakly exothermic and endothermic solid-solid reactions in porous media by counterflow filtration of a gas which carries an active component able to react with an excess of one of the solid reactants. We showed that the interaction between the filtration and gasless combustion zones results in a desired stoichiometric composition of the product for a particular value of the gas flux, which we derived. The temperature of the stoichiometric product of synthesis is determined by the proportions of the solid reactants in the initial mixture similar to adiabatic gasless SHS systems in which the burning temperature can be determined by thermodynamic arguments. For practical applications this means that there is a specific value of the gas flux such that both a desired product composition and a temperature sufficient for self-propagation will be achieved. An estimate of the value of this flux under assumptions that are standard in combustion theory was given.

4. Melting in SHS Waves

Combustion of gasless systems is often accompanied by melting of one or more components of either the initial mixture or of the products, either intermediate or final. This is due to the considerable exothermicity and high combustion temperatures for most mixtures. Melting with subsequent spreading of the melt increases the surface-to-surface contact between reactants which significantly promotes the reaction. Moreover, gasless systems in which melting does not occur may not be able to sustain burning since the surface-to-surface contact between solid particles is not sufficient for an intensive reaction.

We formulated and analyzed a model describing the combustion of porous condensed materials in which a reactant melts and spreads through the pores of the sample [2]. Our model describes the cases when the melt either fills all the pores or when some gas remains in the pores. In each case the melt occupies a prescribed volume fraction of the mixture. We employed both analytical and numerical methods to find uniformly propagating combustion waves, to analyze their stability and to determine behavior in the instability region.

The principal conclusion of our analysis is that the flow of the melted component can result in nonuniform composition of the product. Unlike models which do not take into account the relative motion of the components, this model exhibits a dependence of the structure of the product on the mode of propagation

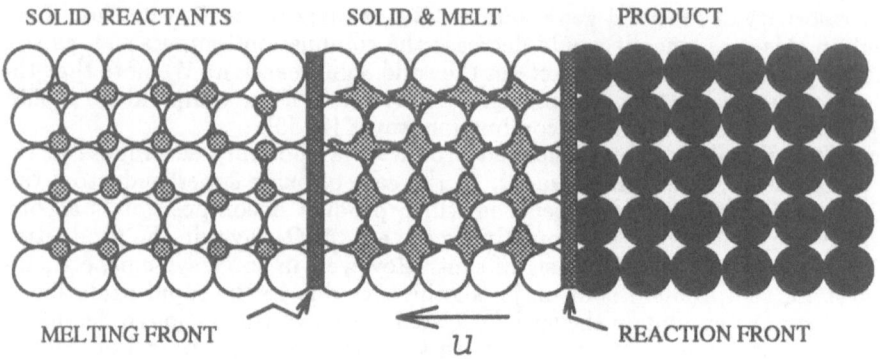

SOLID REACTANTS   SOLID & MELT   PRODUCT

MELTING FRONT u REACTION FRONT

Figure 3:

of the combustion front, in accord with experimental observations.

We proposed a rather simple model of gasless combustion which accounts for melting of one component of the solid mixture, followed by flow of the melt. We considered a sample (Fig. 3) which initially consists of two solid reactants, say S_1 and A_s, of which A_s melts in the combustion wave to form A_l ($A_s \rightarrow A_l$). Melting occurs in a front at which the temperature is equal to the melting temperature. Then the liquid reactant A_l reacts with S_1 to form the product S_2. We assume that neither S_1 nor S_2 melts, and that they form a "hard matrix" for which the volume fraction of the nonmelting components remains constant during the course of the reaction. The melt spreads through either all the pores or only some of them. At each point ahead of the melting front the component A_l occupies a fixed volume fraction (m_{A0}). We assume that at each point behind the melting front where there is melt, it occupies the same prescribed volume fraction (m_{A1}) of the porous medium.

We employed a one dimensional model based on balances of energy and the species masses [2]. The quantities determined in the course of solution are: the distribution of temperature T and effective densities of S_1 and S_2 in the combustion wave, the melt velocity v_A, the velocity v_s of deformation of the solid matrix, the propagation velocity u, the width of the liquid melt layer.

We distinguish two cases, namely the melt and solid deficient cases, depending on which reactant (S_1 or A) is completely consumed in the reaction, thus causing its termination. In particular, in the melt deficient case the region occupied by the melt is the region between the melting and reaction fronts, and there is no melt in the product region. However, in the solid deficient case the melt penetrates the reaction zone, fills at least some of the pores of the sample in the product region, and flows at velocity v_{Ab} either towards the reaction front ($v_{Ab} < 0$) or in the opposite direction ($v_{Ab} > 0$) depending on the effective densities of A_l and S_1, and the stoichiometric coefficient μ. If $v_{Ab} > 0$, the liquid occupies the entire product region. If $v_{Ab} < 0$ the melt fills only part of the burned porous sample. The remaining part of the product remains porous, and contains no liquid (cf. [8]). The ratio of the length of the porous part of the product to the length of that part of the product whose pores are filled with liquid is a constant, which we determine.

The temperature behind the reaction front is always given by the thermodynamic value and does not depend on the direction of the melt flow. The propagation velocity of the combustion wave, as well as the temperature profile in the preheat layer and its length are determined by the same expressions as in the case without melt flow. This result might be considered as a justification of the fact that in the traditional models of gasless combustion the melt flow was not taken into account. However, we emphasize that the identity of results for combustion models with and without melt flow occurs only for uniformly propagating combustion waves. In the nonstationary case the results are drastically different. Filtration of the melt is one of the determining factors for stability of the uniformly propagating wave, for the behavior of the system in the instability region, and for the unsteady behavior seen on the scale of the sample.

We investigated the stability of the uniformly propagating combustion wave, and derived a dispersion relation which was studied both analytically and numerically. In the interesting melt deficient case, in which pulsating combustion results in a layered product, there are two limiting cases in which the dispersion relation can be studied analytically, namely when the distance l between the reaction and melting fronts is either large or small. When l is large and convective heat transfer by the melt is negligibly small, combustion waves in which melt fills the pores completely are more readily destabilized than when the melt fills only a small number of pores. Convective heat transfer stabilizes the uniformly propagating combustion wave for $m \equiv m_{A1}/m_{A0} > 1$ and destabilizes it for $m < 1$. This result can also be formulated in terms of the direction of the melt flow: if the melt flows from the reaction zone toward the melting front, then heat transferred by the flow is stabilizing. If the direction of the flow is opposite, heat transferred by the flow is destabilizing. In a sense, the opposite is true for relatively small values of l. That is, for small l (which can be caused e.g. by melting temperatures close to the reaction temperature) combustion waves in which the melt fills only a small number of pores are more readily destabilized than those in which the pores are completely filled by the melt. Convective heat transfer stabilizes the uniformly propagating combustion wave for $m < 1$ and destabilizes it for $m > 1$. The results of the approximate analysis are in agreement with the results of direct numerical solution of the dispersion relation.

As noted earlier, pulsating combustion waves in the melt deficient case result in a periodic product structure. Instability leads to an oscillation of the distance between the melting and reaction fronts, and to an effective product density which is periodic in time in the coordinate system attached to the front of the (unstable) uniformly propagating wave. Therefore, the density will be periodic in space in the laboratory coordinate system.

5. Smolder waves

Smoldering combustion is important for the study of fire safety. Smoldering itself can cause damage, its products are toxic and it can also lead to the more dangerous gas phase combustion. Examples of porous substances that can sustain smoldering combustion are cotton, dusts, polyurethane foams, thermal insulation materials and wood. We describe smolder combustion through a porous sample when gas is forced into the sample through an end, considering both coflow [9] and counterflow [10] filtration.

We considered a simple one-step reaction in which solid and gaseous reactants produce solid and gaseous products and release heat. This one-step reaction represents the rate limiting step in a more complex reaction scheme. Heat transfer between the solid and the gas is assumed to be sufficiently fast that the gas and solid phases are in local thermodynamic equilibrium. Thus, a single temperature model is used to describe both the solid and gaseous phases. The solid phase is considered to be stationary and nondeforming. Fick's law is used to describe the diffusion of oxygen through the gas. The activation energy is considered to be

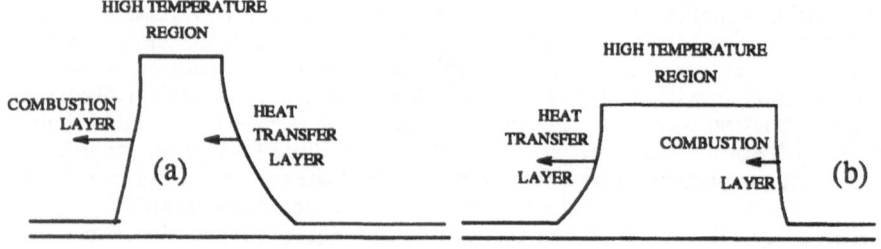

Figure 4:

large so that the reaction occurs in a narrow region. The sample is assumed to be highly porous so that the characteristic length scale of pressure variations is much larger than that for thermal variations. In addition, the ambient pressure is assumed to be larger than the hydrostatic pressure drop over the length of the sample as is generally the case for normal gravity situations.

The model consists of equations describing balances of energy, gas mass, oxygen mass, solid reactant mass, and gas momentum using Darcy's law, as well as an equation of state, and appropriate boundary and initial conditions [9], [10]. Incoming gas flux m^i and the concentration Y^i of the oxidizer in the gas are prescribed at the open end of the sample.

For coflow smoldering we found that on the scale of the sample the temperature profile for a standard solution exhibits three regions in which the temperature is essentially constant (Fig.4a). The sample is cool ahead of the reaction site, hot immediately behind the reaction zone, and cool again where the sample is cooled by the incoming gas. The initial and final temperatures need not be the same if the incoming gas is at a different temperature than the initial sample. Connecting these regions of constant temperature are two transition layers in which the temperature changes rapidly. These two layers, which in general move at different velocities, are referred to as the combustion layer and the heat transfer layer.

In standard solution structures the combustion layer precedes the heat transfer layer. There are two types of standard structure depending on the oxygen mass flux $Y^i m^i$: oxygen controlled and kinetically controlled solutions. Standard oxygen controlled solutions occur when the rate of oxygen supply is slow compared to the reaction rate and the reaction must wait for oxygen to arrive before moving on. This mode of propagation can be identified by the linear dependence of the smolder speed on $Y^i m^i$. When the supply of oxygen is sufficiently high that the rate of consumption limits the speed of propagation the solution is said to be kinetically controlled. Smolder velocity is determined by reaction kinetics. Increasing Y^i affects the smolder velocity weakly through the reaction rate increase due to a higher concentration of oxygen at the reaction site. Varying m^i, however, affects the smolder velocity by changing the burning temperature, which has a large effect on the propagation velocity.

In addition to the standard solution, we also described an inverted solution where the heat transfer layer precedes the combustion layer (Fig.4b). Incoming gas carries the heat rapidly from the combustion layer into the unburned region where it preheats the solid reactant. As in the standard structure, there are two types of solution with the inverted structure. The inverted oxygen controlled solution is similar to the standard oxygen controlled solution in that the smolder velocity is proportional to the oxygen influx $Y^i m^i$, and the burning temperature is affected by Y^i and not by m^i. However, the structure of the wave is very different in that all the heat from the reaction is carried ahead of the reaction front. Kinetically controlled inverted structure solutions occur when the incoming gas absorbs heat sufficiently fast that the solid is cooled before it is completely converted. This solution differs from the standard kinetically controlled solution in its dependence on the oxygen influx because the final depth of conversion of solid and thus the total heat released in the reaction is determined by the oxygen influx. The incoming gas and in particular the incoming oxygen affect the smolder velocity and the burning temperature. Solutions with standard and inverted structures were previously found in [6], [7].

We found that both standard and inverted solutions in forced coflow smolder exhibit super-adiabatic burning temperatures [6], [7]. Super-adiabatic burning temperatures are achieved for forced coflow because the energy released in the reaction remains in the localized high temperature region between the two layers and thus travels with the wave. In the standard structure, we think of the gas extracting heat from the hot solid product, and carrying it toward the reaction site. In this sense, the gas can be said to be preheated, i.e. the gas is heated to the burning temperature before it reaches the combustion layer. In the inverted structure, the gas, which is heated while passing through the reaction zone, heats the solid reactant so that the solid reactant can be said to be preheated. The magnitude of the super-adiabatic effect is determined by the relative velocities of the heat transfer and combustion layers.

Inverted (standard) structure solutions occur for Y^i less (greater) than a critical value, and for a given Y^i, kinetically (oxygen) controlled solutions occur for large (small) incoming mass flux.

For the counterflow smolder problem we find two types of solutions. Gas deficient solutions, in which the oxidizer is consumed in the reaction and solid conversion is incomplete, and solid deficient solutions, in which conversion of the solid reactant is complete.

The most interesting results for the counterflow smolder involve extinction limits. The extinction limits considered are adiabatic extinction limits in the sense that they are not due to heat losses to the external environment. Rather, extinction is due to heat transferred from the reaction to the inert gas passing through it. Extinction occurs when the heat lost to the inert exceeds a critical value.

The model predicts that in the environments where smoldering is usually studied, i.e. at moderate oxygen concentrations and with reaction rates which do not vary too strongly with the degree of conversion, extinction occurs when solid conversion is complete. Extinction can occur within the solid deficient regime or at stoichiometry which corresponds to the limit of the solid deficient regime. In some cases, when the reaction rate exhibits a strong kinetic dependence on the solid fuel, extinction can occur in the gas deficient regime.

This work was supported by DOE grant DE-FG02-87ER25027, NSF grant CTS93-08708 and NASA grant NAG3-1608.

References

[1] Aldushin, A.P.: New results in the theory of filtration combustion. Combust. & Flame **94** (1993) 308-320

[2] Aldushin, A.P., Matkowsky, B.J., Shkadinsky, K.G., Shkadinskaya, G.V., Volpert, V.A.: Combustion of porous samples with melting and flow of reactants. Comb. Sci. Tech. **99** (1994) 313-343

[3] Aldushin, A.P., Matkowsky, B.J., Volpert, V.A.: Interaction of gasless and filtration combustion. Comb. Sci. Tech. **99** (1994) 75-103

[4] Aldushin, A.P., Matkowsky, B.J., Volpert, V.A.: Enhancement of gasless combustion synthesis by counterflow gas filtration. Comb. Sci. Tech. (to appear)

[5] Aldushin, A.P., Merzhanov, A.G.: Theory of filtration combustion. In: *Propagation of Thermal Waves in Heterogeneous Media*, Matros Y.S. (Ed), Nauka, Novosibirsk, 1988, 9-52 (in Russian).

[6] Aldushin, A.P., Seplyarsky, B.S.: Propagation of waves of exothermal reaction in porous medium during gas blow-through. Soviet Phys. - Dokl. **23** (1978) 483-485

[7] Aldushin, A.P., Seplyarsky, B.S.: Inversion of the structure of a combustion wave in a porous medium during blow-through of gas. Soviet Phys. - Dokl. **24** (1979) 928-930

[8] Feng, H.J., Moore, J.J., Wirth, D.C.: Combustion synthesis of ceramic-metal composite materials: the $TiC-Al_2O_3-Al$ system. Metallurgical Transactions A **23** (1992) 2373-2379

[9] Fernandez-Pello, A.C., Matkowsky, B.J., Schult, D.A., Volpert, V.A.: Propagation and extinction of forced counterflow smolder wave. Combust. & Flame (to appear)

[10] Fernandez-Pello, A.C., Matkowsky, B.J., Schult, D.A., Volpert, V.A.: Forced forward smolder combustion. Combust. & Flame (to appear)

[11] Merzhanov, A.G.: Self-propagating high-temperature synthesis: twenty years of search and findings. In: *Combustion and Plasma Synthesis of High-Temperature Materials*, Munir, Z.A., Holt, J.B. (Eds.), VCH, 1990, 1-53

[12] Munir, Z.A., Anselmi-Tamburini, U.: Self-propagating exothermic reactions: the synthesis of high-temperature materials by combustion. Material Science Reports, A Review Journal **3** (1989) 277-365

[13] Shkadinsky, K.G., Shkadinskaya, G.V., Matkowsky, B.J., Volpert, V.A.: Self-compaction or expansion in combustion synthesis of porous materials. Comb. Sci. Tech. **88** (1992) 271-292

[14] Shkadinsky, K.G., Shkadinskaya, G.V., Matkowsky, B.J., Volpert, V.A.: Combustion synthesis of a porous layer, Comb. Sci. Tech. **88** (1992) 247-270

[15] Shkadinsky, K.G., Shkadinskaya, G.V., Matkowsky, B.J., Volpert, V.A.: Combustion of porous samples with deformation of high temperature products. Int'l J. of SHS **1** (1992) 371-391

[16] Shkadinsky, K.G., Shkadinskaya, G.V., Matkowsky, B.J., Volpert, V.A.: Two front traveling waves in filtration combustion, SIAM J. Appl. Math. **53** (1993) 128-140

[17] Shkadinsky, K.G., Shkadinskaya, G.V., Matkowsky, B.J., Volpert, V.A.: Filtration combustion with self deformation, J. Material Synth. & Proces. **1** (1993) 245-274

7. High Mach Numbers

Stability of Reacting Mixing Layers

T.L. Jackson[1]

[1]ICASE, NASA Langley Research Center, Hampton, VA 23681-0001, USA

Abstract. The structure and stability characteristics of reacting, compressible mixing layers are briefly reviewed. In addition, recent unpublished work will also be presented.

Keywords. Stability, mixing layer, reacting

1 Introduction

In recent years there has been renewed interest in understanding the stability characteristics of compressible mixing layers, due in part to the projected use of the scramjet engine for the propulsion of hypersonic aircraft. The study of the stability of these flows is particularly important because experimental and computational results show an increase in the flow stability at high Mach numbers. One effect of this is that the mixing between the fuel and oxidizer may decrease as the Mach number increases, resulting in partial burning and a loss in combustion efficiency. Because of this gain in stability, natural transition may occur at downstream distances which are larger than practical combustor lengths. Therefore, it is desirable to examine techniques which may enhance mixing. Knowledge of these characteristics may allow one, in principle, to control the downstream evolution of such flows. Further discussion of these issues are given in the recent reviews by Beach [1] and Bushnell [2].

The purpose of this article is to briefly review the structure and stability of compressible mixing layers. We shall do this within the context of a three-dimensional mixing layer. For more in-depth reviews, the interested reader is referred to the review articles by Jackson [3] and Grosch [4]. Recent unpublished work will also be presented here.

2 Mean Flow

Consider a three-dimensional reacting compressible mixing layer, with zero pressure gradient, which separates two streams of different speeds and temperatures. Only a brief description will be presented here and complete details can be found in the recent article by Jackson and Grosch [5]. The mean flow at $y = \pm\infty$ is parallel to the (x, z) plane. All three velocity components of the mean flow are non-zero, although they are only a function of the downstream coordinate, x, and the normal coordinate, y, and are independent of the cross stream coordinate, z. We let (U, V, W) be the velocity components in the (x, y, z) directions, respectively, ρ the density, T the temperature, and F_1 and F_2 the mass fractions of the fuel and oxidizer. We also assume that the reaction is single step, irreversible and of the Arrhenius type. All of the variables are non-dimensionalized using the magnitudes of the freestream values at $y = +\infty$. Lengths are referred to some characteristic length scale of the flow. We shall assume that the mean flow is governed by the three-dimensional compressible boundary layer equations. The appropriate set of initial (in x) and boundary (in y) conditions are:

$$T = 1, \quad U = \cos\theta, \quad W = \sin\theta, \quad F_1 = 1, \quad F_2 = 0 \tag{1}$$

for $x = 0$, $y > 0$, and $x > 0$, $y \to \infty$;

$$T = \beta_T, \quad U = \beta_U, \quad W = 0, \quad F_1 = 0, \quad F_2 = \phi^{-1} \tag{2}$$

for $x = 0$, $y < 0$, and $x > 0$, $y \to -\infty$. Note that the fast stream at $y = +\infty$ is moving at an angle θ with respect to the x-axis, with $0° \le \theta \le 90°$. If $\theta = 0°$ there is no crossflow. If $\theta = 90°$ the flow at $+\infty$ is along the z-axis and at $-\infty$ is along the x-axis. The velocity ratio parameter β_U lies in the range $(0, \cos\theta)$. If the temperature ratio parameter β_T is less than one, the gas in the slow stream is relatively cold compared to that in the fast stream, and if β_T is greater than one it is relatively hot. In addition, ϕ is the equivalence ratio defined as the ratio of the mass fraction F_1 in the fast stream to the mass fraction F_2 in the slow stream. If $\phi = 1$, the mixture is stoichiometric, if $\phi > 1$ it is F_1 (Fuel) rich, and if $\phi < 1$, it is F_1 (Fuel) lean. Finally, since the equation for W has the same functional form as the equation for U, we have

$$W = \frac{\sin\theta}{\cos\theta - \beta_U}(U - \beta_U). \tag{3}$$

Note that if $\beta_U = 0$ then W is proportional to U, and by an appropriate rotation of the axes, the mean flow can be reduced to a two-dimensional one, i.e. the angle θ can be scaled out of the problem. Thus we see that the parameter β_U plays a role equivalent to that of the pressure gradient parameter in a three-dimensional boundary layer flow.

Asymptotic and numerical solutions are facilitated if the equations are first transformed into incompressible form by means of the Howarth-Dorodnitzyn transformation

$$Y = \int_0^y \rho\, dy, \quad \hat{V} = \rho V + U \int_0^y \rho_x\, dy. \tag{4}$$

Because the velocity profile U attains a self-similar form at a very small value of x, we seek solutions in terms of the similarity variable for the chemically frozen heat conduction problem

$$\eta = \frac{Y}{2\sqrt{x}}, \quad U = f'(\eta), \quad \hat{V} = x^{-1/2}(\eta f' - f), \quad W = g(\eta). \quad (5)$$

All results are carried out in the (x, η) variables.

3   Ignition and structure

For small x we expect the flow to be chemically frozen. For the special case of Chapman's linear viscosity law $\mu = T$, and unit Prandtl number, the frozen (or inert) solution is given by

$$T_I = 1 - (1 - \beta_T - \lambda)\Psi - \lambda\Psi^2 \quad (6)$$

where

$$\Psi = \frac{\cos\theta - f'}{\cos\theta - \beta_U}, \quad \lambda = \frac{\gamma - 1}{2}M^2(1 - 2\beta_U\cos\theta + \beta_U^2), \quad (7)$$

and $\Psi \in [0, 1]$ for $\beta_U \leq f' \leq \cos\theta$. As x increases, more of the combustible mixes until, at some finite distance downstream of the plate, a thermal explosion occurs characterized by significant departure from the inert. For large Damkohler number, ignition will occur at the maximum of T_I, given by $T_I' = 0$; i.e.

$$[-(1 - \beta_T - \lambda) - 2\lambda\Psi]\Psi' = 0. \quad (8)$$

If the first term is set equal to zero, then the ignition location is given by the implicit relation

$$\Psi_* \equiv \Psi(\eta_*) = -\frac{1 - \beta_T - \lambda}{2\lambda}. \quad (9)$$

Since Ψ_* is required to lie in the range $[0, 1]$, we note the following two cases:

- $\lambda < 1 - \beta_T$ ignition is governed by external heating and the ignition location lies outside the mixing layer; or

- $\lambda > 1 - \beta_T$ ignition is governed by viscous heating and the ignition location lies inside the mixing layer where the classical triple-flame structure emerges downstream and acts as a flameholding device.

Three dimensionality of the mean flow has a significant effect on the ignition location. For example, the case of $\beta_U = \beta_T = 0.5$, ignition will take place within the mixing layer for $M \geq 3.16$ with $\theta = 0^o$, but at an angle of $\theta = 60^o$, ignition will take place for $M \geq 1.83$; also, the ignition location moves upstream towards the splitter plate as the crossflow angle is increased from zero. Asymptotic analysis is used to further describe the ignition regime as a function of the mean flow parameters β_T, β_U, M, and θ, and post-ignition events are investigated numerically [5].

4 Stability

In investigating the stability of mixing layers, it is typical to assume that there exists a local parallel flow about which the governing equations are linearized with respect to spatially and temporally varying disturbances. From this linearization, it is straightforward to calculate either temporal growth rates (assuming fixed spatial wavenumbers) or to calculate spatial growth rates (assuming a fixed temporal frequency). If an instability exists, there is usually a band or bands of wavenumbers (or frequencies) for which there are positive growth rates. These bands are bounded by the neutral modes, whose existence (and phase speeds) can be determined through the Lees and Lin regularity condition, assuming that the phase speeds are subsonic. For modes with phase speeds which are supersonic with respect to either the fast freestream or the slow freestream, the neutral modes must be determined numerically. Another neutral mode, either subsonic or supersonic, can be found in the limit of the wavenumber going to zero. A typical sketch of neutral modes (denoted by the solid curves) in the phase speed-Mach number plane is shown in figure 1. The two dashed curves in the figure correspond to disturbances which are sonic with respect to a freestream and separates subsonic regions from supersonic regions and is useful for classifying modes. That is, modes in the region labeled (i) have phase speeds which are subsonic at both freestreams; modes in region (ii) have phase speeds which are subsonic in the fast stream and supersonic in the slow stream; modes in region (iv) have phase speeds which are supersonic in the fast stream and subsonic in the slow stream; and modes in region (iii) have phase speeds which are supersonic in both freestreams. In this way, modes can be classified as subsonic (region i), fast supersonic (region ii), slow supersonic (region iv), or supersonic-supersonic (region iii) [6]. The regions of instability lie between neighboring neutral modes. For example, unstable modes lies between the neutral modes labeled 1 and 2 (subsonic), 3 and 5 (fast supersonic), and 4 and 7 (slow supersonic). The supersonic-supersonic neutral mode labeled 5 is isolated and has no regions of instability about it.

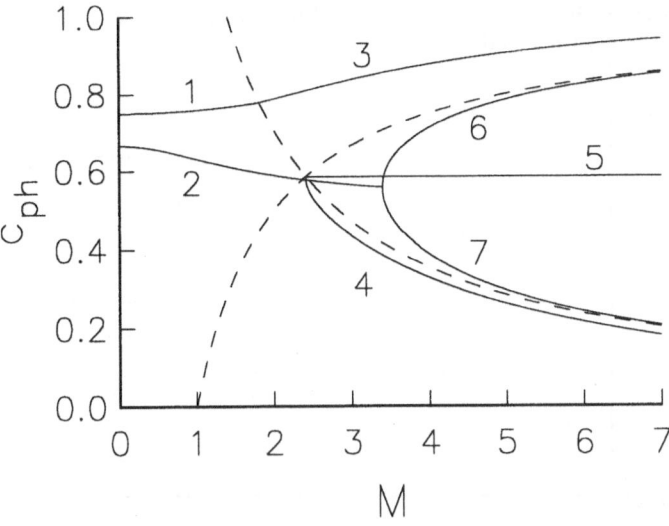

Figure 1. Plot of the neutral phase speeds as a function of Mach number for a particular thermodynamic model and selected mean flow parameters.

The results of previous investigations leads to the conclusion that the temperature profile, which can be significantly altered through external heating or cooling, internal viscous heating as well as exothermic chemical reactions, alters the Lees-Lin regularity condition sufficiently such that an additional pair of unstable modes exist. In the absence of reaction, viscous heating significantly raises the temperature so that at a large enough Mach number, there are three possible neutral modes instead of one. Although these additional modes lie in a region in which the phase speeds would typically be supersonic, significant crossflow can alter the sonic phase speed curves such that all three neutral modes represent a physically realizable subsonic mode. In the case of a reacting mixing layer, a simple one-step exothermic reaction with moderate heat release can easily introduce an extra pair of neutral modes, all of which will be subsonic modes (figure 2). A "flame sheet" analysis can be used to quickly locate these modes, one of which will have a phase speed equal to the flow velocity at the flame sheet location. An extensive study of the spatially evolving reacting mixing layer with finite reaction rate showed that the "flame sheet" results gave accurate values of the phase speeds of the neutral modes as long as the Lees-Lin regularity condition was applied downstream of the ignition point. Further analysis showed that the slow mode may undergo a transition from convective to absolute instability as the heat of reaction increases. It should be noted that although this transition is deemed significant, the backwards propagation of the disturbance, which is the hallmark of an absolute instability, is seen to be (after a wave packet analysis) exceedingly small.

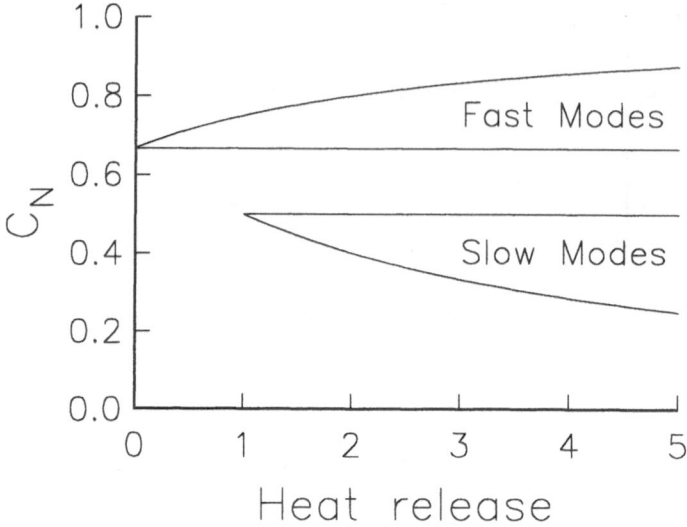

Figure 2. Plot of the neutral subsonic phase speeds as a function of heat release parameter at $M = 0$ for a particular thermodynamic model and selected mean flow parameters.

All previous investigations on the stability of mixing layers, either reacting or non-reacting, have assumed equal molecular weights for the gases above and below the splitter plate. We present here preliminary results on how differing molecular weights for a non-reacting gas can affect the stability characteristics. Complete details will be presented at a later date.

As before, the flow field is perturbed by introducing wave disturbances in the velocity, pressure, temperature and density with amplitudes which are functions of η. The only difference is that the gas law now takes into account different gases. Substitution into the inviscid compressible equations and linearizing yields the compressible Rayleigh's equation for a multi-component gas. Assuming constant thermodynamic properties and the flow to be two-dimensional ($\theta = 0$) with two-dimensional disturbances ($\tilde{\theta} = 0$), the Lees-Lin regularity condition is now given by:

$$S(\eta) = U'' - 2\left(\frac{T'}{T} + \frac{G'}{G}\right)U', \qquad G = \sum_{i=1}^{N} F_i/W_i, \qquad (10)$$

where W_i is the molecular weight of species F_i. As before, zeros of $S(\eta_c)$ correspond to the neutral subsonic phase speeds $c_N = U(\eta_c)$. The growth rates, either spatial or temporal, can then be computed from Rayleigh's equation.

To illustrate how a multi-component gas may alter the stability characteristics, we take as an example a two-component gas with $N = 2$ and let

$$G = \frac{F_1}{W} + F_2, \qquad W = \frac{W_1}{W_2}, \tag{11}$$

where W is the ratio of the molecular weights. We have the following two cases depending on the magnitude of W:

- $\underline{W > 1}$ heavier gas resides in the fast freestream at $\eta = +\infty$ and the lighter gas in the slow freestream at $\eta = -\infty$; or

- $\underline{W < 1}$ lighter gas resides in the fast freestream at $\eta = +\infty$ and the heavier gas in the slow freestream at $\eta = -\infty$.

For the inert gases Ar and He (typical gases used in experiments), we see that W can vary between 0.1 for the $He - Ar$ case, and 9.9 for the $Ar - He$ case. The figures below present the neutral phase speeds and spatial growth rates for various values of W. From these figures we conclude that differing molecular weights can have a significant effect on the phase speeds and growth rates of the disturbances, and should be taken into account when computing stability characteristics of multi-component gases. Finally, we are currently building a code for a multi-component gas with non-constant thermodynamic properties and will present these results elsewhere.

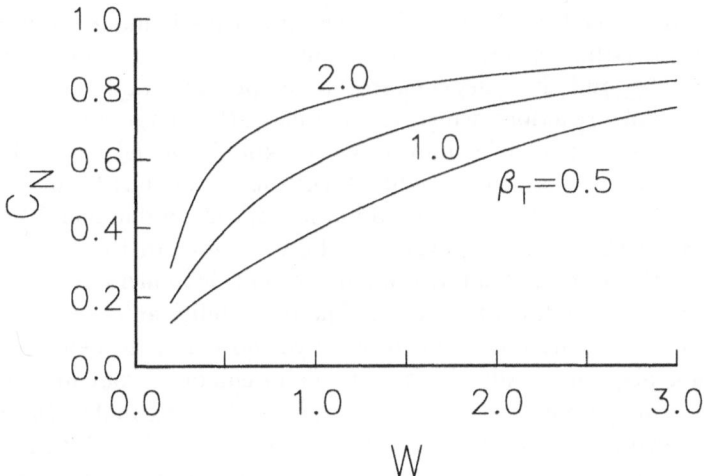

Figure 3. Plot of the neutral phase speeds as a function of W at $M = 0$ and various values of the temperature ratio β_T.

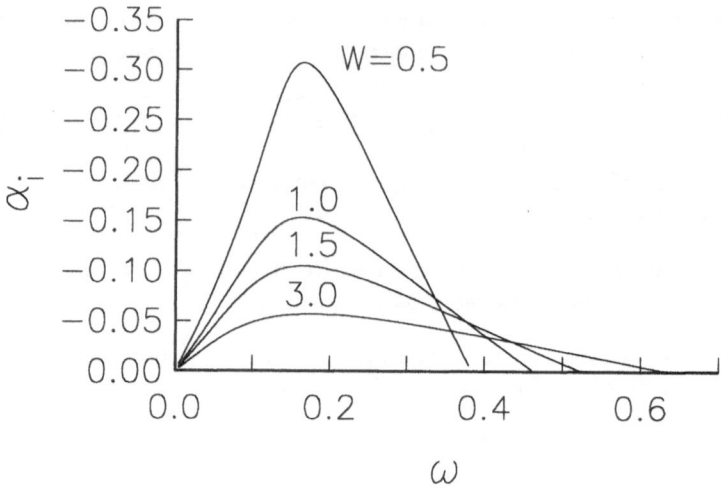

Figure 4. Plot of the maximum growth rates as a function of frequency for $M = 0$, $\beta_T = 1.0$, and $\beta_U = 0.2$.

It is also important to determine how sensitive the results of the stability calculations are to the details of a particular finite rate chemistry model. Past investigations on the stability of reacting mixing layers employed a simplified kinetic model consisting of a one-step, irreversible Arrhenius reaction. The reason for using a simple kinetic model is clear: any realistic modeling of complex kinetics will necessitate a full numerical solution. Then questions arise as to the reaction rates and their dependence on temperature and the relative importance of various reactions which entail thus rather large uncertainties. In a mathematical treatment of combustion, the kinetic model has to be necessarily simple. The one-step irreversible Arrhenius model has been extensively used with significant success in the study of low-speed combustion. This model appears to cover the essential physics of the problem. In the last decade or so, the asymptotic studies of combustion based on this model have significantly enhanced our understanding of ignition, of flame stability and of diffusion flame structure in low subsonic flows. There is as yet no reason to believe that this model will not play an equally significant role in enhancing our understanding of supersonic combustion. In any event, this idealization makes the problem amenable to asymptotic analysis and thus provides a semi-analytical solution. The results can then certainly verify and in turn be verified by full numerical simulations. However, there is still the question of sensitivity of complex kinetics on flow stability characteristics. For example, one such question is how does complex kinetics affect the number of zeros, and hence the number of subsonic neutral modes, of the Lees-Lin regularity condition as compared to the simple one-step model? One way to address this question is to investigate the regularity condition using various reduced mechanisms. In using certain reduced mechanisms, we will not be concerned with the validity of a particular model to high-speed combustion. Preliminary results using the reduced mechanisms given below are shown in figure 5. Note that for each model the number of zeros of the Lees-Lin regularity condition is three, the same as for the simple one-step model. We found at most three zeros as various mean flow parameters where changed. A more complete discussion will be presented elsewhere.

- Birkan and Law

$$F + R_1 \rightarrow 2R_2$$

$$O + R_2 \rightarrow 2R_1$$

$$R_1 + R_2 + M \rightarrow 2P + M$$

- Peters

$$3H_2 + O_2 \leftrightarrow 2H + 2H_2O$$

$$2H + M \leftrightarrow H_2 + M$$

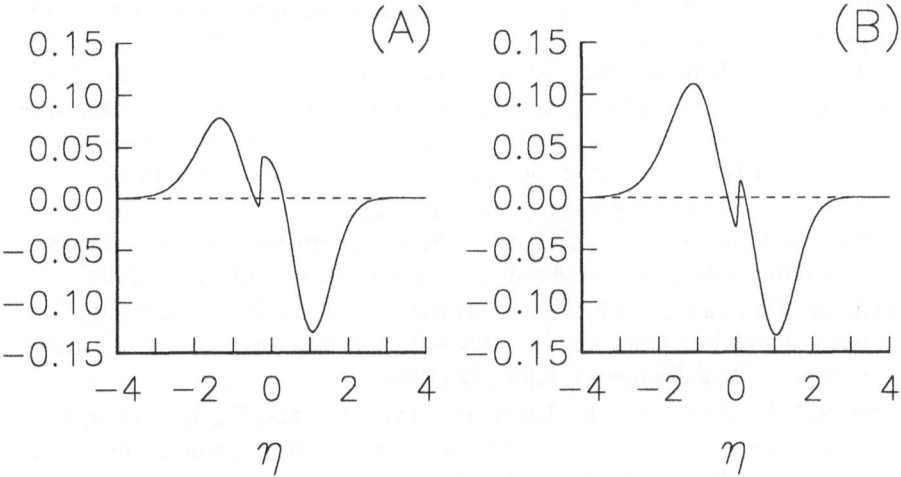

Figure 5. Plot of the Lees-Lin regularity condition $S(\eta)$ for (A) the Birkan and Law model and (B) for Peters model.

5 Conclusions

The structure and stability characteristics of reacting, compressible mixing layers was briefly reviewed. In addition, recent unpublished work on the effects of multi-component gases and various reduced mechanisms on the stability characteristics was also presented.

Many of these concepts can be extended to include, but not limited to: curved mixing layers; flows with non-zero pressure gradients; fuel jets; flow about the triple-deck region at the edge of the splitter plate; or flows with vector or parallel wall fuel injectors.

6 Acknowledgements

The author is greatly indebted to Professors C.E. Grosch (ODU), D.G. Lasseigne (ODU), F.Q. Hu (ODU) and Dr. M.Y. Hussaini (ICASE) for the many years of fruitful scientific interaction and friendship.

7 References

1. Beach, H.L., "Supersonic combustion status and issues", in *Major Research Topics in Combustion*, M.Y. Hussaini, A. Kumar, R.G. Voigt, (eds.), Springler-Verlag, pp. 1-20, (1992).

2. Bushnell, D.M., "Mixing and combustion issues in hypersonic air-breathing propulsion", in *Combustion in High-Speed Flows*, J. Buckmaster, T.L. Jackson, A. Kumar (eds.), Kluwer Academic Publishers, pp. 3-16, (1994).

3. Jackson, T.L., "Stability of laminar diffusion flames in compressible mixing layers", in *Major Research Topics in Combustion*, M.Y. Hussaini, A. Kumar, R.G. Voigt, (eds.), Springler-Verlag, pp. 131-161, (1992).

4. Grosch, C.E., "Reacting compressible mixing layers: structure and stability", in *Combustion in High-Speed Flows*, J. Buckmaster, T.L. Jackson, A. Kumar (eds.), Kluwer Academic Publishers, pp. 131-190, (1994).

5. Jackson, T.L. and Grosch, C.E., "Structure and stability of a laminar diffusion flame in a compressible, three-dimensional mixing layer", *Theoret. Comput. Fluid Dynamics*, 6:89-112 (1994).

6. Grosch, C.E., Jackson, T.L., Klein, R., Majda, A. and Papageorgiou, D.T., "The inviscid discrete eigenvalue spectrum of the compressible mixing layer", unpublished manuscript, (1991).

Ignition and Combustion Characteristics in Supersonic Flow

T. Niioka, Y. Ju, T. Fujimori and K. Takita

Institute of Fluid Science, Tohoku University
Katahira, Sendai 980-77, Japan

Abstract. Numerical and analytical results on ignition and combustion phenomena which are characteristic in supersonic flow are presented. In a supersonic mixing layer of hot air and fuel, the ignition process, which is strongly enhanced by viscous heating at high Mach numbers, is analyzed numerically for hydrogen. Next, extinction of a diffusion flame established in the forward stagnation region of a porous sphere or cylinder in supersonic airflow is solved by large activation energy asymptotics. Lastly, the mechanism of flame-holding supported by shock waves, which are induced into the subsonic region behind a triangular strut, is clarified by a numerical approach.

Keywords. supersonic combustion, ignition, viscous heating, shock heating, flame holding

1. Introduction

In retrospect, we have never fully utilized the advantages of supersonic combustion with a high rate of heat release. A recent important application of such supersonic combustion is the scramjet engine for supersonic or hypersonic vehicles such as the spaceplane. Although supersonic combustion phenomena may not supersede established combustion fundamentals, they offer various very interesting problems to us. Unless the complicated mechanisms of ignition and combustion are clarified and techniques for their control are developed, realization of the scramjet engine will be very difficult, because combustion phenomena change suddenly and drastically due to the intake of air by condition of the scramjet engine in flight.

Of interest in supersonic combustion is that the mass transfer time is very short and therefore the relevant Damköhler number may easily become smaller than unity. Other points of interest are the large amount of heat produced by viscous dissipation, the existence of shock waves or expansion waves, the large total enthalpy, and so on. Supersonic combustion, therefore, holds promise of new findings on the interaction between the combustion reaction and the above-mentioned various phenomena concerned with supersonic flow.

First, the ignition event which takes place in the supersonic mixing layer is numerically solved for hydrogen. When the velocity difference between two parallel flows becomes large, heat produced by viscous dissipation promotes an ignition reaction, and so the ignition point or the starting point of the combustion region moves upstream.

Second, Tsuji's counterflow diffusion flame problem is solved for supersonic airflow. In subsonic flow, when air velocity is sufficiently large, the diffusion flame established in the forward stagnation region is blown off. If airflow is supersonic, however, it has a large total enthalpy and so the airflow temperature becomes high behind the shock wave. It follows that the diffusion flame may once again develop in the forward stagnation region and that the adiabatic flame temperature may increase as the Mach number of airflow increases.

Third, an interesting flameholder is introduced and its mechanisms of flame holding in supersonic flow is discussed. When shock waves are introduced into the subsonic flow region behind a flameholder of a triangular prism, they enable the flame to become anchored. This phenomenon is numerically verified.

2. Ignition in the Supersonic Mixing Layer

As shown in Fig. 1, we consider the reaction process which proceeds in the supersonic mixing layer. Hot airflow and fuel gas flow divided by a smooth splitter plate contact at $x = 0$, diffuse each other and finally ignite to establish a diffusion flame. This configuration involves the most fundamental and important aspects of the scramjet engine. We especially focus our attention on ignition or flame stabilization, which principally determines whether the scramjet combustors can work well at supersonic flow.

In recent years, the reacting supersonic mixing layer has attracted major interest. Jackson and Hussaini [1] asymptotically analyzed the ignition and combustion for one-step reaction kinetics in reacting supersonic flow, and showed that the ignition point moved upstream due to viscous dissipation. Ju and Niioka [2-4] also asymptotically obtained the ignition process of combustible gas for a two-step chemical reaction scheme and graphically illustrated the significant contribution of viscous heating to the ignition reaction.

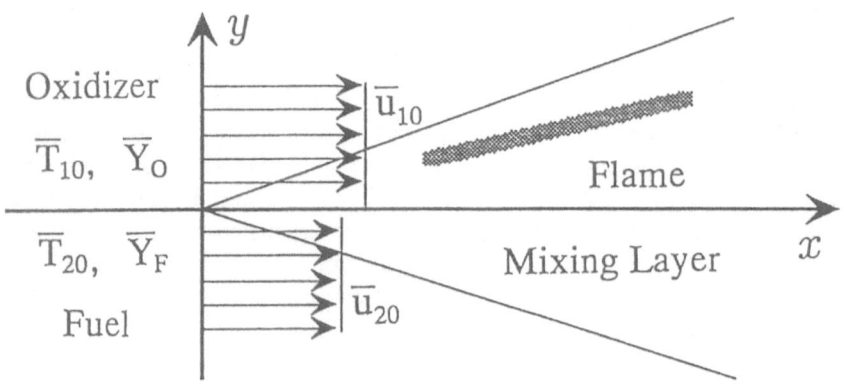

Figure 1 Schematic explanation of ignition problem in parallel supersonic streams.

Silva, Deshaies and Champion [5] numerically calculated the combustion process of hydrogen and the air mixing layer for full kinetics, based on boundary layer equations and constant properties, and reaffirmed an enhancement of ignition by viscous heating. Ju and Niioka solved this problem by use of large activation energy asymptotics for three-step kinetics [6], and then obtained the numerical solutions for the full kinetics of hydrogen and air by using variable properties and the N-S equation [7], the reduced chemistry of the process of hydrogen and air ignition [8], the case of ignition of blended fuel of hydrogen and methane [9], and the asymptotic structure of a diffusion flame in the supersonic mixing layer [10].

Since a recent experiment [11] showed that transition from laminar to turbulent airflow occurred 28 mm downstream of the splitter, which separated Mach 3 and 8 airflow, a laminar flow can be assumed here for simplicity to grasp the original aspect of ignition in the flow system shown in Fig. 1. Two-dimensional and compressible N-S equations were solved for hydrogen and air system with full kinetics of Stahl and Warnatz by the second-order Harten-Yee scheme, improved for reactive real gas. Variable properties were applied to multicomponent gas [7]. The computational domain has a length of 15 cm along the streamwise coordinate and a height of 3 cm along the vertical coodinate. The computational grid has 81 × 61 points.

In Fig. 2, typical temperature distributions for the same initial temperatures and the different Mach numbers of hydrogen and air are shown. Although the ignition point is located around 5 cm downstream in the case of Fig. 2 (a), it shifts to about 12 cm downstream in the case of Fig. 2 (b) when the Mach number of the airflow increases up to 3. In Fig. 2 (c) of the airflow Mach number 4, however, the approach of the ignition location to the splitter plate is seen. Up to ignition around 10 cm downstream, heating of fluid flow due to viscous discipation can apparently be seen. The ignition locations are shown in Fig. 3, with the results of reduced chemistry [8].

Figure 4 shows the case of a confined width of hydrogen injection. The calculation was carried out for velocity of airflow and hydrogen, $U_1 = 1000$ m/s, $U_2 = 550$ m/s, and the temperature of airflow and hydrogen, $T_1 = 1300$ K and $T_2 = 300$ K, respectively. The computational grid increased to 151 × 101 points. The algebraic turbulence model applicable to turbulent wake flows is employed in this calculation. A schematic temperature profile is shown in Fig. 5 for the case of $D = 6$ mm. Since the supersonic mixing layer is extremely thin, there seems to be little interaction between the upper and lower ignition processes. When the width of the hydrogen jet is less than 4 mm, only a slight effect arises, as seen in Fig. 6. Pressure waves are produced from two ignition locations, as shown in Fig. 5, which can also be recognized in Fig. 3.

From the above-mentioned results, it is noted that the mixing in the supersonic shear layer is very slow and so the development of the mixing layer is very slow. To enhance the mixing, Bogdanoff [12] proposed pulsating injectors employing a Hartmann-Sprenger tube. The present authors [7] also calculated the case of supersonic mixing layer with initial disturbances composed of a linear combination of two harmonic components of wave numbers. It was found that ignition is considerably enhanced at high convective Mach numbers by imposing a very small initial perturbation stream. This suggests the possibilities of new advanced mixing techniques for the scramjet engine.

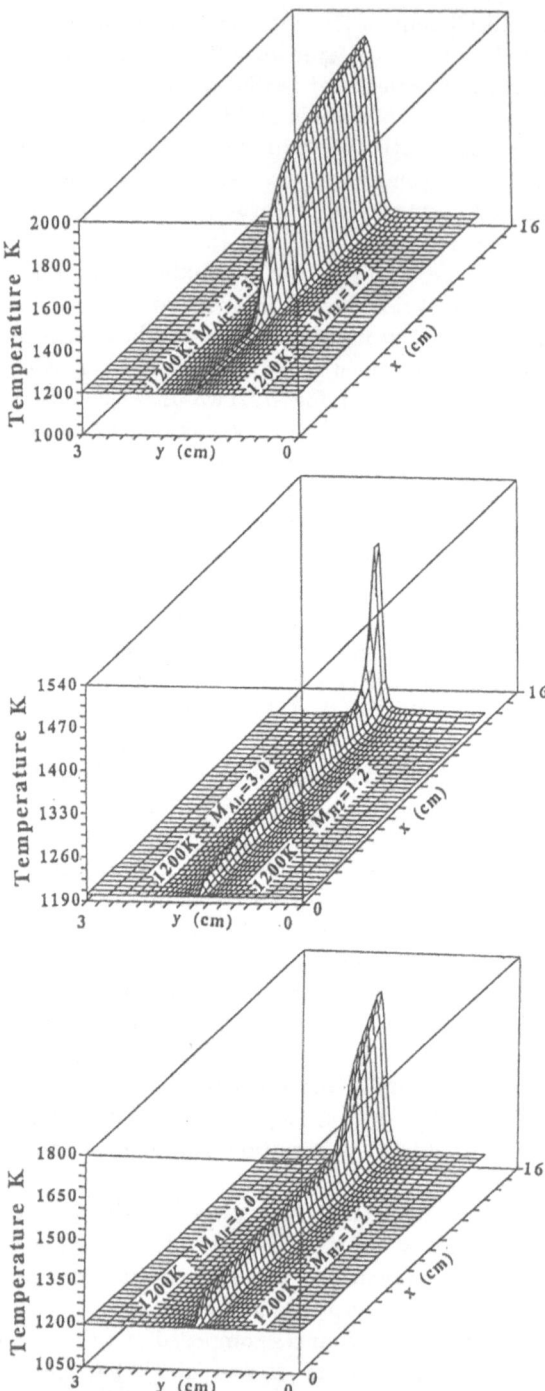

Figure 2 Temperature profiles of ignition process in the case of the same initial temperature $T_{\text{Air}} = T_{H2} = 1200~K$ and $M_{H2} = 1.2$, $(a)M_{\text{Air}} = 1.3$, $(b)M_{\text{Air}} = 3$, $(c)M_{\text{Air}} = 4$.

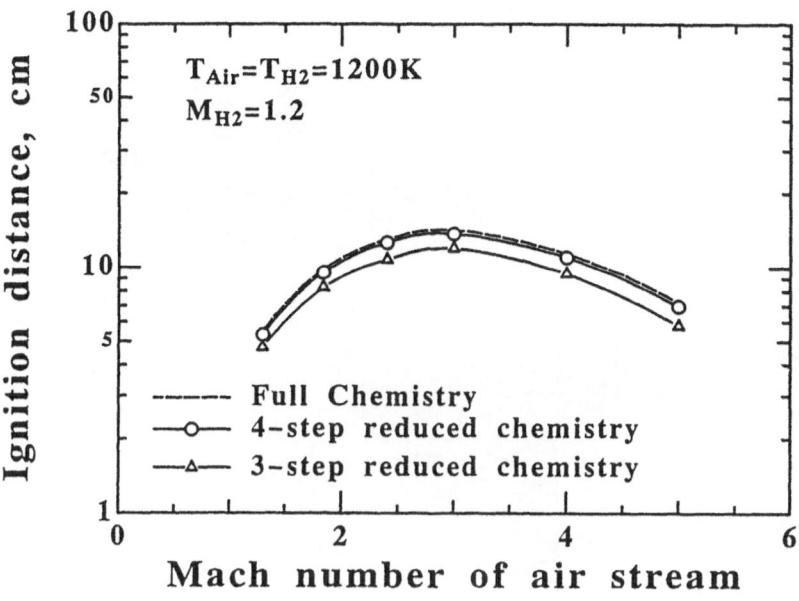

Figure 3 Ignition distance depending upon Mach number of air stream [8].

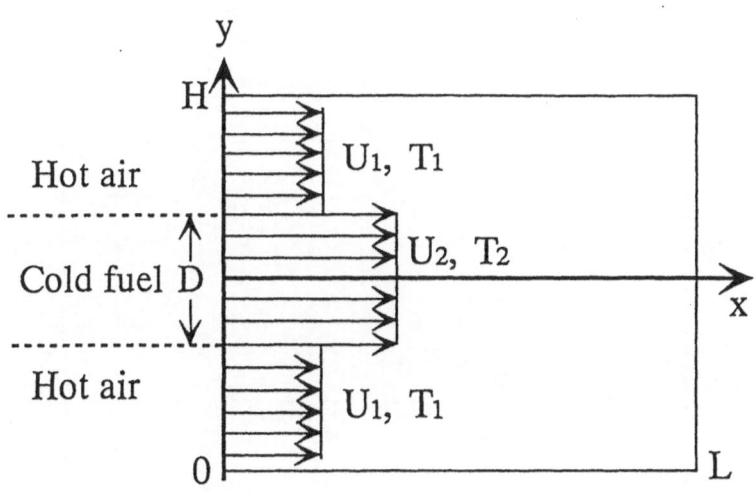

Figure 4 Schematic explanation of ignition in the case of confined hydrogen injection.

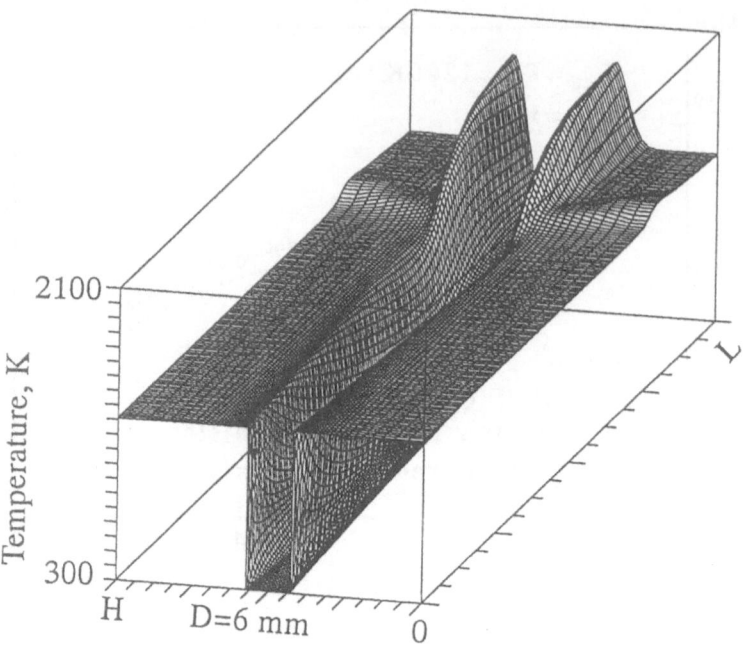

Figure 5 Temperature profile of ignition process for the hydrogen injector width of 6 *mm*.

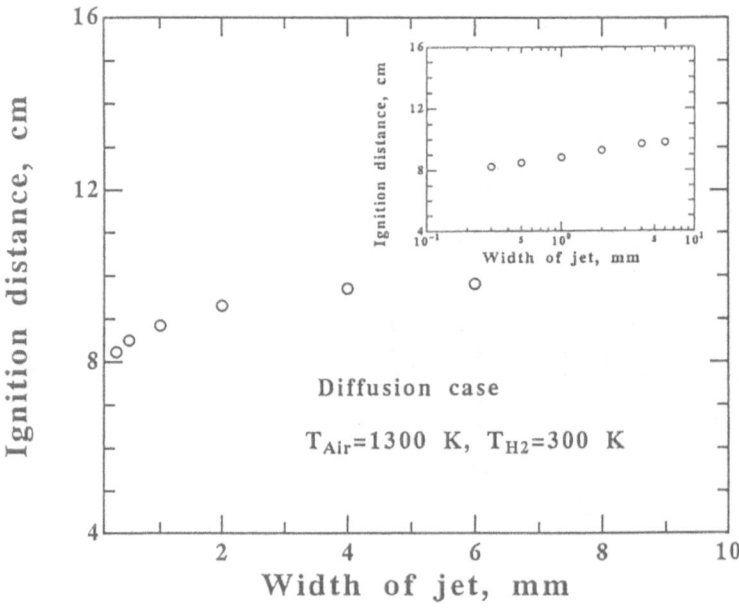

Figure 6 Dependence of ignition distance upon the hydrogen injector width.

3. Counterflow Diffusion Flame in Supersonic Airflow

Counterflow diffusion flames have been studied theoretically and experimentally for many years, because the counterflow configurations are very simple and the flames established in their flow systems offer a very good field in which the interactions between fluid mechanics and chemical kinetics can be observed and analyzed. Especially the so-called Tsuji burner has yielded considerable information on flames [13], and recently detailed or reduced chemical reaction mechanisms, using the counterflow fields, have been investigated by the numerical method.

The most basic point observed from the above-mentioned diffusion flames established in the subsonic counterflow field is that the Damköhler number decreases and extinction takes place when the counterflow velocity increases. As far as we deduce the supersonic case from this, the diffusion flame cannot be held because of the high convective flow velocity. In the case of supersonic flow, however, the total enthalpy becomes large and so a high static temperature is possible in the stagnation point flow behind the shock wave.

Some reports [14-16] have shown that flame extinction never occurs at any large flow velocity when the airflow temperature exceeds a critical point. Therefore, the diffusion flame seems to be established in the stagnation point region of supersonic flow. However, there has been no research done on the counterflow diffusion flames in supersonic airflow, in which the airflow temperature behind the shock wave and the location of the shock wave are estimated from the Mach number of the airflow outside the shock layer.

In the present analysis, as shown in Fig. 7, we analyzed the flow configuration of free-stream supersonic airflow opposing a fuel stream ejected from the surface of a porous sphere or cylinder. To grasp the fundamental characteristics of the diffusion flame, the problem was simplified and analyzed by use of large activation energy asymptotics [17].

For the stagnation point flow field behind the shock wave, inviscid flow, small momentum flow of fuel gas and the radial velocity proportional to the radial distance (x) are assumed to simplify the analysis. Unity Lewis number, constant product of density and thermal conductivity, constant specific heat, and the overall one-step reaction of oxygen and fuel are also presumed. Since the momentum equation becomes independent of the other equations, only equations of energy and mass conservation of species remain to be solved, which can be seen in the former references such as Refs. [18,19].

The location (d) of the shock wave was given by the preceding experimental results [20] and the stretch rate (a) was approximately calculated by the airflow velocity (v_2) behind the shock wave divided by the shock wave location. The airflow temperature behind the shock wave (T_2), which corresponds to the boundary temperature for the diffusion flame, was obtained from the normal shock conditions. Since these calculations yield information on the relation of shock heating and the stretch rate of the flow, the extinction limit may be connected with the conditions (v_1, T_1) of the free-stream supersonic airflow.

The asymptotic analysis for the diffusion flame established in the stagnation point flow region can be made by the same procedure as Liñán [18], and as a result, the energy equation is transformed into a simple ordinary equation [17,18]. The ordinary energy equation was solved in its time-dependent form and so each solution corresponds to a converged value. The values used for calculation are as follows: oxygen concentration of airflow, $Y_0 = 0.233$, fuel concentration $Y_F = 1$, radius of porous material $R_0 = 25\ mm$, stoichiometric mass ratio, $\nu = 0.215$, for hydrogen; activation energy, $E = 30\ kcal/mol$; frequency facfor, $B = 1.3 \times 10^8\ m^3/s/kg$; heat of reaction, $Q = 121\ kJ/g$; and surface temperature of porous material, $T_W = 500\ K$.

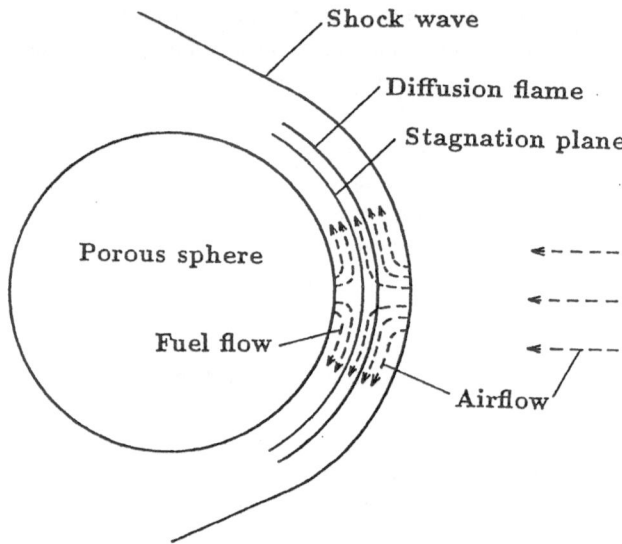

Figure 7 Tsuji problem in supersonic airflow.

Figure 8 shows the dependence of the Damköhler number (δ_0) of this system on the free-stream Mach number (M_1), showing the increase of the Damköhler number with Mach number, for three different temperatures of the free-stream airflow (T_1). This is basically because the airflow temperature increases due to shock heating. Increasing the Mach number (M_1), the airflow velocity (v_2) behind the shock wave does not change much, but the distance (d) between the surface of a porous body and the shock wave decreases. Therefore, the stretch rate ($a \propto v_2/d$) increases rapidly with the Mach number (M_1), and the Damköhler number decreases unless other parameters change due to the increase in the Mach number. Shock heating, however, results in a rise of the adiabatic flame temperature, which increases the Damköhler number as a result.

The non-dimensional maximum flame temperature, $\theta_{max} = T_{max}(c_p Y_F/Q)$, is seen in Fig. 9 , showing the increase of the flame temperature with the Damköhler number. In spite of large differences of the free-stream airflow temperature (T_1), extinction occurs at almost equal Damköhler numbers, because it is a function of

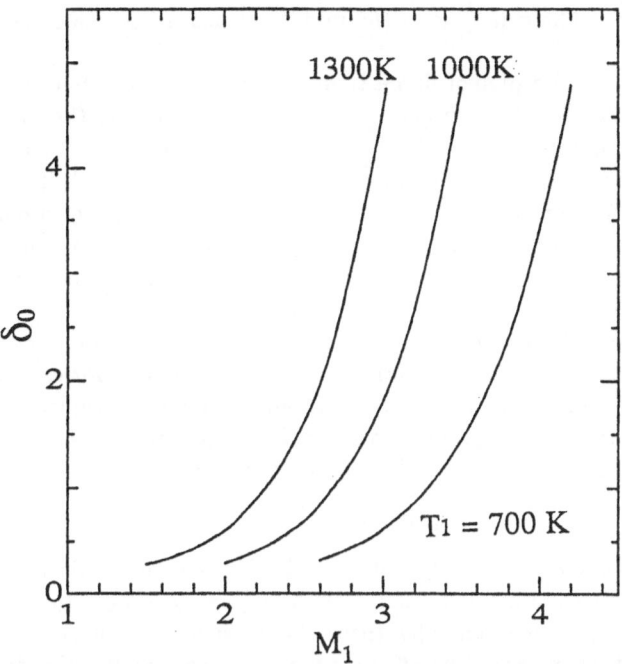

Figure 8 The variation of the Damköhler number with the free-stream Mach number.

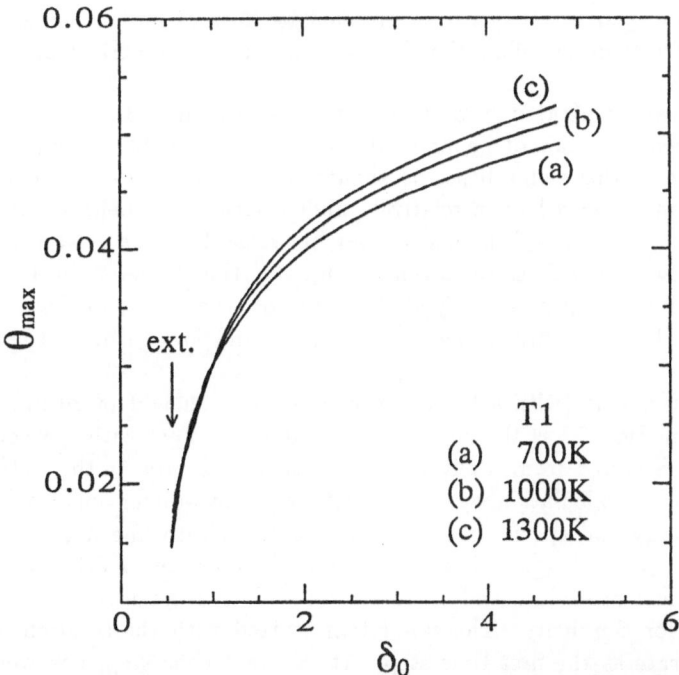

Figure 9 The variation of the nondimensional maximum flame temperature with the Damköhler number.

flame temperature. The solutions of the middle branch of the so-called S curve is not illustrated in Fig. 9.

Figure 10 shows the plot of flame temperature versus the free-stream airflow Mach number for $T_1 = 700\ K$. Since this is the dependence on the Mach number, the gradient at extinction is not infinity and extinction takes place suddenly. In the case of subsonic flow, the flame temperature decreases very slowly (the horizontal axis is logarithmic) when the Mach number increases. In the case of supersonic flow, however, the flame temperature rises very sharply when the Mach number increases, because the adiabatic flame temperature increases sharply with the Mach number.

Lastly, Figure 11 shows the extinction limit curve above which the diffusion flame can become established. Conditions in this extinction limit curve are approximately equivalent to those in scramjet combustor, that is, for example, airflow temperatures (T_1) may be increased up to around 1200 K at the combustor inlet and Mach numbers may be equal to around 2 in the engine. This implies that combustion in the scramjet engine is not necessarily stable.

4. Flame Holding by Shock Waves

There must be shock waves or reflected shock waves not only around the air inlet, but also in the combustor of the scramjet engine, generated by struts, ramps, injectors and so on. Since generally these shock waves or reflected shock waves cannot be avoided, they should be effectively applied to stabilize the flame and/or to enhance the combustion efficiency. Actually, there have been some researches [21,22] on the stabilization of the combustion region developed at the wedge base and assisted by the shock waves, as shown in Fig. 12. However, the flame-holding mechanism of this system has not been clarified because of limited experimental data and the difficulty of modelling the phenomenon, and no numerical analysis has been done.

The shock wave generated in a supersonic flow system and passing through a supersonic free-stream cannot penerate the subsonic flow field, for example the recirculation zone at the wedge base. It should be considered that points where shock waves are introduced have a relatively high pressure and behave like a tiny obstacle placed on the edge of the shear layer between the main stream and the subsonic wake flow. Therefore, as shown in Fig. 12, the shape of the boundary between two streams changes suddenly at the positions where the shock waves were introduced from the shock drivers located on upper and lower side of the wedge strut.

The triangular strut (with a length of 80 mm and a height of 20 mm) with a recess shown in Fig. 12 is the subject of calculation. Fuel with a velocity of Mach number 1 is issued from a slit with a width of 0.2 mm at the bottom of the recess. Numerical calculation was conducted for the two-dimensional N-S equations and the energy and species conservation equations with one-step overall reaction of oxygen and hydrogen. The Harten-Yee second order TVD scheme was used as a numerical procedure and the Baldwin and Lomax algebraic turbulent model was used for simplicity. The simulation started with the solution for the infinite reaction rate as the first time step. At the next time step, however, a finite chemical reaction was introduced and the calculation was continued until a

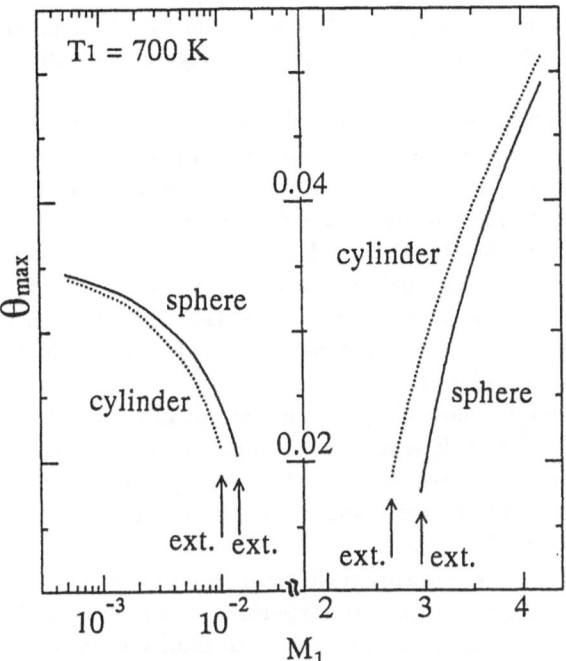

Figure 10 The variation of flame temperature with the free-stream Mach number, for subsonic and supersonic flow.

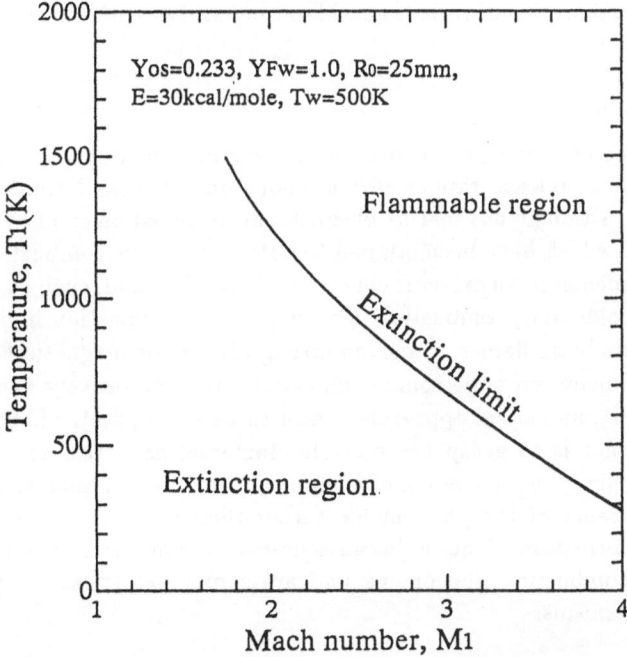

Figure 11 Dependence of extinction limit upon the free-stream conditions.

steady state was attained. Depending upon the flow or the temperature conditions, we often cannot obtain the stable combustion region; we call such a case "blow-off".

Figure 13 shows the streamlines with and without incident shock waves at Mach number 2.5. It should be noted that an enlarged recirculation region develops around the position to which the shock waves reach, and this region suggests that flame-holding may be possible due to the extension of the residence time and/or the feedback of heat. In Fig. 14, the axial velocity change along the centerline of the stream is plotted for various locations (x_s in Fig. 12) of the shock waves. Unless the position of shock waves is far from the wedge base, recirculation flows are strongly formed, that is, the velocity is negative over a wide range, which was also verified by the authors' experiment [23].

Figures 15 and 16 show the reaction rate of oxygen and hydrogen, and the temperature, respectively. It is interesting that reaction is concentrated at the tail of the recirculation zone and is continued for a long period, but that apparent flames cannot be seen behind the wedge and inside the recirculation zone in spite of the significant heat feedback.

The fundamental mechanisms of the flame-holding in the present configuration were grasped and were consistent with experiments. Experimentally, however, there are some curious findings, for example, the rich limit becomes narrower even when air is added to the fuel side. The competition between the characteristic flow time and the chemical reaction time in the recirculation region may be important, as discussed in Ref. [24]. Moreover, the effect of three-dimensional flow seems to be apparent in experiments, although it is difficult to include this effect in the calculation. A combination of various considerations will essential to the clarification of this problem.

Concluding Remarks

Supersonic combustion produces many interesting phenomena, especially the extremely high heat release rate, which is about one thousand times that of an ordinary burner. Although any feature of supersonic combustion can be explained by the fundamentals which have been studied to date, supersonic combustion is a very attractive phenomenon in an extreme condition of reactive fluid, such as combustion in very low Reynolds flow, combustion in microgravity, combustion in supercritical pressure, superadiabatic flames, flames in strong electric or magnetic fields and so on. Since experiments on supersonic combustion are prohibitively expensive and involve great effort, numerical approaches seem to be more practical. However, the most essential point is to grasp the most fundamental aspect of the phenomena concerned. To achive this, modelling of the phenomena and an analytical approach expressing the essence of the phenomena, rather than numerical procedure, seem to yield more information. This is because numerics sometimes lack a wide view of the complex combustion phenomena and are sometimes tricky in complicated combustion mechanisms.

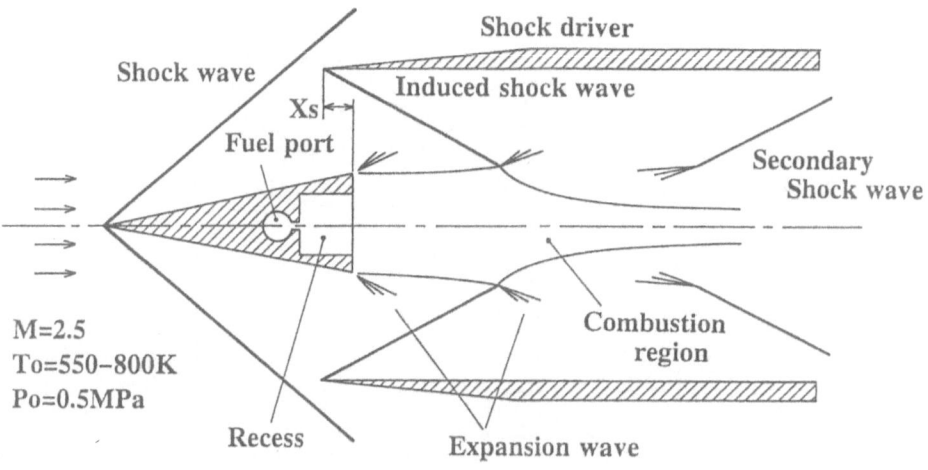

Figure 12 Flame holding by means of induced shock waves.

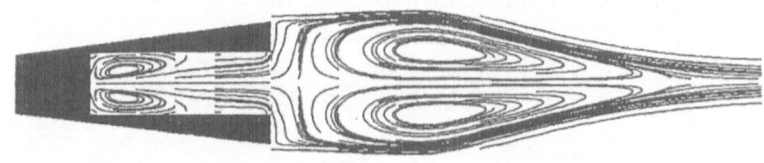

With shock waves

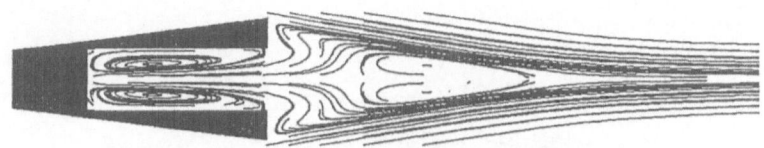

Without shock waves

Figure 13 Stream lines in the case without and with incident shock waves.

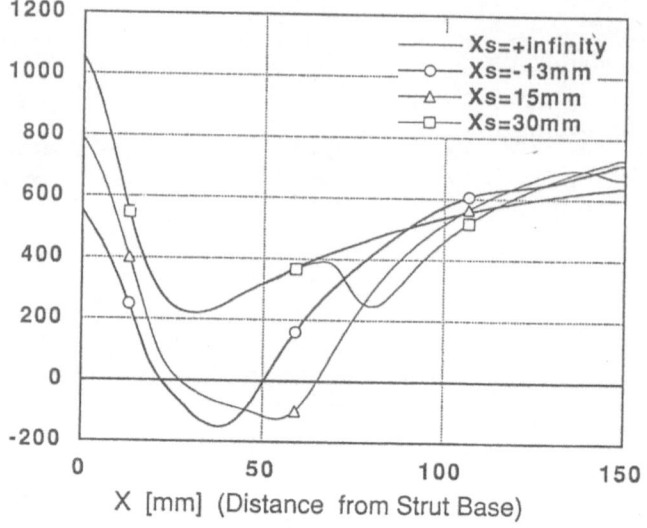

Axial Velocity Profile on the Center Line
Incident Shock Strength P2/P1=1.51, M0=2.5

Figure 14 Axial flow velocity along the centerline.

Figure 15 The profile of the reaction rate.

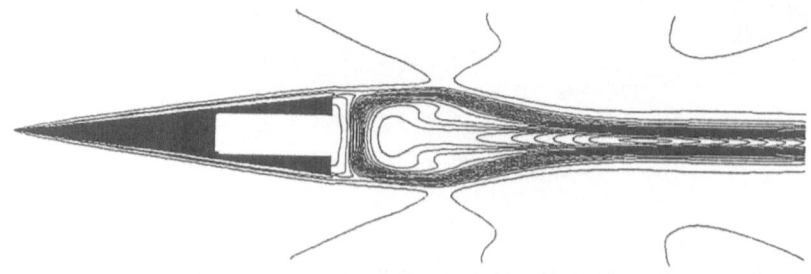

Figure 16 The profile of temperature.

Refernces

[1] Jackson,T.L. and Hussaini,M.Y: Combust. Sci. Tech., Vol.57 (1988), p.129.

[2] Ju,Y., Terada,K. and Niioka,T: Archivum Combustionis, Vol.11 (1991), p.161.

[3] Ju,Y. and Niioka,T: Transport Phenomenon Science and Technology, Higher Education Press, Beijing, 1992, pp.721-726.

[4] Ju,Y. and Niioka,T.: Proceedings of 18th. Int. Symp. on Space and Tech. and Sci., Kagoshima, 1992, p.7.

[5] Silva,L.F.F., Deshaies,B. and Champion,M.: Combust. Sci. Tech., Vol.89 (1993), p.317

[6] Ju,Y. and Niioka,T.: AIAA J. Vol.31 (1993), p.863.

[7] Ju,Y. and Niioka,T.: JSME International Journal, 1994 (in press).

[8] Ju,T. and Niioka,T.: Combust. Flame, Vol.97 (1994), p.423.

[9] Ju,T. and Niioka,T.: Ignition Simulation of Methane/Hydrogen Mixtures in a Supersonic Mixing Layer, Combust. Flame (submitted).

[10] Ju,T. and Niioka,T.: Transactions of Japan Soc.of Aeronautics Space Sci., Vol.37 (1994) p.148.

[11] Demetriades,A. and Yapuncich,F.L.: AIAA J., Vol.30, No.7(1992), p.1795.

[12] Bogdanoff,D.W.: AIAA J., Vol.10 (1994), P.183.

[13] Tsuji,H: Prog. Energy Combust. Sci., Vol.8 (1992), p.93.

[14] Chung,P.M., Fendell,F.E. and Holt,J.F.: AIAA J., Vol.4 (1966), p.1020.

[15] Saitoh,T.: Int. J. Heat Mass Transf., Vol.17 (1974), p.1063.

[16] Darabiha,N. and Candel,S.: Combust. Sci. Tech. Vol.86 (1992), p.67.

[17] Takita,K. and Niioka,T.: 4th ASME/JSME Thermal Eng. Joint Conference, Hawaii, 1995 (to appear).

[18] Liñán,A.: Acta. Astronautica, Vol.1 (1974), p.1007.

[19] Niioka,T. et al.: Combust. Flame, Vol.86 (1991), p.171.

[20] Anderson,Jr.,J.D.: Hypersonic and High Temperature Gas Dynamics, (1989), p. 199, McGraw-Hill

[21] Winterfeld,G.: Investigation of the Stabilization of Hydrogen Diffusion Flames by Means of Flameholders in Supersonic Flow, ESA-TT-347 (Translation of DLR-FB 76-35), 1977.

[22] Guerra,R., Waidmann,W. and Laible,C.: AIAA Paper 91-5102, 1991.

[23] Fujimori,T.,et al.: 31st Japanese Symp. on Combust.,1993, p.129(in Japanese).

[24] Terada,K. et al.: J. Propulsion and Power, 1994 (in press).

Level-Set Techniques Applied to Unsteady Detonation Propagation

D. Scott Stewart[1] *Tariq Aslam*[1] *Jin Yao*[1] *and John B. Bdzil*[2]

[1] Theoretical and Applied Mechanics
   University of Illinois, Urbana, Illinois, 61801, USA
[2] Los Alamos National Laboratory, Los Alamos, New Mexico, 87545, USA

1 Introduction

Here we are concerned with describing the dynamics of multi-dimensional detonation as a self-propagating surface. The detonation shock surface has been shown under certain circumstances to be governed by an intrinsic relation between the normal shock velocity and the local curvature, obtaining a $D_n - \kappa$ relation. Once the initial shock position is given, the subsequent motion of the shock can be determined by solving a scalar partial differential equation (PDE) for the shock position. The ingredients for prediction of the motion of the shock, include the $D_n - \kappa$ relation, determined from theory or experiment, the initial configuration of the shock and confinement boundary conditions. Thus we are also concerned about efficient numerical solution of the scalar PDE in three-dimensions, in cases that include multiply-connected and disjoint shock surfaces. This has led us to consider the level-set techniques of Osher and Sethian [1], which are naturally suited to these problems.

In what follows, we discuss examples of propagating surfaces, from formulations in combustion and heat transfer to which level-set methods apply. In Sect. 3, we discuss the specific example from detonation theory, which summarizes our recent work in [2]. In Sect. 4, we briefly explain the derivation of the $D_n - \kappa$ relation, in the context of detonation and mention some recent extensions of the theory, that includes shock acceleration terms [3]. These new results can all be summarized as a replacement of the $D_n - \kappa$ relation, by a relation of the form $F(\dot{D}_n, D_n, \kappa) = 0$ where $\dot{D}_n$ is the acceleration of the detonation shock along its normal. Importantly, the resulting equation is hyperbolic in character as opposed to parabolic, for a simple $D_n - \kappa$ relation. Finally we indicate the interesting new features of the dynamics that can be observed in the detonation shock surface evolution, and comment on their relevance to the formation of sustained detonation cells.

2 Examples of Propagating Surfaces

Theory for propagating surfaces arise naturally from discussions of phase transformation, that involve a jump in enthalpy across the surface. Examples include solidification and the Stefan problem, flame propagation and detonation propagation. In the first case, the surface is the boundary between solid and liquid; in the second case the flame surface, in the last case the detonation shock surface. In all three cases, the actual surface is not a material surface, but a phase surface through which material passes. The surface is assumed to separate the two phases (called here, *burnt* and *unburnt*), and the normal unit vector is defined to be positive in the direction of the unburnt material. At each point on the surface, the normal velocity is designated D_n and the local total curvature (the sum of the principle curvatures) is designated by κ. Further, κ is assumed to be positive when the surface is convex, relative to a normal pointing towards the unburnt material.

Next we delineate between two types of propagating surfaces that can be treated successfully with the level-set techniques; *Not Self-Propagating Surfaces (NSPS)* and *Self-Propagating Surfaces (SPS)*. We distinguish these two cases as follows. We define a surface that is *Not Self-Propagating* to be one that requires information normal to the surface to define the normal velocity D_n. For an NSPS one includes relations of the form $F(D_n, \kappa, \mathbf{x}, t, n_+) = 0$, where F generally depends on the curvature, the spatial position of the wave, time, and the values of quantities on one side (here the burnt side) of the surface. The slowly-varying hydrodynamic limit of a flame, described in [4], is an excellent example of an NSPS.

In contrast, we define a *Self-Propagating Surface* that only requires information defined in the surface to determine normal velocity D_n. So for an SPS one includes relations of the form $F(D_n, \kappa, \mathbf{x}, t) = 0$, or $F(D_n, \kappa, \mathbf{x}, t, \dot{D}_n) = 0$, where F generally depends on the curvature, the spatial position of the wave, time and possibly the self-acceleration of the surface, in its normal direction. Examples of SPS with D_n of the form $D_n = G(\kappa)$ include the simple Markstein flame, (see [5]), or the simplest version of the $D_n - \kappa$ relation obtained from Detonation Shock Dynamics (DSD); $D_n = D_{CJ} - \alpha\kappa$, where D_{CJ} and α are positive constants. As we mention in Sect. 4 the acceleration term $\dot{D}_n$ also arises naturally in the description of weakly-curved detonation and enlarges the dynamics that is considered in the DSD-theory.

3 Level-Set Methods: Tools for Computing the Dynamics of Interfaces

Here we outline the level-set method and explain its application and utility as a tool for computing the dynamics of propagating interfaces. First, notice that an interface (or surface) is a subset of lower dimensionality than the space that it travels in. The level-set technique solves for a field function $\psi(\mathbf{x}, t)$ that depends on physical space and time, and the field identifies surfaces of constant values of ψ. The surface $\psi(\mathbf{x}, t) = 0$, is typically identified with the surface of physical interest. Therefore, the computational task involves computing a field in space-time, and one then exhibits the surface of interest by searching for the special surface $\psi = 0$.

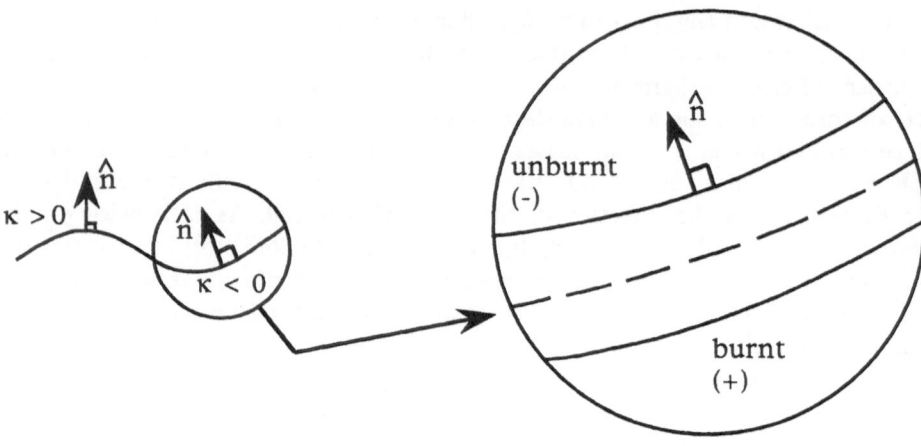

Fig. 1. Schematic of a propagating surface. The outward normal points toward "unre-acted" material. The blow up indicates a layer, within the structure of the surface that is has physics that may control its propagation, like a reaction-zone

This imbedding method is in contrast to what are sometimes known as *surface methods*. Differential representations of the surface, are based on surface param-eterizations, while discrete representations of the surface often include marker particles in the surface or finite elements. With surface methods, one represents the surface of physical interest by a representation of the same dimension. For example, in two-dimensions, the detonation shock locus is a space-curve in the (x,y)-plane and a numerical discretization represents the shock as a 1D array. For a 3D application the shock surface is a 2D space-surface and the discretization is represented by a 2D array.

While numerical methods based on surface parameterization can be very ef-fective for many problems, and can yield results with high accuracy, they also have substantial numerical and logical problems as the geometric complexity of the underlying problem increases. If the surface rapidly expands or contracts, markers must be added or removed. Surface markers can cross and the stability and accuracy of the method can be lost. The logical complexity of the pro-gramming, for a surface parameterization method can be overwhelming if one considers problems that have surfaces that are disjoint or multiply-connected.

It might seem that additional computation is required for the level-set tech-niques, since they solve for a field in the dimension of the physical space. One compensates for that by using an efficient, high-accuracy numerical method, that is logically simple to program; a point that was made dramatically in [1]. For our applications, we have found that the advantages of the logical simplicity of implementation of the level-set methods, easily compensates for any perceived increase in computational cost due to working in a higher dimensional space .

3.1 Detonation Shock Dynamics

Detonation Shock Dynamics (DSD) is a name that we use to describe a collection of results derived from an asymptotic theory that describes the evolution of a multi-dimensional, curved detonation. The detonation shock is supported by a combustion reaction-zone that trails behind the shock, and the radius of curvature of the detonation shock is assumed to be large when compared to the reaction-zone thickness. Most of the results ([6], [7], [8]) that have been developed so far, assume that the speed of the detonation is close to its, plane, Chapman-Jouguet (CJ) value. In particular, the theoretical results give explicit expressions for the $D_n - \kappa$ relation for an explosive material, described by the Euler equations, with a specified equation of state and reaction rate law. The work mentioned in [3], and in Sect. 4, extends this to include $\dot{D}_n$.

The theory of DSD suggests that the detonation shock, in some regimes, propagate according to a material specific evolution law. This theoretical suggestion has provided the motivation to verify the assertion experimentally in explosive systems. Fig. 2, shows a facsimile of the experimentally determined $D_n - \kappa$ curve for a condensed explosive PBX9502. For positive curvature, the experiments were conducted by Davis and Bdzil of Los Alamos National Laboratory (LANL), [9], and for those of negative curvature, the experiments were conducted by Hull of LANL, [10]. The two sets of experiment were carried out in quite different geometries. Davis and Bdzil's experiments were for round sticks of explosive of different diameters, ignited at the bottom. Hull's experiments were generated by an entirely different sort of experiment, where two, separated point detonations were ignited far within a block of the explosive and the waves then eventually merged to form a single detonation shock. Importantly, the combined data of the two separate experiments share the Chapman-Jouguet (CJ) value for the detonation velocity at zero curvature, and have the same slope $dD_n/d\kappa$ where they join.

The Bdzil-Davis reduction of the experimental data for PBX9502, for the positive curvature side, also indicates an extinction point; defined here as a maximum value of positive curvature, beyond which the $D_n - \kappa$ relation may not be continued. Under certain assumptions, theory also shows a similar property for $D_n - \kappa$ curves.

Without further explanation or assumption in this section, we will assume that we have a $D_n - \kappa$ relation that describes the motion of a detonation shock for some range of normal velocities and curvatures, such as the ones mentioned above. A $D_n - \kappa$ relation then can be assumed to have the form

$$D_n = D_{CJ} - \alpha(\kappa). \tag{1}$$

where α is a function of κ. The $D_n - \kappa$ relation based on intrinsic description corresponds to a SPS, in the sense defined in Sect. 2. If $\alpha = 0$, one is lead to a Huygens' construction for the motion of the shock surface. In the presence of non-zero α, one can propagate the surface by a modified Huygens' construction. If D_n is a monotonically decreasing function of the curvature, then the underlying dynamics of the surface are those of a parabolic partial differential equation.

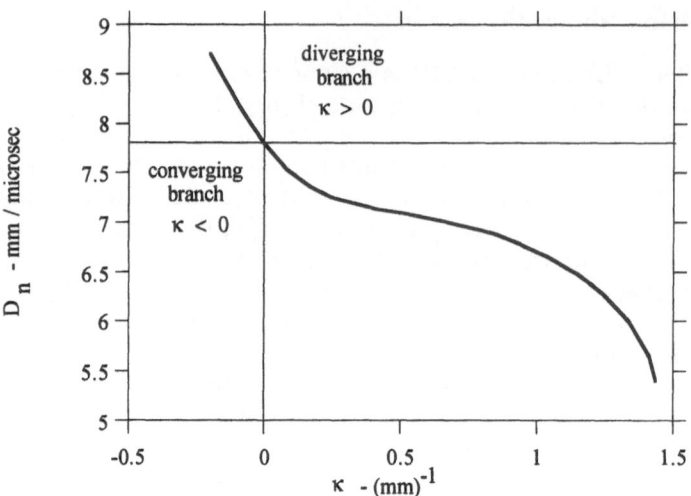

Fig. 2. Calibrated $D_n - \kappa$ response of condensed explosive PBX9502

Indeed under certain assumption the shock slope of the shock surface is shown to obey Burgers' equation.

3.2 Level-Set Formulation

Next we turn to the level-set technique as a way to solve for the motion of the surface, given this specific example of DSD. It is assumed that there is a field $\psi(\mathbf{x}, t)$ that will define level surfaces of the form, $\psi(\mathbf{x}, t) = \text{constant}$. The *shock* location for all time, will be defined as the special surface $\psi(\mathbf{x}, t) = 0$. The initial location of the shock will be associated with the locus $\psi(\mathbf{x}, 0) = 0$.

The ψ function obeys the *level-set equation* which is derived as follows. On any level curve $\psi(\mathbf{x}, t) = \text{constant}$, the time derivative of ψ in a frame, traveling with the curve is zero, i.e.

$$\frac{d\psi}{dt} = \frac{\partial \psi}{\partial t} + \frac{d\mathbf{x}}{dt} \cdot \boldsymbol{\nabla}\psi = 0, \tag{2}$$

where the derivative $d\mathbf{x}/dt \equiv \mathbf{D}$, is the pointwise velocity of the surface. By using the definition of the normal $\hat{\mathbf{n}} = \boldsymbol{\nabla}\psi/|\boldsymbol{\nabla}\psi|$ and noticing that $\mathbf{D} \cdot \boldsymbol{\nabla}\psi$ can be rewritten as $D_n|\boldsymbol{\nabla}\psi|$, the above equation, now referred to as the *level-set equation*, can be restated as

$$\frac{\partial \psi}{\partial t} + D_n|\boldsymbol{\nabla}\psi| = 0. \tag{3}$$

If D_n is a constant (the Huygens' construction), then the level-set equation is a Hamilton-Jacobi equation. If D_n is a function of the curvature, then the level-set equation is a Hamilton-Jacobi-like equation. Importantly, the type of the equation is controlled by the highest order derivative that appears. For example in the current context, if D_n is a monotonically decreasing function of the curvature, then the level-set equation is at most first order in time, is second order in space, and is classified as a parabolic partial differential equation (PDE).

To illustrate the level-set PDE more completely in the form used for DSD applications, one needs the Cartesian expression for the curvature. The curvature is generally represented as $\kappa = \boldsymbol{\nabla} \cdot \hat{\mathbf{n}}$, which in 2D reduces to

$$\kappa = \frac{\psi_{xx}\psi_y^2 - 2\psi_{xy}\psi_x\psi_y + \psi_{yy}\psi_x^2}{\left(\psi_x^2 + \psi_y^2\right)^{3/2}}. \tag{4}$$

The PDE for ψ in a Cartesian frame, is wholly prescribed once the function $D_n(\kappa)$ is given. The initial data for ψ can be generated as follows. At time $t = 0$, define $\psi(\mathbf{x}, 0) = 0$ to be the initial shock position. Note that one could have more than one closed surface identifying initial shocks. Then the remainder of the initial data for the field is defined by setting the value of ψ at any point (x,y,z) equal to the minimum signed distance to the detonation shock. Fig. 3 shows an example of the level-set function ψ defined initially (as the minimum distance function) and at a later time. The problem considered is two cylindrically expanding shocks that are at first separated and then merge.

The numerical solution of the PDE with given initial data and subject to boundary conditions, generates an approximation to the field $\psi(\mathbf{x}, t)$, and the location of the shock is then simply found by searching for the level surface $\psi = 0$. The surface $\psi = 0$ is easily determined by recording when ψ passes through zero. A tabular function of crossing times, $t_{cross}(x, y, z)$, is found from the computation. The shock location at given time is simply a contour of constant crossing time.

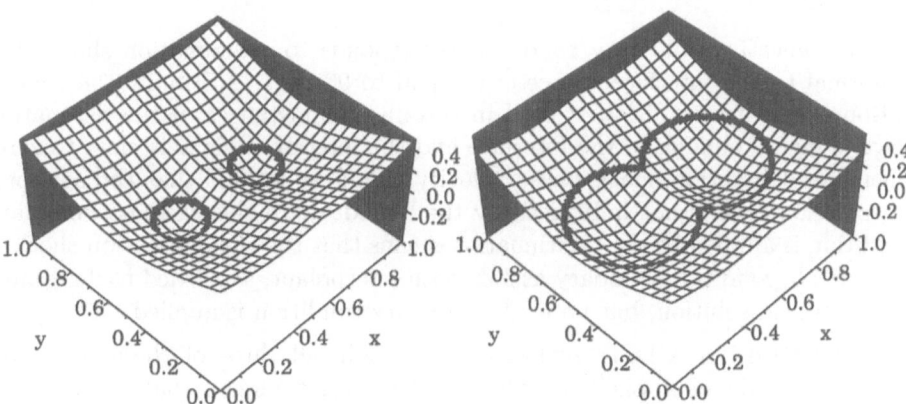

Fig. 3. The level-set function ψ defined initially (as the minimum signed distance function) and at a later time, for the example of two cylindrically expanding shocks that are at first separated and then merge

3.3 Boundary Conditions

We give a brief summary of the boundary conditions that are applied in the DSD application. A detailed description of the boundary conditions and their rationale is found in [2]. The need for boundary conditions comes from the application. The problems of interest in explosive design mostly involve domains of finite size, and the collision of the detonation shock with confinement boundaries. In typical explosive systems, one places the unreacted explosive in a container. After having been ignited, the detonation sweeps through the system and the detonation shock intersects the interfaces. Often the confinement is a thin layer of metal which separates the explosive products from the ambient atmosphere.

The boundary conditions that have been considered so far, are motivated by analysis that models the interaction of the detonation shock with the confinement boundaries, and include three simple categories: i) shock-edge angle boundary conditions, ii) reflective boundary conditions, and iii) continuation boundary conditions. The shock-edge boundary condition was put forward by Bdzil in [11], and used later in [6]. It states that in certain instances, the angle that the shock makes with the interface is fixed, and that the fixed angle is a material constant for an explosive/confining material pair. Let the outward normal of the detonation shock at the edge be represented as $\hat{n}_{edge}$, and let the outward normal of the interface, where the detonation shock and the interface intersect, be represented as $\hat{n}_{mat}$. The interior angle between those two direction vectors such that $\cos\omega = \hat{n}_{edge} \cdot \hat{n}_{mat}$, is some fixed value $\omega = \omega_c$. For example, the angle for a PBX9502 explosive with a particular material as edge confinement is a fixed number; a typical value for explosive detonated without confinement is 45 degrees.

The reflective boundary condition corresponds to a detonation shock that is normal to the interface, hence ω is equal to 90 degrees. Finally the continuation boundary condition is used in certain circumstances, if the detonation wave is highly oblique to the interface and the interior angle ω is close to zero. Then the detonation shock phase velocity would be so fast that the supporting reaction-zone is not influenced by the boundary. In this case, no boundary condition is applied at all. Continuation means that the the detonation shock is extended beyond the boundary as a smooth interpolant, as needed to determine the numerical solution, but no angle boundary condition is applied.

In practice, for a DSD explosive application, all three of these boundary conditions might be applied according to the interior angle ω that is monitored at the edge interface. One of the most important points to stress is that all of the boundary conditions described above, are at most functions of the derivatives of ψ. Thus a level curve propagated according to the $D_n - \kappa$ relation, will evolve only according to data developed in its own surface. The boundary conditions that have been considered for the DSD applications, do not change this property and thus one is lead to a class of problems in finite domains that can be solved consistently using level-set techniques.

3.4 The Recipe for DSD Application Using Level-Set Methods

The recipe for using level-set methods for the DSD application can then be summarized as follows. 1) Determine the initial detonation shock locations and designate them as $\psi(\mathbf{x}, 0) = 0$. 2) Define the ψ field everywhere at time $t = 0$ say, by setting ψ equal to the minimum signed distance to the detonation shock. 3) Solve the level-set equation for the ψ field. 4) At the boundary, apply the boundary condition for each level curve, as if it were the physical shock of interest. 5) Find the physical shock at any time by searching for $\psi(x, y, z, t) = 0$.

3.5 The Numerical Methods

We give a brief description of an efficient general numerical method for solving the level-set equation for the DSD application, on a fixed, Eulerian, finite difference grid. For the interior algorithm, we follow Osher and Sethian, [1]. The normal velocity D_n is explicitly written as $D_{CJ} - \alpha(\kappa)$, where if the second term was absent, then one solves only the Huygens' construction. The update of ψ is split into a Huygens' advection followed by a diffusive correction. The Huygens' advection uses a second-order ENO scheme. The diffusive correction, due to the curvature terms in $\alpha(\kappa)$ is approximated by central differencing. The boundary conditions are implemented with central differences and are second-order accurate. The reader is referred to [2] for more details. Suffice it to say that the advantage of the ENO-based schemes for the advection is the simplicity of implementation and accuracy of the results.

4 Asymptotic Theory

Here we summarize the asymptotic theory that is developed in [3]. A standard mathematical model of explosive materials is adopted which describes a compressible Euler fluid with an ideal equation of state, and Arrhenius form for the reaction rate r,

$$e = \frac{p}{\rho}\frac{1}{\gamma - 1} - Q\lambda, \quad r(p, \rho, \lambda) = k(1 - \lambda)^\nu e^{-E/(p/\rho)}, \tag{5}$$

where e is the specific internal energy, ρ is the density, p is the pressure, λ is the reaction progress variable, γ is the polytropic exponent, Q is the heat of combustion and k, ν and E are respectively the pre-multiplying rate constant, the depletion factor and the activation energy. The laboratory-based velocity will be represented by $\mathbf{u}$. From here on, we adopt the notation convention that a quantity with a $\tilde{(\,)}$ refers to a dimensional quantity and the quantities without a tilde are dimensionless quantities, and scaled with respect to the dimensional unit. In particular, the length, velocity and time scales are given by $\tilde{\ell}_{rz}, \tilde{D}_{CJ}$ and $\tilde{\ell}_{rz}/\tilde{D}_{CJ}$, respectively. The length $\tilde{\ell}_{rz}$, is taken to be the characteristic 1D, steady reaction-zone length. The density scale is $\tilde{\rho}_0$ and pressure scale is $\tilde{\rho}_0\tilde{D}_{CJ}^2$. Consequently the sound speed, reaction rate, curvature and heat of combustion appear as $c = \tilde{c}/\tilde{D}_{CJ}$, $r = \tilde{r}\tilde{\ell}_{rz}/\tilde{D}_{CJ}$, $\kappa = \tilde{\kappa}\tilde{\ell}_{rz}, q = \tilde{Q}/\tilde{D}_{CJ}^2 = 1/[2(\gamma^2 - 1)]$.

(The plane, steady, CJ detonation velocity in the strong shock approximation is given by, $\tilde{D}_{CJ}^2 = 2(\gamma^2 - 1)\tilde{Q}$).

The jumps across the lead detonation shock are determine by the equation of state and the upstream state. We will assume that the upstream state is quiescent with $\mathbf{u} = 0$, density ρ_0 and ambient pressure p_0. For convenience, we will assume that the lead detonation shock is sufficiently strong so that the strong shock approximation holds. The normal shock relations (in the strong shock approximation) for an ideal gas moving into an ambient atmosphere, reduce to

$$U_n = -\frac{\gamma - 1}{\gamma + 1}D_n, \ \ u_t = 0, \ \ \rho = \frac{\gamma + 1}{\gamma - 1}, \ \ p = \frac{2}{\gamma + 1} \ D_n^2, \ \ \lambda = 0, \ \ \text{at} \ \ n = 0, \quad (6)$$

where the $n-$ and $t-$ subscripts respectively refer to the normal and tangential components of the velocity in the shock-attached frame.

4.1 Intrinsic, Shock-Attached Coordinates and Governing Equations

In order to make the analysis tractable, it is essential to write the equations of motion in a suitable form. Given that the material derivative is given by $D/Dt \equiv \partial/\partial t + \mathbf{u} \cdot \nabla$, then the Euler equations, with reaction, are given by $D\rho/Dt + \rho\nabla \cdot \mathbf{u} = 0$, $\rho\,D\mathbf{u}/Dt + \nabla p = 0$, $De/Dt + p\,Dv/Dt = 0$, where $v \equiv 1/\rho$, and $D\lambda/Dt = r(p, \rho, \lambda)$.

Intrinsic, shock-attached coordinates, are used to describe curved, time-evolving detonation waves. We restrict the formulas shown here to 2D to simplify the presentation; the results apply equally well in 3D. The shock surface can be represented quite generally in terms of laboratory-fixed coordinates $\mathbf{x} = (x, y)$ by a function $\psi(\mathbf{x}, t) = 0$. This equation constrains the lab-coordinate position vectors in the surface to $\mathbf{x} = \mathbf{x}_s(x, y, t)$. The shock surface can also be represented by a surface parameterization $\mathbf{x} = \mathbf{x}_s(\xi, t)$, where ξ measures length along the shock curve, relative to the reference point $\mathbf{x}_s(0, t)$. The outward normal (in the direction of the unreacted explosive) and unit tangent vector in the shock surface, (which form a local basis) are given by $\hat{\mathbf{n}} = \nabla\psi/|\nabla\psi|$, $\hat{\mathbf{t}} = \partial\mathbf{x}_s/\partial\xi$. The total shock curvature is given by $\kappa(\xi, t) = \nabla \cdot \hat{\mathbf{n}}$. Finally, the intrinsic coordinates are related to the laboratory coordinates by the change of variable given by $\mathbf{x} = \mathbf{x}_s(\xi, t) + n\,\hat{\mathbf{n}}(\xi, t)$, where the variables n, ξ are respectively, the distance measured in the direction of the normal to the shock wave, and the arc- length measured along the shock curve.

Next the equations of motion are transformed to this shock-attached, intrinsic frame, i.e. from (x,y,t)-coordinates to (n, ξ, t) coordinates. In particular we note, that the normal shock velocity and curvature are only functions of ξ and t, i.e. $D_n = D_n(\xi, t)$ and $\kappa = \kappa(\xi, t)$. The relevant normal velocity that appears subsequently is $U_n = u_n - D_n$. The manipulations of the transformation are lengthy but straightforward and the transformed equations have a direct correspondence to the Euler equations. Importantly, the curvature appears explicitly in the transformed equations.

For the transformed equations, we retain only the explicit time dependence and the first curvature effects and write down a set of approximate equations to analyze, that are valid under the assumption that $|\kappa| \ll 1$. Consistent with the normal shock relations, for a shock propagating into a quiescent material, we neglect u_ξ in this analysis and take it effectively to be zero. The equations are then written in a quasi-conservative form as

$$\frac{\partial(\rho U_n)}{\partial n} = -\kappa\rho(U_n + D_n) - \rho_{,t}, \tag{7}$$

$$\frac{\partial(\rho U_n^2 + p)}{\partial n} = -\kappa\rho U_n(U_n + D_n) - \rho_{,t}U_n - \rho(U_{n,t} + D_{n,t}), \tag{8}$$

$$\frac{\partial}{\partial n}\left(\frac{1}{2}U_n^2 + \frac{c^2}{\gamma - 1} - q\lambda\right) = -(U_{n,t} + D_{n,t})$$
$$- \frac{1}{U_n}\left(\frac{1}{\gamma - 1}\frac{p_{,t}}{\rho} - \gamma\frac{p}{\rho^2}\frac{\rho_{,t}}{\gamma - 1} - q\lambda_{,t}\right). \tag{9}$$

The rate equation can be written as

$$\frac{\partial\lambda}{\partial n} - \frac{r}{U_n} = -\frac{\lambda_{,t}}{U_n}. \tag{10}$$

An auxiliary equation, referred to as the *master equation*, can be written

$$. \ (c^2 - U_n^2)\frac{\partial U_n}{\partial n} = qr(\gamma - 1) - \kappa c^2(U_n + D_n) + U_n(U_{n,t} + D_{n,t}) - vp_{,t}. \tag{11}$$

Note that the intrinsic coordinate, time derivative appearing above is $(\ )_{,t} = (\partial/\partial t)_{n,\xi} + (\partial\xi/\partial t)_\mathbf{x}(\partial/\partial\xi)_{n,t}$, where $(\partial\xi/\partial t)_\mathbf{x}$ is the rate of change of arclength along the shock. Importantly, $(\partial\xi/\partial t)_\mathbf{x}$ is independent of n to the order being considered. Reinterpreted in Cartesian coordinates, the operator $(\ )_{,t}$ is simply the time-rate of change in the shock-normal direction (see references [2] and [3])

$$(\ )_{,t} = \left(\frac{\partial}{\partial t}\right)_\mathbf{x} + D_n\hat{n}\cdot\nabla. \tag{12}$$

The analysis proceeds the assumption that the right-hand side of the structure equations (7) - (10) are in some sense uniformly small and can be approximated by a quasi-steady, plane solution. One applies the shock boundary conditions, (6) at $n = 0$ and attempts to generate a uniform solution throughout the reaction-zone behind the shock.

4.2 The Generalized CJ Conditions

The master equation (11) exhibits the special character of the sonic point that generates a condition that can be used, under appropriate circumstances, to generate the eigenvalue relation between curvature and the normal detonation speed, and the self-acceleration, $\dot{D}_n$.

Suppose the flow has a sonic locus such that

$$\eta = c^2 - U_n^2 = 0, \tag{13}$$

then equation (11) is satisfied at that point, in general, only if, the right-hand side, vanishes simultaneously at that point, i.e.

$$qr(\gamma - 1) - \kappa c^2(U_n + D_n) + U_n(U_{n,t} + D_{n,t}) - vp_{,t} = 0. \tag{14}$$

The pair of conditions (13, 14), called the sonic and the thermicity conditions respectively, taken together are called the *generalized CJ-conditions*, after Wood and Kirkwood, [13].

4.3 The Method of Successive Approximation

The problem outlined above, for quasi-steady, near-CJ, curved detonation, in the absence of explicit time-dependent terms, has been solved by a layer analysis, in [6], [7], [13], [8]. However in [3] we used a technique that is equivalent and perhaps simpler, and is based on an integral formulation rather than the differential formulation.

For the purpose of generating the corrections we assume that the detonation velocity and the state corresponds to a quasi-steady, 1D solution, plus a correction,

$$D_n = D + \kappa D', \tag{15}$$

and

$$U_n = -D\frac{\gamma - \ell}{\gamma + 1} + \kappa U', \quad v = \frac{\gamma - \ell}{\gamma + 1} + \kappa v', \quad p = D^2\frac{1 + \ell}{\gamma + 1} + \kappa p', \tag{16}$$

where $\ell = \sqrt{1 - \lambda/D^2}$. The quasi-steady 1D solution referred to is exhibited in the relations shown above by setting $\kappa = 0$. To keep notation to a minimum, a * subscript refers to the first approximation for the fluid state and a prime superscript refers to the correction to that approximation, e.g., $U_n = U_*(\ell, D) + \kappa U'$. We represent the leading order approximation to D_n, $(D_n)_*$, where it would appear, by a plain D. All that is assumed for now, in the various expansions (illustrated by the expansion for U_n) is that the correction term $\kappa U' \sim o(U)$ as $\kappa \to 0$. The resulting integral equations, shown below are further simplified by using the first approximation in the integrals. Integrating (7 - 9) with respect to n, yields

$$\rho U_n + D_n = \int_0^n [-\kappa \rho_*(U_* + D) - \rho_{*,t}]d\tilde{n}, \tag{17}$$

$$\rho U_n^2 + p - D_n^2 = -\int_0^n [(\rho_* - 1)D_{,t} - \kappa D(U_* + D)]d\tilde{n}, \tag{18}$$

$$\frac{1}{2}U_n^2 + \frac{c^2}{\gamma - 1} - q\lambda - \frac{1}{2}D_n^2 = \int_0^n [-\frac{p_{*,t}}{D} - (1 + \frac{D}{U_*})D_{,t}]d\tilde{n}. \tag{19}$$

The source terms in these equations are evaluated by switching the order of differentiation and integration, since ( )$_{,t}$ is independent of n, and then evaluating the resulting integrals using the substitution $dn = (U_*/r_*)d\lambda$. Since $\lambda(n, D)$, care must be exercised to remove the contribution from the integration limit when ( )$_{,t}$ is applied to the result. From the resulting expressions, one can evaluate the approximate state at the CJ-point, by setting $\lambda = \lambda_{CJ}$. These formulas then represent a correction of the Rankine-Hugoniot jump relations for the state at the generalized-CJ point,

$$(\rho U_n)_{CJ} = -D_n + \kappa I_1 D^2 + J_1 D_{,t}, \tag{20}$$

$$(\rho U_n^2)_{CJ} + p_{CJ} = D_n{}^2 - \kappa I_2 D^3 + I_1 DD_{,t}, \tag{21}$$

$$\frac{1}{2}(U_n^2)_{CJ} + \frac{c_{CJ}^2}{\gamma - 1} - q\lambda_{CJ} = \frac{D_n{}^2}{2} - (I_1 + J_2)DD_{,t}, \tag{22}$$

where the reaction rate integrals I_1, I_2, J_1, J_2 are defined by

$$I_1 = \frac{1}{(\gamma + 1)} \int_0^{\lambda_{CJ}} \frac{(1 + \ell)}{r}d\lambda, \quad I_2 = \frac{1}{(\gamma + 1)^2} \int_0^{\lambda_{CJ}} [\frac{(\gamma - \ell)(1 + \ell)}{r}]d\lambda, \tag{23}$$

$$I_3 = \frac{1}{(\gamma + 1)^2} \int_0^{\lambda_{CJ}} \frac{\ell(\gamma - \ell)}{r}d\lambda, \quad I_4 = \int_0^{\lambda_{CJ}} \frac{\ell}{r}d\lambda, \tag{24}$$

$$J_1 = \frac{1}{\gamma} \frac{d(DI_4)}{dD}, \quad J_2 = -\frac{1}{D^2} \frac{d(D^3 I_3)}{dD}. \tag{25}$$

The formal algebraic solution of (20) - (22) are subject to the sonic constraint that $c^2 = U_n^2$, determines the state $\rho_{CJ}, (U_n)_{CJ}, p_{CJ}$ and a condition on the speed D_n, in the same way as is obtained for the simplest case of a steady, plane, CJ wave. The result for U_n, and the sonic condition $c^2 = U_n^2$ can then be used in (22) to obtain a condition between $D_{,t}, D_n, \kappa$ and λ_{CJ}, which is given by:

$$D_n^2 - \lambda_{CJ} + \gamma^2 \left\{ \frac{[D_n^2 - \kappa I_2 D^3 + I_1 DD_{,t}]^2}{[D_n - \kappa I_1 D^2 - J_1 D_{,t}]^2} - D_n^2 \right\}$$
$$+ 2(\gamma^2 - 1)(I_1 + J_2)DD_{,t} = 0, \tag{26}$$

which can be further simplified by retaining only the first correction in $O(\kappa)$ and $O(D_{,t})$, which are assumed small to obtain the reduced $(D_{,t}, D_n, \kappa, \lambda_{CJ})$ relation

$$D_n^2 - \lambda_{CJ} + 2\kappa\gamma^2(I_1 - I_2)D_n^3 + 2D_n\dot{D}_n[(\gamma^2 - 1)(I_1 + J_2) + \gamma^2(I_1 + J_1)] = 0, \quad (27)$$

where we have replaced D by D_n and $D_{,t}$ by $\dot{D}_n$.

In most respects, (27) is the key result and holds generally for slowly varying, weakly-curved detonation that have an embedded sonic locus in their structure. The evolution equation is obtained once λ_{CJ} is estimated, which follows from consideration of the thermicity condition (14).

4.4 Large Activation Energy

In the general case, the quantities, I_1, I_2, J_1 and J_2 are functions of D_n and $\dot{D}_n$. Thus it is generally difficult to write down the $\dot{D}_n - D_n - \kappa$ relation in very simple terms. For the purpose of illustration, we focus on the case of large activation energy, which follows our work in [8]. In this case, the reaction-zone structure is assumed to be that of an induction-zone, followed by an exponentially thin reaction-zone. It follows that we can assume that λ_{CJ} is exponentially close to one. Further we assume that D_n is close to one, and that quasi-steady time variation in the induction zone is due to the motion of the shock, and that $\dot{D}_n$ and κ are small and of the same order. Equation (27) can be further simplified to

$$D_n = 1 - \gamma^2(I_1 - I_2)\kappa - [\gamma^2(I_1 + J_1) + (\gamma^2 - 1)(I_1 + J_2)]\dot{D}_n. \quad (28)$$

The characteristic reaction-zone length is estimated in terms of the induction zone length scale, $\tilde{\ell}_{rz} = \tilde{k}^{-1}\tilde{D}_{CJ}exp[\theta/c_s^2]/\theta$, and thus the reaction rate is expressed as

$$r = \frac{(1 - \lambda)^\nu}{\theta}e^{\theta/c_s^2 - \theta/c^2}. \quad (29)$$

For $\gamma < 2$, the rate term is exponentially large outside the induction zone, hence the values of the rate integrals I_1, I_2, J_1, J_2 only depend on their behavior in the induction zone.

Consideration of the induction zone allows for calculation of the temperature (or sound speed squared) perturbation in the zone, in terms of the curvature and the slow acceleration of the detonation and small depletion, and obtains the estimate for c^2,

$$c^2 = c_s^2 + \alpha\lambda + \frac{c_s^4}{\theta}\ln\left\{\frac{\theta\kappa\beta}{\alpha}\left(1 - e^{-\alpha\lambda\theta/c_s^4}\right) + exp[\frac{(\gamma + 1)^2}{\gamma(\gamma - 1)}\theta(D_n - 1)]\right\}, \quad (30)$$

where $c_s^2 = [2\gamma(\gamma - 1)]/(\gamma + 1)^2$ and

$$\alpha = \frac{\gamma(3-\gamma)}{2(\gamma+1)^2}, \quad \kappa\beta = 4\left(\frac{\gamma(\gamma-1)^3}{(\gamma+1)^4}\kappa + 2\frac{\gamma(\gamma-1)(\gamma-2)}{(\gamma+1)^3}\frac{\dot{D}_n}{\theta}\right). \tag{31}$$

All that remains is the integral asymptotics, which can be summarized as follows. For large θ, the dominant contributions to the integrals are close to the shock, where $\ell = 1$. It follows that $J_1 \sim 0$ and $I_1(\gamma+1)/2 \sim (I_1-I_2)(\gamma+1)^2/4 \sim -J_2(\gamma+1)^2/[4(\gamma^2-1)] \sim I$, where

$$I = \int_0^1 \frac{1}{r}d\lambda \sim \int_0^\infty e^{-\theta(c^2-c_s^2)/c_s^4}dz \text{ with } z = \lambda\theta. \tag{32}$$

In turn, I can be estimated using the approximation for c^2 in the reaction rate r, as

$$I = \frac{c_s^4}{\theta\kappa\beta}[\ell n(\sigma) - \ell n(\sigma - \theta\kappa\frac{\beta}{\alpha})], \tag{33}$$

where

$$\sigma - \theta\frac{\kappa\beta}{\alpha} = exp\left(\frac{(\gamma+1)^2}{\gamma(\gamma-1)}\theta(D_n-1)\right). \tag{34}$$

Now we substitute these various results back into (28) to obtain the explicit evolution equation

$$\kappa\beta = \frac{\alpha}{\theta}e^{(2/c_s^2-\beta/\mu)\theta(D_n-1)}(1 - e^{(\beta/\mu)\theta(D_n-1)}), \tag{35}$$

where

$$\kappa\mu = c_s^4[\frac{4\gamma^2}{(\gamma+1)^2}\kappa + \frac{2\gamma(4\gamma-3)}{\gamma+1}\frac{\dot{D}_n}{\theta}]. \tag{36}$$

Note that when $\dot{D}_n$ is absent, then

$$\kappa = \frac{\alpha}{\theta\beta}e^{b\theta(D_n-1)}(1 - e^{a\theta(D_n-1)}), \tag{37}$$

where

$$a = \frac{\beta}{\mu}\Big|_{\dot{D}_n=0} = \frac{(\gamma+1)^2(\gamma-1)}{4\gamma^3}, \quad b = \frac{2}{c_s^2} - a = \frac{(\gamma+1)^3(3\gamma-1)}{4\gamma^3(\gamma-1)}, \tag{38}$$

and

$$\frac{\beta}{\alpha}\Big|_{\dot{D}_n=0} = \frac{8(\gamma-1)^3}{(3-\gamma)(\gamma+1)^2}, \tag{39}$$

which agree with the steady ($\dot{D}_n = 0$) $D_n - \kappa$ relation established in [8]. Fig. 4 shows two representations of the $\dot{D}_n - D_n - \kappa$ relation in the limit of large

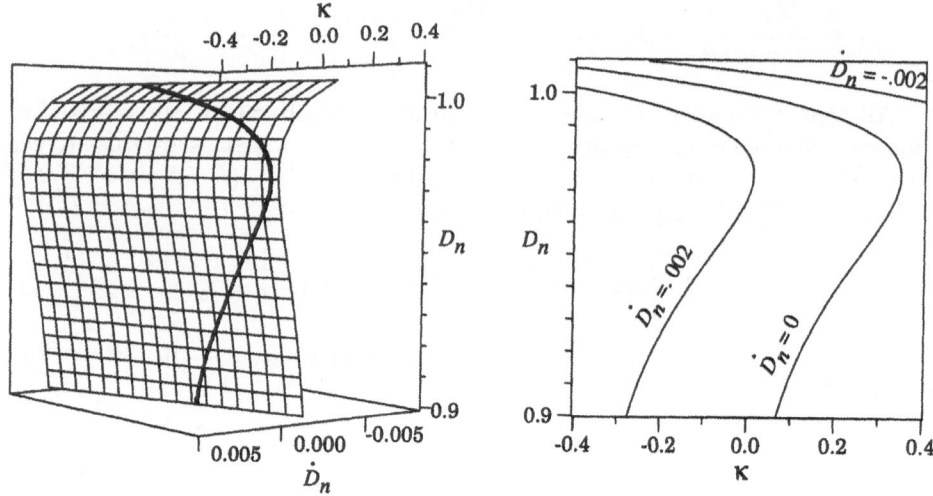

Fig. 4. Two representations of the $\dot{D}_n - D_n - \kappa$ relation in the limit of large activation energy, plotted for $\theta = 2, \gamma = 1.2$. The left plot shows a three-dimensional representation of the surface in the $\dot{D}_n, D_n, \kappa$ - space, with the imbedded curve, $\dot{D}_n =$ shown. The right plot shows $D_n - \kappa$ curves taken at different values of $\dot{D}_n$.

activation energy. The left plot shows a three-dimensional representation of the surface in the $\dot{D}_n, D_n, \kappa$ - space, and the right plot shows $D_n - \kappa$ curves taken at different values of $\dot{D}_n$.

It can be shown, that the evolution (35) is hyperbolic. In addition for regions where $\partial D_n / \partial \kappa |_{\dot{D}_n} < 0$, the dispersion relation that is developed from a frozen coefficient analysis at a point on the surface, corresponds to stable growth of disturbances, superimposed on the solution to the evolution equation, while for $\partial D_n / \partial \kappa |_{\dot{D}_n} > 0$, corresponds to unstable growth.

5 The Dynamics of a $\dot{D}_n - D_n - \kappa$ Relation

Given that the asymptotic analysis suggests that the shock surface evolves according to a $\dot{D}_n - D_n - \kappa$ relation, we discuss some of the changes to the numerics that are required in the level-set formulation, and illustrate some simple aspects of the changes in behavior that are observed from the dynamics of a $D_n - \kappa$ relation.

5.1 Numerical Methods

As in the original level-set method, if one considers the surface to be SPS, but one that obeys a relation of the type $F(\dot{D}_n, D_n, \kappa) = 0$, it is still the case that the level-set equation holds, i.e.

$$\left(\frac{\partial \psi}{\partial t}\right)_{\mathbf{x}} + D_n |\boldsymbol{\nabla}\psi| = 0. \tag{40}$$

But now instead of having $D_n(\kappa)$, we generate from $F = 0$, an expression for the acceleration of the shock in the normal direction in terms of the normal velocity and curvature, $\dot{D}_n(D_n, \kappa)$. This dependence on $\dot{D}_n$ is equivalent to $F(\dot{D}_n, D_n, \kappa) = 0$ being an additional PDE rather than the algebraic constraint considered in Sect. 3. From (12) it is clear that this additional equation is then

$$\left(\frac{\partial D_n}{\partial t}\right)_{\mathbf{x}} + D_n \hat{\mathbf{n}} \cdot \boldsymbol{\nabla} D_n = \dot{D}_n(D_n, \kappa), \tag{41}$$

where $\hat{\mathbf{n}} = \boldsymbol{\nabla}\psi/|\boldsymbol{\nabla}\psi|$. The structure of the the operators in (40) and (41) is the same.

Equations (40) and (41) are a set of two, coupled, PDE's that must be solved simultaneously, for the evolution of the shock surface, for a given $\dot{D}_n - D_n - \kappa$ relation. Notice that not only the initial position of the shock is needed, but also its initial velocity, as well. This 2-PDE reformulation of the level-set method for DSD can be used when $F(D_n, \kappa) = 0$, by treating $F(D_n, \kappa)/\epsilon$, where $0 < \epsilon \ll 1$, as a source term in (41).

5.2 Numerical Examples

Here we demonstrate numerically the differences between the evolution of a wave front governed by a $D_n - \kappa$ relation and a $\dot{D}_n - D_n - \kappa$ relation. The computational domain is $0 \leq x \leq 5$ and $0 \leq y \leq 1$, with continuation boundary conditions at $x = 0, 5$ and perfectly reflecting boundary conditions at $y = 0, 1$. We run two experiments, where for both, the initial location of the wave is given by $x = .2(1 - cos(2\pi y))$, or equivalently a $\psi(x, y, 0) = x - .2(1 - cos(2\pi y))$. Experiment (a) corresponds to the numerical solution of a $D_n - \kappa$ relation given by $D_n = 1 - .05\kappa$. While experiment (b) corresponds to the numerical solution of the $\dot{D}_n - D_n - \kappa$ relation, given by $\dot{D}_n = -.025(D_n - 1) - .5\kappa$. In addition, for experiment (b) we assume that the initial velocity distribution is given by $D_n(x, y, 0) = 1$. Both relations admit $D_n = 1$ as the steady, plane, solution.

The results of the numerical experiment are shown in Fig. 5. The lines are contours of the crossing table $t_{cross}(x, y)$ (i.e., location of waves at time intervals of 0.2), while the grey scale indicates the value of the detonation normal velocity as the wave crosses a node point. Experiment (a) shows how the initial cosine wave smoothly evolves into a flat CJ wave, as expected for a $D_n - \kappa$ relation. In contrast, experiment (b) starts out with smooth data, but in a short time the level-set function (and hence the wave shape) forms cusps and D_n itself becomes discontinuous. As the wave evolves further, these discontinuities reflect off the walls and exhibit the characteristic cell-like pattern, often found in detonation smoke foil records. Amazingly enough, even the qualitative shape of the traces of the triple points on the smoke foil record, at the junction of the intersecting shock waves is reproduced. The dynamics of the motions of the cusps are governed by the nonlinear hyperbolic PDE that corresponds to the $\dot{D}_n - D_n - \kappa$

relation of experiment (b). While experiment (b) does not exhibit self-sustained cells (by construction the dynamics of the evolution are dissipative), the necessary ingredients to construct a theory of detonation cells based on an intrinsic evolution equation, are now available.

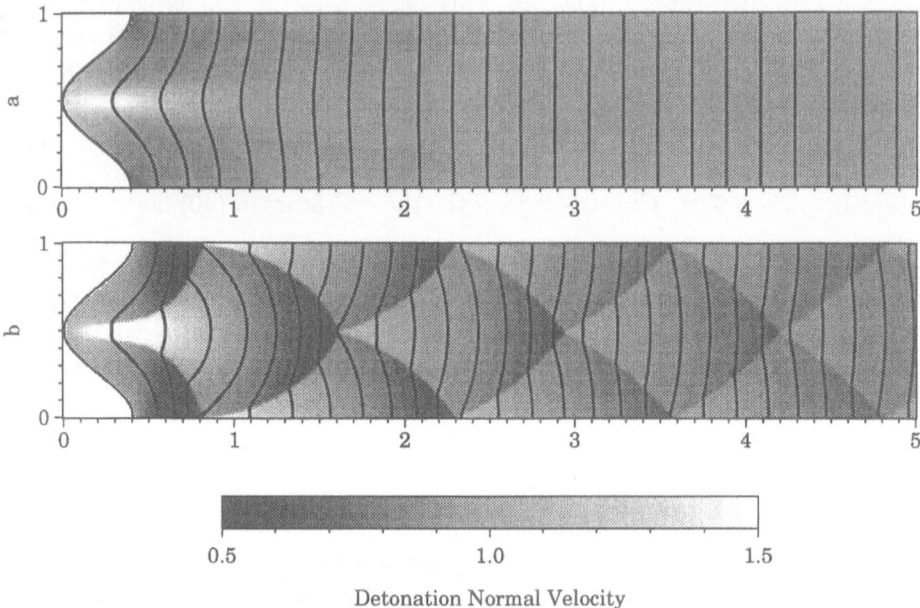

Fig. 5. An example of the comparison between the $D_n - \kappa$ relation and $\dot{D}_n - \kappa$ relation

Acknowledgments

This work has been supported by the United States Air Force, Wright Laboratory, Armament Directorate, Eglin Air Force Base, F08630-92- K0057, and with computing resources from the National Center for Supercomputing Applications (NCSA). Tariq Aslam has partially been supported by an AASERT grant by AFOSR, Summer of 1994. D. S. Stewart had travel support from the National Science Foundation. John Bdzil is supported by the US Department of Energy.

References

1. Osher, Stanley and Sethian, James, A., "Fronts Propagating with Curvature Dependent Speed: Algorithms Based on Hamilton-Jacobi Formulations", *Journal of Computational Physics*, 79, 12-49 (1988)
2. Aslam, T., Bdzil, J., and Stewart, D. S., "The Level-Set Method Applied to Modeling Detonation Shock Dynamics", to be submitted for publication.

3. Yao, Jin and Stewart, D. S., "On the Dynamics of Detonation", to be submitted for publication.
4. Matalon, M. and Matkowsky, M., "Flames as Gasdynamic Discontinuities", *J. Fluid Mech.*, vol. 124, pp., 239-259 (1982)
5. Buckmaster, J. D. and Ludford, G. S. S., *Theory of Laminar Flames*, Cambridge University Press, (1982), page 208.
6. Stewart, D. S. and Bdzil, J. B., "The Shock Dynamics of Stable Multi- Dimensional Detonation", *Combustion and Flame*, 72, 311-323 (1988).
7. Klein, R. and Stewart, D. S., "The relation between curvature and rate state-dependent detonation velocity", *SIAM Journal of Applied Mathematics*, Vol. 53, No. 5, pp. 1401-1435, (1993).
8. Yao, Jin and Stewart, D. S., "On the Normal Detonation Shock Velocity Curvature Relationship for Materials with Large Activation Energy.", to appear in *Combustion and Flame*, Oct. (1994).
9. Bdzil, J. B., Davis, W. C. and Critchfield, R. R. "Detonation Shock Dynamics (DSD) 'Calibration for PBX 9502' ", , *Proceedings of the Tenth Symposium (International) on Detonation*, Boston, Mass, 1993, to appear.
10. Private communication.
11. Bdzil, J. B., "Steady State Two-Dimensional Detonation", *Journal of Fluid Mechanics*, Vol. 108, (1981), pp. 185-226.
12. Wood, W. W. and Kirkwood, J. G. "Diameter effects in condensed explosives: The relation between velocity and radius of curvature", *Journal of Chemical Physics*, 22: 1920-1924 (1954).
13. Klein, R., "On the Dynamics of Weakly-Curved Detonation", in *Dynamical Issues in Combustion Theory*, Fife, P., Linan, A. and Williams, F. A., eds., pp. 127- 166, IMA volumes in Mathematics and Application, Springer-Verlag (1991).

Springer-Verlag
and the Environment

We at Springer-Verlag firmly believe that an international science publisher has a special obligation to the environment, and our corporate policies consistently reflect this conviction.

We also expect our business partners – paper mills, printers, packaging manufacturers, etc. – to commit themselves to using environmentally friendly materials and production processes.

The paper in this book is made from low- or no-chlorine pulp and is acid free, in conformance with international standards for paper permanency.

Lecture Notes in Physics

For information about Vols. 1–408
please contact your bookseller or Springer-Verlag

New Series m: Monographs